THEORY OF
MAGNETIC RESONANCE

THEORY OF MAGNETIC RESONANCE
Second Edition

Charles P. Poole, Jr., and Horacio A. Farach
Department of Physics
University of South Carolina, Columbia

A Wiley-Interscience Publication
JOHN WILEY & SONS
New York · Chichester · Brisbane · Toronto · Singapore

QC
762
P65
1987

Library of Congress Cataloging-in-Publication Data:

Poole, Charles P.
  Theory of magnetic resonance.

  "A Wiley-Interscience publication."
  Includes index.
  1. Magnetic resonance.   I. Farach, Horacio A.
II. Title.

QC762.P65 1986   538'.36   86-11013
ISBN 0-471-81530-6

Printed in the United States of America

10 9 8 7 6 5 4 3 2 1

To Our Children

CPP to   *Kathleen*
          Charles
          Michael
          Mary Ellen
          Elizabeth

HAF to   *Leticia*
          Sylvia
          Cecilia
          Horacio
          Laura
          Martin

PREFACE

In the first edition of this book the theory of magnetic resonance was developed using a uniform formalism based on the direct product matrix expansion technique. This approach stresses the overall unity of the various branches of the subject. After presenting some mathematical background material, the first edition treated a prototype two-spin $(\frac{1}{2}-\frac{1}{2})$ Hamiltonian which was then specialized to the two particular cases of electron spin resonance and nuclear magnetic resonance. Succeeding chapters discussed the various intricacies of this simple Hamiltonian and then expanded it to include more spins, higher spins, and other Hamiltonian terms.

Since the appearance of the 1976 edition a number of aspects of magnetic resonance have become increasingly important to research workers, such as various types of double resonance and dynamic polarization, and specialized topics like spin labels, saturation transfer, and Fourier transform methods. Many of these topics can be treated by means of the matrix expansion technique, and this has the advantage of showing clearly their relationships to the main body of magnetic resonance theory.

This edition retains the matrix expansion background chapter and most of the basic prototype two-spin $(\frac{1}{2}-\frac{1}{2})$ Hamiltonian treatment since these are essential for an understanding of the remainder of the work. The extensions to higher spins and to anisotropies have been considerably shortened to make room for the new topics. The applications to atomic spectra, crystal field theory, and Mössbauer resonance have also been curtailed. The various types of double resonance and dynamic polarization have been developed as extensions of the same prototype two-spin $(\frac{1}{2}-\frac{1}{2})$ Hamiltonian that was the basic building block of several chapters in the first edition. The concluding chapters are on spin labels and Fourier transform NMR.

One of us (H.A.F.) lectured on the material in the spin label chapter at the International Summer School of Theoretical Physics in Trieste during the summers of 1982 and 1984 and the other (C.P.P.) summarized the material in the double resonance chapters at the 9th Waterloo Nuclear Magnetic Resonance Summer Institute in June 1985. The authors wish to thank Dr. Deming Li of Shanxi University for his careful editing of the first eleven chapters of this book.

CHARLES P. POOLE, JR.
HORACIO A. FARACH

Columbia, South Carolina
August 1986

PREFACE TO THE
FIRST EDITION

On several occasions from 1966 to 1969 we taught courses in magnetic resonance. During this period lecture notes were written that emphasized the overall unity of the various branches of the subject. This volume is an enlarged version of those notes. References were added which (1) extend the presentation to more specialized cases and (2) introduce the student to the literature. The extensive bibliography to conference proceedings, books, and review articles found in Appendix III should constitute an adequate introduction to the research literature. Also, a set of problems was compiled, which should help the student to acquire a better grasp of the subject matter.

In this book the theory was developed using a uniform formalism that stresses the unity of magnetic resonance. The various subfields such as electron spin resonance, nuclear magnetic resonance, quadrupole spectroscopy, and the Mössbauer effect are derived as special cases of the general theory. The direct product matrix expansion technique was employed throughout, and it constitutes the unifying feature of the book.

The material treated here presupposes a knowledge of the fundamentals of quantum mechanics. Such a background is usually acquired in a standard quantum-mechanics course. The specialized matrix techniques used throughout the text are reviewed in Chapter 2 and applied immediately thereafter. The general theory of Chapters 2 and 3 is applied and extended in later chapters, some of which are more specialized and also more practical than the earlier ones.

The formalism adopted here is well suited for calculating energies and transition probabilities, and this task constitutes the major emphasis of the text. The phenomena that are treated are static in nature and are usually studied in steady-state (CW) experiments. Dynamic phenomena such as relaxation time mechanisms are best treated by other formalisms such as the Bloch equations or the density matrix technique. These topics were treated in detail in a companion volume, *Relaxation in Magnetic Resonance*.

A mastery of the material presented here will prepare the reader to consult advanced treatises such as Abragam's *The Theory of Nuclear Magnetism*, Abragam and Bleaney's *Electron Paramagnetic Resonance of Transition Ions*, and Emsley, Feeney, and Sutcliffe's two-volume work, *High Resolution Nuclear Magnetic Resonance*.

The authors wish to thank Mr. Sung Il Jon and Drs. Robert C. Nicklin and Milton P. Stombler for their helpful comments on the manuscript.

CHARLES P. POOLE, JR.
HORACIO A. FARACH

Columbia, South Carolina
April 1972

CONTENTS

THEORY OF
MAGNETIC RESONANCE

1

INTRODUCTION

1-1 MAGNETIC RESONANCE THEORIES

There are several different starting points from which one may construct a theory of magnetic resonance. (1) In the classical phenomenological approach one writes the Bloch differential equations for the time dependence of the bulk magnetization and then solves these equations for various initial conditions.[1,2] (2) In the density matrix technique one calculates the bulk magnetization quantum mechanically by summing over the diagonal elements.[3,4] Time-dependent effects arise from the equation of motion of the density matrix operator.[5] (3) If the secular equation viewpoint is adopted, one writes the Hamiltonian matrix and diagonalizes it to obtain the energy levels and eigenfunctions. The first two methods are useful for relaxation studies;[6-9] the third is preferable for computing positions and intensities of spectral lines.

This book emphasizes the calculation of the energy levels and transition probabilities of various spin systems. As a result, the third or secular viewpoint is emphasized throughout. In Chapter 2 the mathematical background for the direct product formulation is developed to facilitate the construction of Hamiltonian matrices in subsequent chapters. The Cartesian coordinate angular momentum operators J_x and J_y have been found to be more convenient to use than their raising and lowering counterparts, J^+ and J^-.

1-2 THE HAMILTONIAN

The energy of an atom or radical containing unpaired electrons and nuclei with nonzero spins may be expressed in terms of the Hamiltonian operator:

$$\mathcal{H} = \mathcal{H}_{el} + \mathcal{H}_{CF} + \mathcal{H}_{LS} + \mathcal{H}_{SS} + \mathcal{H}_{Ze} + \mathcal{H}_{HF} + \mathcal{H}_{Zn} + \mathcal{H}_{II} + \mathcal{H}_{Q} \qquad (1\text{-}1)$$

Each of the terms in this Hamiltonian will be defined and discussed in turn. The first three terms constitute the atomic Hamiltonian, and the last six form the spin Hamiltonian.

The electronic Hamiltonian operator $\mathcal{H}_{el}$ is the sum of the kinetic energy of each electron, $mv_i^2/2 = p_i^2/2m$, the potential energy of each electron relative to the nuclei, $-z_n e^2/r_{ni}$, and the interelectronic repulsion energies, e^2/r_{ij}:

$$\mathcal{H}_{el} = \sum_i \frac{p_i^2}{2m} - \sum_{i,n} \frac{z_n e^2}{r_{ni}} + \sum_{i>j} \frac{e^2}{r_{ij}} \tag{1-2}$$

where the index n is a sum over the nuclei and i, j are sums over the electrons. The electronic energy has the magnitude 10^4–10^5 cm^{-1}, and so it is in the optical region of the spectrum, many orders of magnitude larger than the Zeeman energies.

The crystal field term $\mathcal{H}_{CF}$ shifts and splits the electronic energy. This crystal field arises from the electrostatic charges of the ions in ionic compounds and from the chemical bonds in covalent compounds. It is essentially a Stark effect resulting from an electric potential of the type

$$V = \sum_{i,j} \frac{Q_j}{r_{ij}} \tag{1-3}$$

where the summation is over the Q_j ionic charges and the i electrons. The potential of Eq. 1-3 is given in the point charge approximation, and ordinarily only nearest-neighbor ligands are taken into account. A more realistic potential would contain integrals over the charge distributions. For highly symmetric configurations of nearest neighbors the potential of Eq. 1-3 assumes particularly simple forms, as will be shown in Chapter 10.

The spin–orbit interaction for an atom may be written

$$\mathcal{H}_{LS} = \lambda \vec{L} \cdot \vec{S} \tag{1-4}$$

where λ is the spin–orbit coupling constant, and $\vec{L}$ and $\vec{S}$ are, respectively, the orbital and the spin angular momenta. The spin–orbit energy, $H_{LS} \sim 10^2$ cm^{-1}, is much smaller than typical crystal field splittings for the first transition series and exceeds crystal field splittings for the rare earths.

The lowest order spin–spin interaction is of the form

$$\mathcal{H}_{SS} = D[S_z^2 - \tfrac{1}{3}S(S+1)] + E(S_x^2 - S_y^2) \tag{1-5}$$

This energy (0–10 cm^{-1}) is frequently of the same order of magnitude as the electronic Zeeman energy, in which case the spectrum becomes strongly angular dependent. For axial symmetry $E = 0$, and the spin–spin term is easier to handle.

The electronic Zeeman term

$$\mathcal{H}_{Ze} = \beta \vec{H} \cdot (\vec{L} + 2\vec{S})$$
$$= \beta \vec{S} \cdot \vec{\vec{g}} \cdot \vec{H} \tag{1-6}$$

has a magnitude of about 0.3 cm^{-1} at x band (10^{10} Hz) and is sometimes anisotropic. The nuclear Zeeman term

$$\mathcal{H}_{Zn} = -\sum_i g_{ni} \beta_n \vec{H} \cdot \vec{I}_i \tag{1-7}$$

is three orders of magnitude less and is usually isotropic. These are the main interactions in electron spin resonance and nuclear magnetic resonance, respectively. Equation 1-6 can be written with a summation for several spins.

The hyperfine interaction

$$\mathcal{H}_{HF} = \vec{S} \sum_i \cdot \vec{\vec{A}}_i \cdot \vec{I}_i \tag{1-8}$$

and the nuclear spin–spin interaction

$$\mathcal{H}_{II} = \sum_{i>j} \vec{I}_i \cdot \vec{\vec{J}}_{ij} \cdot \vec{I}_j \tag{1-9}$$

are of the same form. The former is summed only once, since one electron spin typically couples to several nuclei. The hyperfine coupling constants A_i are frequently anisotropic, while most NMR studies deal with isotropic spin–spin coupling constants. The Hamiltonian terms of Eqs. 1-8 and 1-9 are responsible for the structure that is observed in magnetic resonance spectra.

The main quadrupolar energy term is

$$\mathcal{H}_Q = \frac{e^2 Q}{4I(2I-1)} \left(\frac{\partial^2 V}{\partial z^2} \right) [3I_z^2 - I(I+1) + \eta(I_x^2 - I_y^2)] \tag{1-10}$$

where Q is the quadrupole moment, $\partial^2 V/\partial z^2$ is the field gradient, and η is the asymmetry parameter:

$$\eta = \frac{\partial^2 V/\partial x^2 - \partial^2 V/\partial y^2}{\partial^2 V/\partial z^2} \tag{1-11}$$

The quadrupole energy may be observed directly, or its influence on magnetic resonance spectra may be studied, as discussed in Chapter 8. For axial symmetry $\eta = 0$, and the Hamiltonian is simplified.

1-3 ATOMIC HAMILTONIAN

The high-energy terms, $\mathcal{H}_{el}$, $\mathcal{H}_{CF}$, and $\mathcal{H}_{LS}$, are much greater in energy than the main Zeeman terms, and therefore they are handled separately in the calculation. First the secular equation is solved with these three interactions alone. Then the influence of these terms on the spin Hamiltonian is calculated. When the spin–orbit coupling interaction is much less than the crystal field splitting, ΔE_{CF}, the wavefunction of the ground crystal field state will contain admixtures of the next higher state of the order of $\lambda / \Delta E_{CF}$.

As a result of this admixture of higher crystal field states into the ground state, the g-factor of a transition-metal ion deviates from its free electron value by a term that is proportional to the spin–orbit coupling constant and inversely proportional to the crystal field splittings. In the particular case of Cu^{2+} in a tetragonal crystalline field Polder[10] showed that

$$g_{\parallel} = 2 - \frac{8\lambda}{E_3 - E_1}$$

$$g_{\perp} = 2 - \frac{2\lambda}{E_4 - E_1}$$

(1-12)

where E_1, E_2, E_3, and E_4 are the four levels into which the crystal field splits the $3d^9$, ^2D configuration, and E_1 is the ground state of the group. For Cu^{2+}, $\lambda = -852$ cm^{-1}, and the energy denominators are more than 10 times this value; hence the g-factors of Cu^{2+} vary from 2.15 to 2.4. Formulas similar in form to these relations may be obtained for other transition-metal ions in lattice sites of various symmetries.

The crystal field splitting between the ground orbital level E_1 and the next excited orbital level E_2 helps to determine the spin–lattice relaxation time T_1 produced by the direct and Raman processes, and typical formulas for $S = \frac{1}{2}$ are[11]

$$T_1 = \frac{10^4 (E_2 - E_1)^4}{\lambda^2 H^4 T} \quad \text{sec} \quad \text{(Direct process)}$$

$$T_1 = \frac{10^4 (E_2 - E_1)^6}{\lambda^2 H^2 T^7} \quad \text{sec} \quad \text{(Raman process)}$$

(1-13)

where E_1, E_2, and λ are in the units of reciprocal centimeters, H is in gauss, and T is the absolute temperature.

1-4 SPIN HAMILTONIAN

Most of this book is devoted to solving the spin Hamiltonian for the eigenenergies and eigenfunctions. Ordinarily, only two or three of the total spin Hamiltonian terms

$$\mathcal{H}_{\text{spin}} = \mathcal{H}_{Ze} + \mathcal{H}_{HF} + \mathcal{H}_{SS} + \mathcal{H}_Q + \mathcal{H}_{Zn} + \mathcal{H}_{II}$$

(1-14)

will be present at a time. The direct products involving Pauli spin matrices, $\vec{\sigma}_i / 2$, and their higher spin analogs are employed for the calculations with this Hamiltonian.

The fields of electron spin resonance, nuclear magnetic resonance, nuclear quadrupole resonance, and Mössbauer resonance will be treated by the same mathematical formalism. Typical spin Hamiltonians for these cases are

$$
\begin{array}{ll}
\mathcal{H} = \mathcal{H}_{Ze} + [\mathcal{H}_{HF} + \mathcal{H}_{SS}] & \text{(ESR)} \\
\mathcal{H} = \mathcal{H}_{Zn} + [\mathcal{H}_{II}] & \text{(NMR)} \\
\mathcal{H} = \mathcal{H}_{Q} + [\mathcal{H}_{Zn}] & \text{(Quadrupole)} \\
\mathcal{H} = \mathcal{H}_{\text{nuclear}} + [\mathcal{H}_{Zn} + \mathcal{H}_{Q}] & \text{(Mössbauer)}
\end{array}
\qquad (1\text{-}15)
$$

Many spectra are singlets without structure, in which case only the first term in each Hamiltonian is observed. The term $\mathcal{H}_{\text{nuclear}}$ of the Mössbauer Hamiltonian is a nuclear transition typically several hundred kiloelectron volts in energy, and therefore it is not really part of a "spin Hamiltonian." Structure in ESR ordinarily arises from the hyperfine term alone. In Mössbauer resonance one ordinarily observes either $\mathcal{H}_{Zn}$ or $\mathcal{H}_{Q}$, although the two may be present together.

1-5 EQUIVALENT OPERATORS

Some authors write the spin Hamiltonian for an unimpaired spin $\vec{S}$ interacting with a nucleus $\vec{I}$ in the following manner:

$$
\mathcal{H} = \vec{S} \cdot \vec{\vec{g}} \cdot H - \beta_N \vec{H} \cdot \vec{\vec{g}}_N \cdot \vec{I} + \vec{S} \cdot \vec{\vec{T}} \cdot \vec{I}
$$

$$
+ \sum_{\substack{k=0 \\ \text{even}}}^{2S} \sum_{q=-k}^{k} B_k^q O_{k(S)}^q + \sum_{\substack{n=0 \\ \text{even}}}^{2I} \sum_{m=-n}^{n} C_n^m O_{n(I)}^m
\qquad (1\text{-}16)
$$

where the first three terms on the right-hand side, namely, the electronic Zeeman, nuclear Zeeman, and hyperfine ones, are treated in the usual manner. The first is linear in $\vec{S}$, the second is linear in $\vec{I}$, and the third is linear in both. Higher-order cross-terms (quadratic in both, etc.) are possible.

The O_k^q operators are referred to as equivalent operators and have the form shown in Table 10-2. Abragam and Bleaney[12] tabulate these matrix elements. For $O_{k(S)}^q$ the operators J_i are replaced by S_i, and for $O_{n(I)}^m$ operators J_i are replaced by I_i. The raising and lowering operators may be transformed to their Cartesian analogs in the following manner:

$$
\tfrac{1}{2}(J_+^2 + J_-^2) = J_x^2 - J_y^2
$$

$$
\tfrac{1}{2}(J_+^3 + J_-^3) = J_x^3 - J_x J_y^2 - J_y J_x J_y - J_y^2 J_x
$$

$$
\tfrac{1}{2}(J_+^4 + J_-^4) = J_x^4 + J_y^4 - J_x^2 J_y^2 - J_y^2 J_x^2 - J_x J_y^2 J_x
$$

$$
\qquad\qquad - J_y J_x^2 J_y - J_x J_y J_x J_y - J_y J_x J_y J_x
$$

$$
(1\text{-}17)
$$

The coefficients k and n must be nonzero even integers, while q and m are any positive or negative integers within the range

$$-k \leq q \leq k$$
$$-n \leq m \leq n \tag{1-18}$$

The coefficients B_k^q and C_n^m assume definite values for specific interactions. Several correspondences are written down for reference purposes. The D and E zero field splitting interactions are

$$D[S_z^2 - \tfrac{1}{3}S(S+1)] = B_2^0 O_{2(S)}^0 \tag{1-19}$$
$$E(S_x^2 - S_y^2) = B_2^2 O_{2(S)}^2 \tag{1-20}$$

which show that

$$D = 3B_2^0, \qquad E = B_2^2 \tag{1-21}$$

The cubic zero field splitting which exists for $S > \tfrac{3}{2}$ has the form

$$\tfrac{1}{6}a[S_x^4 + S_y^4 + S_z^4 - \tfrac{1}{5}S(S+1)(3S^2 + 3S - 1)] = B_4^0 O_{4(S)}^0 + B_4^4 O_{4(S)}^4 \tag{1-22}$$

For the fourfold axis coordinate system one has

$$\tfrac{1}{6}a = 20B_4^0 = 4B_4^4 \tag{1-23}$$

and for the threefold ones

$$\frac{1}{6}a = -\frac{40}{3}B_4^0 = -\frac{800\sqrt{2}}{3}B_4^3 \tag{1-24}$$

The zero field F term, which exists for axial distortions when $S > \tfrac{3}{2}$, has the form

$$\frac{F}{180}\{35S_z^4 - 30S(S+1)S_z^2 + 25S_z^2$$

$$+ 3S(S+1)[S(S+1) - 2]\} = B_4^0 O_{4(S)}^0 \tag{1-25}$$

so that

$$F = 180B_4^0 \tag{1-26}$$

The axial electric quadrupole term is

$$\frac{e^2qQ}{8I(2I-1)}[3I_z^2 - I(I+1)] = C_2^0 O_{2(I)}^0 \tag{1-27}$$

so that

$$\frac{e^2qQ}{8I(2I-1)} = C_2^0 \tag{1-28}$$

The lower symmetry quadrupole interaction may be written

$$\frac{e^2qQ\eta}{8I(2I-1)}\,(I_x^2 - I_y^2) = C_2^2 O_{2(I)}^2 \tag{1-29}$$

to give

$$\frac{e^2qQ\eta}{8I(2I-1)} = C_2^2 \tag{1-30}$$

Therefore the asymmetry parameter, η, is defined by the ratio

$$\eta = \frac{C_2^2}{C_2^0} \tag{1-31}$$

These relations should be useful in transforming some parameters in the literature to our formalism. Unfortunately, other types of coefficient such as b_k^q occur in the literature; examples of these are

$$\begin{aligned} b_2^0 &= D \\ b_2^2 &= 3E \\ b_4^0 &= \tfrac{1}{3}F \end{aligned} \tag{1-32}$$

One must be very careful of definitions in comparing literature references.

Another completely different way to express the spin Hamiltonian is to adopt the irreducible tensor (spherical tensor) formalism, which is discussed in Section 11-12 of the first edition for the specific case of quadrupole interactions. It may be employed for other Hamiltonian terms also.

1-6 EFFECTIVE SPIN HAMILTONIAN

The actual interactions of electronic spins in transition-series ions are quite complex. Fortunately, the complicated expressions occurring in accurate theoretical formulations can frequently be simplified by the use of an effective spin S and a Hamiltonian written in terms of the actual or effective spin.

As an example of the origin of a simplified spin Hamiltonian consider an atomic Hamiltonian with the spin–orbit, spin–spin, and electronic Zeeman interactions, respectively, as follows:

$$\mathcal{H} = \lambda \vec{L} \cdot \vec{S} - \rho[(\vec{L} \cdot \vec{S})^2 + \tfrac{1}{2}(\vec{L} \cdot \vec{S}) - \tfrac{1}{3}L(L+1)S(S+1)] + \beta \vec{H} \cdot (k\vec{L} + g_s\vec{S}) \tag{1-33}$$

where $\vec{L}$ and $\vec{S}$ are the orbital and spin angular momenta, respectively, and the various terms arise from summations over individual electrons with $\vec{l}_i$ and $\vec{s}_i$ angular momenta:

$$\lambda \vec{L} \cdot \vec{S} = \sum_i (\zeta_i \vec{l}_i \cdot \vec{s}_i)$$

$$k\vec{L} + g_s \vec{S} = \sum_i (k\vec{l}_i + g_s \vec{s}_i) \tag{1-34}$$

The spin–orbit coupling constant $|\lambda| = |\zeta/2S|$, and $k \le 1$ takes into account bonding electrons. For the usual case of spin–orbit and crystal field interactions that greatly exceed the Zeeman term, the orbital operators may be averaged out by taking matrix elements between the ground and higher crystal field levels. The crystal field can "quench" the orbital motion of the Zeeman term by quantizing it along the crystal field direction, so that only S can interact with the magnetic field. For complete quenching this gives the Zeeman splitting $g_s \beta \vec{H} \cdot \vec{S}$, where $g_s = 2.0023$ for a free electron. Ordinarily, only partial quenching results, and one obtains a g-factor that may be anisotropic.

The above considerations permit the electronic interaction to be written

$$H = \vec{S} \cdot \vec{\vec{D}} \cdot \vec{S} + \beta \vec{H} \cdot \vec{\vec{g}} \cdot \vec{S} \tag{1-35}$$

where $\vec{\vec{D}}$ is a function of the orbital matrix elements with excited crystal field states and the spin–orbit parameter, and $\vec{\vec{g}}$ takes into account the partial quenching of $\vec{L}$.

Sometimes a group of close-lying electronic levels is widely separated from all upper levels. The behavior of such a group can be represented by an effective spin S such that the total number of levels in the group is $2S + 1$. Treatises on the electron spin resonance of transition-metal ions frequently employ the concept of effective spin. The nature of the parameters $\vec{\vec{D}}$ and $\vec{\vec{g}}$ in a Hamiltonian of the type of Eq. 1-35 written for an effective spin takes into account the specific origin of the group of levels. Other books[12-15] should be consulted for further details (cf. also Section 10-13).

REFERENCES

1. F. Bloch, *Phys. Rev.*, **70**, 460 (1946).

2. N. Bloembergen, E. M. Purcell, and R. V. Pound, *Phys. Rev.*, **73**, 679 (1948).

3. C. P. Slichter, *Principles of Magnetic Resonance*, Harper & Row, New York, 1963.

4. A. Abragam, *The Principles of Nuclear Magnetism*, Clarendon Press, Oxford, 1961.

5. A. G. Redfield, *Adv. Magn. Reson.* **1**, 1 (1965).

6. C. P. Poole, Jr. and H. A. Farach, *Relaxation in Magnetic Resonance*, Academic Press, New York, 1971.

7. K. J. Standley and R. A. Vaughan, *Electron Spin Relaxation Phenomena in Solids*, Plenum Press, New York, 1970.

8. D. Wolf, *Spin Temperature and Nuclear Spin Relaxation in Matter: Basic Principles and Applications*, Oxford University Press, London, 1976.

9. L. T. Muus and P. W. Atkins, eds., *Electron Spin Relaxation in Liquids*, Plenum, New York, 1972.

10. D. Polder, *Physica*, **9**, 709 (1942).

11. R. Kronig, *Physica*, **6**, 33 (1939).

12. A. Abragam and B. Bleaney, *Electron Paramagnetic Resonance of Transition Ions*, Clarendon Press, Oxford, 1970.

13. J. W. Orton, *Electron Paramagnetic Resonance*, Iliffe Books, London, 1968.

14. J. E. Harriman, ed., *Theoretical Foundations of Electron Spin Resonance*, Academic Press, New York, 1978.

15. J. E. Wertz and J. R. Bolton, *Electron Spin Resonance*, Chapman & Hall, New York, 1986.

MATHEMATICAL AND QUANTUM-MECHANICAL BACKGROUND

2-1 INTRODUCTION

The general approach to be followed in this volume is the setting up of secular equations by means of the direct product (inner product) matrix expansion technique. This may be regarded as a direct product counterpart to Condon and Shortley's formulation.[1] The present chapter outlines the mathematical and quantum-mechanical properties of matrices and angular momentum operators and thereby prepares the way for an understanding of the material in subsequent chapters. Those who already possess the proper background may conveniently omit most of the chapter, although it is strongly recommended that they peruse Sections 2-6 and 2-7.

2-2 TYPES OF MATRICES

Quantum-mechanical operators will be in the form of Hermitian matrices, and the secular equations will constitute Hermitian matrices. Before discussing unitary and Hermitian matrices in detail it is best to define several other types of square matrices. Some of these definitions will be in the form of an example of a 2×2 or 3×3 matrix.

The unit matrix has the form

$$\vec{\vec{I}} = \begin{pmatrix} 1 & 0 & 0 \\ 0 & 1 & 0 \\ 0 & 0 & 1 \end{pmatrix} \qquad (2\text{-}1)$$

and multiplication by the unit matrix leaves another matrix $\vec{\vec{M}}$ unaffected:

$$\vec{\vec{M}}\vec{\vec{I}} = \vec{\vec{I}}\vec{\vec{M}} = \vec{\vec{M}} \tag{2-2}$$

The reciprocal $\vec{\vec{M}}^{-1}$ of a matrix has the property

$$\vec{\vec{M}}\vec{\vec{M}}^{-1} = \vec{\vec{M}}^{-1}\vec{\vec{M}} = \vec{\vec{I}} \tag{2-3}$$

We shall give below the form of this reciprocal for several types of matrix.
   A matrix $\vec{\vec{M}}$,

$$\vec{\vec{M}} = \begin{pmatrix} a & b & c \\ d & e & f \\ g & h & i \end{pmatrix} \tag{2-4}$$

has a transpose, $\tilde{\vec{\vec{M}}}$,

$$\tilde{\vec{\vec{M}}} = \begin{pmatrix} a & d & g \\ b & e & h \\ c & f & i \end{pmatrix} \tag{2-5}$$

a complex conjugate $\vec{\vec{M}}^*$,

$$\vec{\vec{M}}^* = \begin{pmatrix} a^* & b^* & c^* \\ d^* & e^* & f^* \\ g^* & h^* & i^* \end{pmatrix} \tag{2-6}$$

and an adjoint $\vec{\vec{M}}^+$ corresponding to the transpose complex conjugate matrix:

$$\vec{\vec{M}}^+ = \tilde{\vec{\vec{M}}}^* = \begin{pmatrix} a^* & d^* & g^* \\ b^* & e^* & h^* \\ c^* & f^* & i^* \end{pmatrix} \tag{2-7}$$

A symmetric matrix is one which equals its transpose:

$$\vec{\vec{M}} = \tilde{\vec{\vec{M}}} = \begin{pmatrix} a & b & c \\ b & e & f \\ c & f & i \end{pmatrix} \tag{2-8}$$

An antisymmetric matrix is the negative of its transpose:

$$\vec{\vec{A}} = -\tilde{\vec{\vec{A}}} = \begin{pmatrix} 0 & b & c \\ -b & 0 & f \\ -c & -f & 0 \end{pmatrix} \tag{2-9}$$

An orthogonal matrix is one with real matrix elements whose reciprocal is its transpose:

$$\vec{A}^{-1} = \vec{A} = \begin{pmatrix} a & d & g \\ b & e & h \\ c & f & i \end{pmatrix} \tag{2-10}$$

It has the property that its determinant is ± 1:

$$|A| = |A^{-1}| = \pm 1 \tag{2-11}$$

A unimodular matrix has the determinant $+1$. A 3×3 proper rotation matrix is unimodular, while an improper rotation (i.e., rotation-inversion) matrix has a determinant of -1. An example of the former is

$$\vec{A} \begin{pmatrix} \cos\theta & -\sin\theta & 0 \\ \sin\theta & \cos\theta & 0 \\ 0 & 0 & 1 \end{pmatrix} \tag{2-12}$$

A real matrix contains only real matrix elements and hence is identical with its complex conjugate:

$$\vec{A} = \vec{A}* \tag{2-13}$$

A purely imaginary matrix contains only imaginary matrix elements and hence equals the negative of its complex conjugate:

$$\vec{A} = -\vec{A}* = \begin{pmatrix} a & b & c \\ d & e & f \\ g & h & i \end{pmatrix} = \begin{pmatrix} -a* & -b* & -c* \\ -d* & -e* & -f* \\ -g* & -h* & -i* \end{pmatrix} \tag{2-14}$$

A Hermitian matrix is identical with its adjoint:

$$\vec{A} = \vec{A}^{+} = \begin{pmatrix} a & b & c \\ b* & e & f \\ c* & f* & i \end{pmatrix} \tag{2-15}$$

where the diagonal matrix elements, a, e, and i, are real. An anti-Hermitian matrix is equal to the negative to its adjoint:

$$\vec{A} = \vec{A}^{+} = \begin{pmatrix} a & b & c \\ -b* & e & f \\ -c* & -f* & i \end{pmatrix} \tag{2-16}$$

where the diagonal elements, a, e, and i, are purely imaginary (e.g., $a* = -a$).

2-3 UNITARY MATRICES

A unitary matrix $\vec{U}$ is one whose inverse equals its adjoint:

$$\vec{U} = \begin{pmatrix} a & b & c \\ d & e & f \\ g & h & i \end{pmatrix} \tag{2-17}$$

$$\vec{U}^{-1} = \vec{U}^{+} = \begin{pmatrix} a^* & d^* & g^* \\ b^* & e^* & h^* \\ c^* & f^* & i^* \end{pmatrix} \tag{2-18}$$

A unitary matrix has the property that the absolute value of its determinant is unity. We only consider the special cases where the determinant is ± 1:

$$\begin{vmatrix} a & b & c \\ d & e & f \\ g & h & i \end{vmatrix} = \pm 1 \tag{2-19}$$

Since it is tedious to evaluate the relations between the elements of a 3×3 unitary matrix, the principles are illustrated by means of a 2×2 unitary matrix with the general form

$$\vec{U} = \begin{pmatrix} \alpha & \beta \\ \gamma & \delta \end{pmatrix} \tag{2-20}$$

The number of independent matrix elements may be reduced by carrying out the operation

$$\vec{U}^{-1}\vec{U} = \vec{U}\vec{U}^{-1} = \begin{pmatrix} 1 & 0 \\ 0 & 1 \end{pmatrix} \tag{2-21}$$

as follows:

$$\begin{pmatrix} \alpha & \beta \\ \gamma & \delta \end{pmatrix}\begin{pmatrix} \alpha^* & \gamma^* \\ \beta^* & \delta^* \end{pmatrix} = \begin{pmatrix} \alpha\alpha^* + \beta\beta^* & \alpha\gamma^* + \beta\delta^* \\ \alpha^*\gamma + \beta^*\delta & \gamma\gamma^* + \delta\delta^* \end{pmatrix} \tag{2-22}$$

$$\begin{pmatrix} \alpha^* & \gamma^* \\ \beta^* & \delta^* \end{pmatrix}\begin{pmatrix} \alpha & \beta \\ \gamma & \delta \end{pmatrix} = \begin{pmatrix} \alpha\alpha^* + \gamma\gamma^* & \alpha^*\beta + \delta\gamma^* \\ \alpha\beta^* + \delta^*\gamma & \beta\beta^* + \delta\delta^* \end{pmatrix} \tag{2.23}$$

From the diagonal elements of the two matrices on the right, one finds

$$\alpha\alpha^* + \beta\beta^* = 1$$
$$\alpha\alpha^* + \gamma\gamma^* = 1$$
$$\delta\delta^* + \beta\beta^* = 1$$
$$\delta\delta^* + \gamma\gamma^* = 1$$
$$\tag{2-24}$$

and so

$$\alpha\alpha^* = +\delta\delta^*$$
$$\beta\beta^* = +\gamma\gamma^*$$
$$\tag{2-25}$$

From the off-diagonal components one deduces

$$\frac{\alpha}{\beta} = -\left(\frac{\delta}{\gamma}\right)^* \tag{2-26a}$$

$$\frac{\alpha}{\gamma} = -\left(\frac{\delta}{\beta}\right)^* \tag{2-26b}$$

Let the determinant be +1:

$$\begin{vmatrix} \alpha & \beta \\ \gamma & \delta \end{vmatrix} = \alpha\delta - \beta\gamma = 1 \tag{2-27a}$$

$$\begin{vmatrix} \alpha^* & \gamma^* \\ \beta^* & \delta^* \end{vmatrix} = \alpha^*\delta^* - \beta^*\gamma^* = 1 \tag{2-27b}$$

From Eqs. 2-26a and 2-27a it follows that

$$\beta = -\gamma^*, \qquad \alpha = \delta^* \tag{2-28}$$

to give, for the most general 2×2 unitary matrix with determinant +1,

$$\vec{U} = \begin{pmatrix} \alpha & -\gamma^* \\ \gamma & \alpha^* \end{pmatrix} \tag{2-29}$$

The corresponding matrix with determinant -1 is

$$\vec{U} = \begin{pmatrix} \alpha & \gamma^* \\ \gamma & -\alpha^* \end{pmatrix} \tag{2-30}$$

We will have occasion to use the +1 case only for matrices that diagonalize a Hermitian matrix. The Pauli spin matrices, on the other hand, are unitary matrices with a determinant -1.

One should note that a real unitary matrix is equivalent to an orthogonal matrix. Also the matrix elements of a unitary matrix are related in such a way that every row is orthogonal to the complex conjugate of every other row, and each row is normalized to unity relative to the complex conjugate of itself. The same applies to the columns. For example, these rules applied to unitary matrix 2-17 give

$$ad^* + be^* + cf^* = 0 \tag{2-31}$$

$$aa^* + bb^* + cc^* = 1 \tag{2-32}$$

The same orthonormality properties apply to orthogonal matrices in which all matrix elements are real.

2-4 HERMITIAN MATRICES

In this book we are interested in finding the eigenvalues of Hermitian matrices. A Hermitian matrix may be diagonalized by a unitary transformation, and the transformation matrix provides the coefficients for the eigenfunctions of the system. This will become clear in Chapter 3 when the $I_1 = I_2 = \frac{1}{2}$ spin system is worked out in detail. At this time it is helpful to illustrate the principles by an example.

The Pauli spin matrix, $\vec{\sigma}_y$,

$$\vec{\sigma}_y = \begin{pmatrix} 0 & -i \\ i & 0 \end{pmatrix} \tag{2-33}$$

is Hermitian and has the eigenvalues λ,

$$\begin{vmatrix} 0 - \lambda & -i \\ i & 0 - \lambda \end{vmatrix} = 0 \tag{2-34}$$

to give

$$\lambda^2 - 1 = 0$$
$$\lambda = \pm 1 \tag{2-35}$$

The eigenvectors of the unitary matrix $\vec{U}$ which diagonalizes $\vec{\sigma}_y$ by $\vec{U}^{-1}\vec{\sigma}_y\vec{U} = \vec{\lambda}$,

$$\begin{pmatrix} \alpha^* & \gamma^* \\ -\gamma & \alpha \end{pmatrix}\begin{pmatrix} 0 & -i \\ i & 0 \end{pmatrix}\begin{pmatrix} \alpha & -\gamma^* \\ \gamma & \alpha^* \end{pmatrix} = \begin{pmatrix} \lambda_1 & 0 \\ 0 & \lambda_2 \end{pmatrix} \tag{2-36}$$

are found from the relations $(\vec{\sigma}_y\vec{U} = \vec{U}\vec{\lambda})$

$$\begin{pmatrix} 0 & -i \\ i & 0 \end{pmatrix}\begin{pmatrix} \alpha \\ \gamma \end{pmatrix} = \lambda\begin{pmatrix} \alpha \\ \gamma \end{pmatrix} \tag{2-37}$$

Inserting the eigenvalue $\lambda = +1$ and multiplying out Eq. 2-37 gives

$$-i\gamma = \alpha \tag{2-38}$$

subject to the normalization condition

$$\alpha\alpha^* + \gamma\gamma^* = 1 \tag{2-39}$$

with a possible result

$$\alpha = \frac{1}{\sqrt{2}}, \qquad \gamma = \frac{i}{\sqrt{2}} \tag{2-40}$$

since $\alpha\alpha^* = \gamma\gamma^*$ in this instance. The resulting unitary matrix has the explicit form

$$\vec{\tilde{U}} = \begin{pmatrix} \dfrac{1}{\sqrt{2}} & \dfrac{i}{\sqrt{2}} \\ \dfrac{i}{\sqrt{2}} & \dfrac{1}{\sqrt{2}} \end{pmatrix} \tag{2-41}$$

The unitary transformation for a higher-order Hermitian matrix may be found by a similar procedure, but the manipulations become quite tedious. It is wise, therefore, to make use of a computer.

In this book we are concerned with the addition,

$$\begin{pmatrix} A & B \\ C & D \end{pmatrix} + \begin{pmatrix} a & b \\ c & d \end{pmatrix} = \begin{pmatrix} A+a & B+b \\ C+c & D+d \end{pmatrix} \tag{2-42}$$

multiplication,

$$\begin{pmatrix} A & B \\ C & D \end{pmatrix}\begin{pmatrix} a & b \\ c & d \end{pmatrix} = \begin{pmatrix} Aa+Bc & Ab+Bd \\ Ca+Dc & Cb+Dd \end{pmatrix} \tag{2-43}$$

and direct product expansion,

$$\begin{pmatrix} A & B \\ C & D \end{pmatrix} \times \begin{pmatrix} a & b \\ c & d \end{pmatrix} = \begin{pmatrix} Aa & Ab & Ba & Bb \\ Ac & Ad & Bc & Bd \\ Ca & Cb & Da & Db \\ Cc & Cd & Dc & Dd \end{pmatrix} \tag{2-44}$$

of square matrices. The first two operations are defined for matrices of the same order, $n \times n$, and produce a matrix of that same order. The direct product expansion, on the other hand, forms an $(mn) \times (mn)$ matrix from an $m \times m$ and an $n \times n$ matrix. Column vectors may be added,

$$\begin{pmatrix} A \\ B \end{pmatrix} + \begin{pmatrix} a \\ b \end{pmatrix} = \begin{pmatrix} A+a \\ B+b \end{pmatrix} \tag{2-45}$$

and expanded as direct products,

$$\begin{pmatrix} A \\ B \end{pmatrix} \times \begin{pmatrix} a \\ b \end{pmatrix} = \begin{pmatrix} Aa \\ Ab \\ Ba \\ Bb \end{pmatrix} \tag{2-46}$$

A square matrix times a column vector gives another column vector, and similarly for the reciprocal operator:

$$\begin{pmatrix} A & B \\ C & D \end{pmatrix}\begin{pmatrix} a \\ b \end{pmatrix} = \begin{pmatrix} Aa+Bb \\ Ca+Db \end{pmatrix} \tag{2-47}$$

$$(a^*b^*)\begin{pmatrix} A & B \\ C & D \end{pmatrix} = (Aa^* + Cb^* \quad Ba^* + Db^*) \tag{2-48}$$

and the simple scalar product is

$$(a^* \quad b^*)\begin{pmatrix} A \\ B \end{pmatrix} = Aa^* + Bb^* \tag{2-49}$$

These operations are easily generalized to higher dimensions.

2-5 ANGULAR MOMENTUM OPERATORS

Before proceeding to the matrix formulation of angular momentum operators, it is appropriate to summarize their principal operator properties.

An angular momentum $\vec{J}$ has a degeneracy of $2J + 1$, with its magnetic quantum number m assuming the integral or half-integral values within the range $-J \le m \le J$. The z component of the angular momentum operates on the ket vector $|m\rangle$ in accordance with the relation

$$J_z |m\rangle = m|m\rangle \tag{2-50}$$

and thus the ket vector $|m\rangle$ is an eigenfunction of the operator J_z. The x and y components may be expressed as linear combinations of raising and lowering operators:

$$J_x = \tfrac{1}{2}(J^+ + J^-)$$
$$J_y = \frac{1}{2i}(J^+ - J^-) \tag{2-51}$$

$$J^+ = J_x + iJ_y$$
$$J^- = J_x - iJ_y \tag{2-52}$$

The raising and lowering operators transform the ket vector $|m\rangle$ into a new ket in the following manner:

$$J^+|m\rangle = \sqrt{(J - m)(J + m + 1)}|m + 1\rangle$$
$$J^-|m\rangle = \sqrt{(J + m)(J - m + 1)}|m - 1\rangle \tag{2-53}$$

These expansions may be utilized to evaluate matrix elements of the type $\langle m|I_x|m'\rangle$ with the aid of the orthogonality relations

$$\langle m|m'\rangle = \delta_{mm'} \tag{2-54}$$

The ket vector $|m\rangle$ is also an eigenfunction for the square of the total angular momentum:

$$J^2|m\rangle = J(J + 1)|m\rangle \tag{2-55}$$

where we say that $J(J + 1)$ is the eigenvalue of the operator J^2.

2-6 ANGULAR MOMENTUM MATRICES

In order to employ the direct product method for spins with magnitudes greater than 2, it is necessary to know the corresponding spin matrices. These are easily generated from Eqs. 2-50 and 2-53, the former giving the diagonal matrix elements for J_z and the latter the matrix elements adjacent to the diagonal for J_x and J_y. All other matrix elements vanish. We are content to write down the matrices for low magnitudes of J, and then to point out a convenient mnemonic method for constructing them.

The various matrices for $J = \frac{1}{2}$, 1, and $\frac{3}{2}$ have the following explicit forms, where J may be any orbital, spin, or total angular momentum:

$$
\text{Unit Matrix} \rightarrow \begin{pmatrix} 1 & 0 \\ 0 & 1 \end{pmatrix} \rightarrow \begin{pmatrix} 1 & 0 & 0 \\ 0 & 1 & 0 \\ 0 & 0 & 1 \end{pmatrix} \rightarrow \begin{pmatrix} 1 & 0 & 0 & 0 \\ 0 & 1 & 0 & 0 \\ 0 & 0 & 1 & 0 \\ 0 & 0 & 0 & 1 \end{pmatrix}
\tag{2-56}
$$

$$
\vec{J}_x \rightarrow \frac{1}{2}\begin{pmatrix} 0 & 1 \\ 1 & 0 \end{pmatrix} \rightarrow \frac{1}{2}\begin{pmatrix} 0 & \sqrt{2} & 0 \\ \sqrt{2} & 0 & \sqrt{2} \\ 0 & \sqrt{2} & 0 \end{pmatrix} \rightarrow \frac{1}{2}\begin{pmatrix} 0 & \sqrt{3} & 0 & 0 \\ \sqrt{3} & 0 & 2 & 0 \\ 0 & 2 & 0 & \sqrt{3} \\ 0 & 0 & \sqrt{3} & 0 \end{pmatrix}
\tag{2-57}
$$

$$
\vec{J}_y \rightarrow \frac{1}{2}\begin{pmatrix} 0 & -i \\ i & 0 \end{pmatrix} \rightarrow \frac{1}{2}\begin{pmatrix} 0 & -i\sqrt{2} & 0 \\ i\sqrt{2} & 0 & -i\sqrt{2} \\ 0 & i\sqrt{2} & 0 \end{pmatrix}
$$

$$
\rightarrow \frac{1}{2}\begin{pmatrix} 0 & -\sqrt{3}i & 0 & 0 \\ \sqrt{3}i & 0 & -2i & 0 \\ 0 & 2i & 0 & -\sqrt{3}i \\ 0 & 0 & \sqrt{3}i & 0 \end{pmatrix}
\tag{2-58}
$$

$$
\vec{J}_z \rightarrow \frac{1}{2}\begin{pmatrix} 1 & 0 \\ 0 & -1 \end{pmatrix} \rightarrow \begin{pmatrix} 1 & 0 & 0 \\ 0 & 0 & 0 \\ 0 & 0 & -1 \end{pmatrix} \rightarrow \frac{1}{2}\begin{pmatrix} 3 & 0 & 0 & 0 \\ 0 & 1 & 0 & 0 \\ 0 & 0 & -1 & 0 \\ 0 & 0 & 0 & -3 \end{pmatrix}
\tag{2-59}
$$

$$
\vec{J}^2 \rightarrow \frac{3}{4}\begin{pmatrix} 1 & 0 \\ 0 & 1 \end{pmatrix} \rightarrow 2\begin{pmatrix} 1 & 0 & 0 \\ 0 & 1 & 0 \\ 0 & 0 & 1 \end{pmatrix} \rightarrow \frac{15}{4}\begin{pmatrix} 1 & 0 & 0 & 0 \\ 0 & 1 & 0 & 0 \\ 0 & 0 & 1 & 0 \\ 0 & 0 & 0 & 1 \end{pmatrix}
\tag{2-60}
$$

$$\vec{J}^+ \to \begin{pmatrix} 0 & 1 \\ 0 & 0 \end{pmatrix} \to \begin{pmatrix} 0 & \sqrt{2} & 0 \\ 0 & 0 & \sqrt{2} \\ 0 & 0 & 0 \end{pmatrix} \to \begin{pmatrix} 0 & \sqrt{3} & 0 & 0 \\ 0 & 0 & 2 & 0 \\ 0 & 0 & 0 & \sqrt{3} \\ 0 & 0 & 0 & 0 \end{pmatrix} \tag{2-61}$$

$$\vec{J}^- \to \begin{pmatrix} 0 & 0 \\ 1 & 0 \end{pmatrix} \to \begin{pmatrix} 0 & 0 & 0 \\ \sqrt{2} & 0 & 0 \\ 0 & \sqrt{2} & 0 \end{pmatrix} \to \begin{pmatrix} 0 & 0 & 0 & 0 \\ \sqrt{3} & 0 & 0 & 0 \\ 0 & 2 & 0 & 0 \\ 0 & 0 & \sqrt{3} & 0 \end{pmatrix} \tag{2-62}$$

Each matrix has $2J + 1$ rows and an equal number of columns. The unit matrix has ones along the diagonal and zeros elsewhere. The matrix for J_z is diagonal with elements $J, J - 1, J - 2, \ldots, -J$ from upper left to lower right. The matrix for J^2 equals the unit matrix times $J(J + 1)$ and is easily calculated from the expression

$$\vec{J}^2 = \vec{J}_x^2 + \vec{J}_y^2 + \vec{J}_z^2 \tag{2-63}$$

by matrix multiplication ($\vec{J}_x\vec{J}_x$, etc.) followed by matrix addition. These matrices are rather straightforward to form and should present no difficulty for even higher spins ($J > \frac{3}{2}$).

The x, y, and z component matrices obey the usual commutation law:

$$\vec{J}_i\vec{J}_j - \vec{J}_j\vec{J}_i = [\vec{J}_i, \vec{J}_j] = \vec{J}_k\sqrt{-1} \quad (i, j, k \text{ cyclic}) \tag{2-64}$$

as is easily proved by carrying out the matrix multiplication. For example, when $J = 1$ the commutation law takes the form

$$[\vec{J}_x, \vec{J}_y] = i\vec{J}_z \tag{2-65}$$

$$\frac{1}{4}\left[\begin{pmatrix} 0 & \sqrt{2} & 0 \\ \sqrt{2} & 0 & \sqrt{2} \\ 0 & \sqrt{2} & 0 \end{pmatrix}\begin{pmatrix} 0 & -i\sqrt{2} & 0 \\ i\sqrt{2} & 0 & -i\sqrt{2} \\ 0 & i\sqrt{2} & 0 \end{pmatrix} \right.$$

$$\left. - \begin{pmatrix} 0 & -i\sqrt{2} & 0 \\ i\sqrt{2} & 0 & -i\sqrt{2} \\ 0 & i\sqrt{2} & 0 \end{pmatrix}\begin{pmatrix} 0 & \sqrt{2} & 0 \\ \sqrt{2} & 0 & \sqrt{2} \\ 0 & \sqrt{2} & 0 \end{pmatrix} \right] = i \begin{pmatrix} 1 & 0 & 0 \\ 0 & 0 & 0 \\ 0 & 0 & -1 \end{pmatrix} \tag{2-66}$$

One should note that the matrix for $\vec{J}^2$ automatically has the eigenvalue $J(J + 1)$:

$$\begin{pmatrix} \vec{J}^2 \\ \text{Matrix} \end{pmatrix} = J(J + 1)\begin{pmatrix} \text{Unit} \\ \text{Matrix} \end{pmatrix} \tag{2-67}$$

The Pauli spin matrices, $\vec{\sigma}_i$, are double the spin-$\frac{1}{2}$ matrices:

$$\vec{\sigma}_x = \begin{pmatrix} 0 & 1 \\ 1 & 0 \end{pmatrix}, \qquad \vec{\sigma}_y = \begin{pmatrix} 0 & -i \\ i & 0 \end{pmatrix}, \qquad \vec{\sigma}_z = \begin{pmatrix} 1 & 0 \\ 0 & -1 \end{pmatrix} \qquad (2\text{-}68)$$

and they have some special properties not shared by the higher-order matrices. For example, they anticommute in pairs:

$$\vec{\sigma}_i \vec{\sigma}_j + \vec{\sigma}_j \vec{\sigma}_i = 0 \qquad (2\text{-}69)$$

for any pair $i, j = x, y, z$. In addition,

$$\vec{\sigma}_x^2 = \vec{\sigma}_y^2 = \vec{\sigma}_z^2 = \begin{pmatrix} 1 & 0 \\ 0 & 1 \end{pmatrix} \qquad (2\text{-}70)$$

which means that $\frac{1}{4}\Sigma_i \sigma_i^2$ also satisfies Eq. 2-67, as expected.

The matrices for J_x and J_y and their raising and lowering operator counterparts (2-52) are less familiar for high spins, and the following mnemoic method is useful for their formation. The only nonzero elements occur adjacent to the diagonal and have values obtained from the following triangle:

$$
\begin{array}{ll}
J = \frac{1}{2} & \sqrt{1} \\
J = 1 & \sqrt{2}\;\sqrt{2} \\
J = \frac{3}{2} & \sqrt{3}\;\sqrt{2\cdot2}\;\sqrt{3} \\
J = 2 & \sqrt{4}\;\sqrt{3\cdot2}\;\sqrt{2\cdot3}\;\sqrt{4} \\
J = \frac{5}{2} & \sqrt{5}\;\sqrt{4\cdot2}\;\sqrt{3\cdot3}\;\sqrt{2\cdot4}\;\sqrt{5} \\
J = 3 & \sqrt{6}\;\sqrt{5\cdot2}\;\sqrt{4\cdot3}\;\sqrt{3\cdot5}\;\sqrt{2\cdot5}\;\sqrt{6} \\
J = \frac{7}{2} & \sqrt{7}\;\sqrt{6\cdot2}\;\sqrt{5\cdot3}\;\sqrt{4\cdot4}\;\sqrt{3\cdot5}\;\sqrt{2\cdot6}\;\sqrt{7}
\end{array}
\qquad (2\text{-}71)
$$

The row for a general spin J is

$$\sqrt{(2J)\cdot(1)}\;\sqrt{(2J-1)\cdot(2)}\;\sqrt{(2J-2)\cdot(3)}$$
$$\sqrt{(2J-3)\cdot(4)}\cdots\sqrt{(1)\cdot(2J)} \qquad (2\text{-}72)$$

One should note that triangle 2-71 is symmetrical about its vertical axis. A closer inspection reveals several obvious methods for generating successive rows for higher spins.

The matrix for J_x is real and symmetric and contains, adjacent to its diagonal, 1/2 times the numbers in the corresponding row of triangle 2-71. For example, the $J = \frac{5}{2}$ matrix has the sequence $\sqrt{5}/2$, $\sqrt{2}$, $\frac{3}{2}$, $\sqrt{2}$, $\sqrt{5}/2$ on either side of its diagonal. The matrix for J_y is purely imaginary and is obtained from J_x by multiplying each element to the left below the diagonal by $i = \sqrt{-1}$, and each element to the right above the diagonal by $-i$. The raising operator, J^+, contains sequence 2-72 only immediately to the right above the diagonal, while the lowering operator, J^-, contains the same numerical sequence only immediately below and to the left of the diagonal. An inspection of the examples in Eqs. 2-57 to 2-62 will clarify these rules.

2-7 ADDITION OF ANGULAR MOMENTUM

The total angular momentum in a system is the vector sum of all other angular momenta:

$$\vec{J} = \Sigma \vec{J}_i \qquad (2\text{-}73)$$

In this section we treat the case of two angular momenta,

$$\vec{J} = \vec{J}_1 + \vec{J}_2 \qquad (2\text{-}74)$$

for which the following values of J are allowed:

$$|J_1 - J_2| \le J \le J_1 + J_2 \qquad (2\text{-}75)$$

Since the situation is discussed in standard books on quantum mechanics, only the application of the direct product method to this problem is presented here.

There are two important representations for $\vec{J}$, the $\vec{J}'$ representation, which uses the basis functions $|m_1 m_2\rangle'$, and the J'' representation, which uses the basis functions $|Jm\rangle''$. The Hamiltonian which we treat in the text has principal eigenvalue representations $| \rangle'''$, in which the Hamiltonian $\mathcal{H}'''$ is diagonal. The quantum-mechanical problem ordinarily consists of writing the Hamiltonian in the primed $\mathcal{H}'$ representation and transforming it to its diagonal form.

It is also possible to start with the H'' or zero field representation. This section illustrates both the primed and the double-primed representation for angular momentum matrices. All operators have dimensionality

$$(2J_1 + 1)(2J_2 + 1) = \Sigma \, (2J + 1) \qquad (2\text{-}76)$$

where the summation is over the range of J values given in Eq. 2-75. The specific example of $J_1 = \frac{1}{2}$, $J_2 = \frac{1}{2}$ with dimensionality 4 is discussed in detail.

One begins by taking the direct product sum:

$$\vec{J}' = \vec{J}'_x \hat{i} + \vec{J}'_y \hat{j} + \vec{J}'_z \hat{k}$$

$$= \vec{J}_1 \times \vec{1}_2 + \vec{1}_1 \times \vec{J}_2 \qquad (2\text{-}77)$$

$$= (\vec{J}_{1x} \times \vec{1}_2 + \vec{1}_1 \times \vec{J}_{2x})\hat{i} + (\vec{J}_{1y} \times \vec{1}_2 + \vec{1}_1 \times \vec{J}_{2y})\hat{j}$$

$$+ (\vec{J}_{1z} \times \vec{1}_2 + \vec{1}_1 \times \vec{J}_{2z})\hat{k}$$

where $\vec{1}_1$ and $\vec{1}_2$ are unit matrices in the J_1 and J_2 spaces. The components J'_i have the following explicit forms for $J_1 = J_2 = \frac{1}{2}$:

$$\vec{J}'_x = \frac{1}{2}\left[\begin{pmatrix} 0 & 1 \\ 1 & 0 \end{pmatrix} \times \begin{pmatrix} 1 & 0 \\ 0 & 1 \end{pmatrix} + \begin{pmatrix} 1 & 0 \\ 0 & 1 \end{pmatrix} \times \begin{pmatrix} 0 & 1 \\ 1 & 0 \end{pmatrix} \right] \qquad (2\text{-}78)$$

$$= \frac{1}{2} \begin{pmatrix} 0 & 1 & 1 & 0 \\ 1 & 0 & 0 & 1 \\ 1 & 0 & 0 & 1 \\ 0 & 1 & 1 & 0 \end{pmatrix}' \tag{2-79}$$

$$\vec{J}'_j = \frac{1}{2} \begin{pmatrix} 0 & -i & -i & 0 \\ i & 0 & 0 & -i \\ i & 0 & 0 & -i \\ 0 & i & i & 0 \end{pmatrix}' \tag{2-80}$$

$$\vec{J}'_z = \begin{pmatrix} 1 & 0 & 0 & 0 \\ 0 & 0 & 0 & 0 \\ 0 & 0 & 0 & 0 \\ 0 & 0 & 0 & -1 \end{pmatrix}' \tag{2-81}$$

The ket vectors, $|m_1 m_2\rangle'$, corresponding to this primed representation are expressed in terms of the $|m_i\rangle$ kets:

$$|\tfrac{1}{2}\rangle = \begin{pmatrix} 1 \\ 0 \end{pmatrix}, \qquad |-\tfrac{1}{2}\rangle = \begin{pmatrix} 0 \\ 1 \end{pmatrix} \tag{2-82}$$

by the direct product expansion:

$$|\tfrac{1}{2} \tfrac{1}{2}\rangle' = \begin{pmatrix} 1 \\ 0 \end{pmatrix} \times \begin{pmatrix} 1 \\ 0 \end{pmatrix} = \begin{pmatrix} 1 \\ 0 \\ 0 \\ 0 \end{pmatrix}'$$

$$|\tfrac{1}{2} -\tfrac{1}{2}\rangle' = \begin{pmatrix} 1 \\ 0 \end{pmatrix} \times \begin{pmatrix} 0 \\ 1 \end{pmatrix} = \begin{pmatrix} 0 \\ 1 \\ 0 \\ 0 \end{pmatrix}'$$

$$\tag{2-83}$$

$$|-\tfrac{1}{2} \tfrac{1}{2}\rangle' = \begin{pmatrix} 0 \\ 1 \end{pmatrix} \times \begin{pmatrix} 1 \\ 0 \end{pmatrix} = \begin{pmatrix} 0 \\ 0 \\ 1 \\ 0 \end{pmatrix}'$$

$$|-\tfrac{1}{2} -\tfrac{1}{2}\rangle' = \begin{pmatrix} 0 \\ 1 \end{pmatrix} \times \begin{pmatrix} 0 \\ 1 \end{pmatrix} = \begin{pmatrix} 0 \\ 0 \\ 0 \\ 1 \end{pmatrix}'$$

Thus we have all the eigenfunctions and angular momentum operators in the $|m_1 m_2\rangle'$ or primed representation. The solutions of most of the eigenvalue problems occurring throughout the text are arrived at by writing the Hamiltonian H' in this representation.

The second basic representation might be referred to as the total J, zero field, or double-primed representation with the operators J'', ket vectors $|Jm\rangle''$, and Hamiltonian H''. This representation separates the angular momentum operators into individual operators for each magnitude of angular momentum. For the present example the 4×4 matrix for each J_i is decomposed into a 3×3 one for $J = 1$ and a 1×1 one for $J = 0$:

$$\begin{pmatrix} 4 \times 4 \\ \text{Matrix} \end{pmatrix}' \rightarrow \begin{pmatrix} \begin{array}{c|ccc} 1 \times 1 & 0 & 0 & 0 \\ \hline 0 & & & \\ 0 & & 3 \times 3 & \\ 0 & & & \end{array} \end{pmatrix}'' \tag{2-84}$$

with an analogous change for the ket vector:

$$(1 \times 4)' \rightarrow \begin{pmatrix} 1 \times 1 \\ 1 \times 3 \end{pmatrix}'' \tag{2-85}$$

The generalization to higher spins is obvious.

The transformation between the two representations is carried out by means of the Clebsch–Gordan coefficient[2-6] matrix, $\vec{\vec{C}}$, and its inverse (transpose), $\vec{\vec{C}}^{-1}$:

$$\vec{\vec{C}}^{-1}(\quad)' = (\quad)'' \tag{2-86}$$

$$\vec{\vec{C}}(\quad)'' = (\quad)' \tag{2-87}$$

$$\vec{\vec{C}}\vec{\vec{J}}''\vec{\vec{C}}^{-1} = \vec{\vec{J}}' \tag{2-88}$$

$$\vec{\vec{C}}^{-1}\vec{\vec{J}}'\vec{\vec{C}} = \vec{\vec{J}}'' \tag{2-89}$$

where $\vec{\vec{C}}$ is given by

$$\vec{\vec{C}} = (\langle \tfrac{1}{2}\tfrac{1}{2}m_1 m_2 | Jm\rangle) = \begin{array}{c} \\ \langle \tfrac{1}{2} \ \tfrac{1}{2}| \\ \\ \langle \tfrac{1}{2} \ -\tfrac{1}{2}| \\ \\ \langle -\tfrac{1}{2} \ \tfrac{1}{2}| \\ \\ \langle -\tfrac{1}{2} \ -\tfrac{1}{2}| \end{array} \begin{array}{cccc} |0\,0\rangle & |1\,1\rangle & |1\,0\rangle & |1-1\rangle \\ \begin{pmatrix} 0 & 1 & 0 & 0 \\ \dfrac{1}{\sqrt{2}} & 0 & \dfrac{1}{\sqrt{2}} & 0 \\ -\dfrac{1}{\sqrt{2}} & 0 & \dfrac{1}{\sqrt{2}} & 0 \\ 0 & 0 & 0 & 1 \end{pmatrix} \end{array} \tag{2.90}$$

for the case at hand. This matrix is real and unitary by definition. The matrix $\vec{\vec{C}}$ provides the algebraic transformation

$$|Jm\rangle = \sum_{m_1,m_2} \langle J_1 J_2 m_1 m_2 | Jm \rangle |m_2 m_2\rangle \tag{2-91a}$$

and its inverse

$$|m_1 m_2\rangle = \sum_{Jm} \langle Jm | J_1 J_2 m_1 m_2 \rangle |Jm\rangle \tag{2-91b}$$

where $\langle Jm | J_1 J_2 m_1 m_2 \rangle$ is the transverse Clebsch–Gordan matrix element corresponding to $\vec{C}^{-1}$. As an example of the transformation consider the case

$$\vec{C}^{-1} \vec{J}_y' \vec{C} = \vec{J}_y'' \tag{2-92}$$

which has the explicit form

$$\frac{1}{2}\begin{pmatrix} 0 & \frac{1}{\sqrt{2}} & -\frac{1}{\sqrt{2}} & 0 \\ 1 & 0 & 0 & 0 \\ 0 & \frac{1}{\sqrt{2}} & \frac{1}{\sqrt{2}} & 0 \\ 0 & 0 & 0 & 1 \end{pmatrix} \begin{pmatrix} 0 & -i & -i & 0 \\ i & 0 & 0 & -i \\ i & 0 & 0 & -i \\ 0 & i & i & 0 \end{pmatrix}' \begin{pmatrix} 0 & 1 & 0 & 0 \\ \frac{1}{\sqrt{2}} & 0 & \frac{1}{\sqrt{2}} & 0 \\ -\frac{1}{\sqrt{2}} & 0 & \frac{1}{\sqrt{2}} & 0 \\ 0 & 0 & 0 & 1 \end{pmatrix}$$

$$= \begin{pmatrix} 0 & 0 & 0 & 0 \\ 0 & 0 & -\frac{i}{\sqrt{2}} & 0 \\ 0 & \frac{i}{\sqrt{2}} & 0 & -\frac{i}{\sqrt{2}} \\ 0 & 0 & \frac{i}{\sqrt{2}} & 0 \end{pmatrix}'' \tag{2-93}$$

$$(\vec{C}^{-1})(\vec{J}_{1y} \times \vec{J}_{2y})'(\vec{C}) = \begin{pmatrix} J_y & & & \\ J=0 & 0 & 0 & 0 \\ \hline 0 & & J_y & \\ 0 & & J=1 & \\ 0 & & & \end{pmatrix}'' \tag{2-94}$$

The ket vector $|\frac{1}{2} - \frac{1}{2}\rangle$ tranforms as follows:

$$
\begin{pmatrix}
0 & \frac{1}{\sqrt{2}} & -\frac{1}{\sqrt{2}} & 0 \\
1 & 0 & 0 & 0 \\
0 & \frac{1}{\sqrt{2}} & \frac{1}{\sqrt{2}} & 0 \\
0 & 0 & 0 & 1
\end{pmatrix}'
\begin{pmatrix}
0 \\
1 \\
0 \\
0
\end{pmatrix}'
=
\begin{pmatrix}
\frac{1}{\sqrt{2}} \\
0 \\
\frac{1}{\sqrt{2}} \\
0
\end{pmatrix}''
\tag{2-95}
$$

The $\vec{J}''_x$ and $\vec{J}''_z$ matrices for $J_1 = J_2 = \frac{1}{2}$ are explicitly

$$
\vec{J}''_x =
\begin{pmatrix}
0 & 0 & 0 & 0 \\
0 & 0 & \frac{1}{\sqrt{2}} & 0 \\
0 & \frac{1}{\sqrt{2}} & 0 & \frac{1}{\sqrt{2}} \\
0 & 0 & \frac{1}{\sqrt{2}} & 0
\end{pmatrix}''
\tag{2-96}
$$

$$
\vec{J}''_z =
\begin{pmatrix}
0 & 0 & 0 & 0 \\
0 & 1 & 0 & 0 \\
0 & 0 & 0 & 0 \\
0 & 0 & 0 & -1
\end{pmatrix}''
\tag{2-97}
$$

Equation 2-90 gives the Clebsch–Gordan coefficient transformation matrix $\vec{C}$ for the case $J_1 = J_2 = \frac{1}{2}$. The general form of this matrix for arbitrary spins may be expressed symbolically by

$$
\vec{C}_{(J_1 J_2)} = (\langle J_1 J_2 m_1 m_2 | J m \rangle)
\tag{2-98}
$$

and for several specific spin cases $\vec{C}$ is given by the following matrices:

$$
\vec{C}_{(1/2\ 1)} =
\begin{array}{c}
\\
\langle m_1 m_2| \\[4pt]
\langle \tfrac{1}{2}\ 1| \\[6pt]
\langle \tfrac{1}{2}\ 0| \\[6pt]
\langle \tfrac{1}{2}\ -1| \\[6pt]
\langle -\tfrac{1}{2}\ 1| \\[6pt]
\langle -\tfrac{1}{2}\ 0| \\[6pt]
\langle -\tfrac{1}{2}\ -1|
\end{array}
\begin{array}{cccccc}
|Jm\rangle\ \ |\tfrac{1}{2}\tfrac{1}{2}\rangle & |\tfrac{1}{2}-\tfrac{1}{2}\rangle & |\tfrac{3}{2}\tfrac{3}{2}\rangle & |\tfrac{3}{2}\tfrac{1}{2}\rangle & |\tfrac{3}{2}-\tfrac{1}{2}\rangle & |\tfrac{3}{2}-\tfrac{3}{2}\rangle \\[4pt]
\left(\begin{array}{cccccc}
0 & 0 & 1 & 0 & 0 & 0 \\[6pt]
-\dfrac{1}{\sqrt{3}} & 0 & 0 & \sqrt{\tfrac{2}{3}} & 0 & 0 \\[6pt]
0 & -\sqrt{\tfrac{2}{3}} & 0 & 0 & \dfrac{1}{\sqrt{3}} & 0 \\[6pt]
\sqrt{\tfrac{2}{3}} & 0 & 0 & \dfrac{1}{\sqrt{3}} & 0 & 0 \\[6pt]
0 & \dfrac{1}{\sqrt{3}} & 0 & 0 & \sqrt{\tfrac{2}{3}} & 0 \\[6pt]
0 & 0 & 0 & 0 & 0 & 1
\end{array}\right)
\end{array}
\tag{2-99}
$$

$$
\vec{C}_{(1\ 1/2)} =
\begin{array}{c}
\\
\langle m_1 m_2| \\[4pt]
\langle 1\ \tfrac{1}{2}| \\[6pt]
\langle 1\ -\tfrac{1}{2}| \\[6pt]
\langle 0\ \tfrac{1}{2}| \\[6pt]
\langle 0\ -\tfrac{1}{2}| \\[6pt]
\langle -1\ \tfrac{1}{2}| \\[6pt]
\langle -1\ -\tfrac{1}{2}|
\end{array}
\begin{array}{cccccc}
|Jm\rangle\ \ |\tfrac{1}{2}\tfrac{1}{2}\rangle & |\tfrac{1}{2}-\tfrac{1}{2}\rangle & |\tfrac{3}{2}\tfrac{3}{2}\rangle & |\tfrac{3}{2}\tfrac{1}{2}\rangle & |\tfrac{3}{2}-\tfrac{1}{2}\rangle & |\tfrac{3}{2}-\tfrac{3}{2}\rangle \\[4pt]
\left(\begin{array}{cccccc}
0 & 0 & 1 & 0 & 0 & 0 \\[6pt]
\sqrt{\tfrac{2}{3}} & 0 & 0 & \dfrac{1}{\sqrt{3}} & 0 & 0 \\[6pt]
-\dfrac{1}{\sqrt{3}} & 0 & 0 & \sqrt{\tfrac{2}{3}} & 0 & 0 \\[6pt]
0 & \dfrac{1}{\sqrt{3}} & 0 & 0 & \sqrt{\tfrac{2}{3}} & 0 \\[6pt]
0 & -\sqrt{\tfrac{2}{3}} & 0 & 0 & \dfrac{1}{\sqrt{3}} & 0 \\[6pt]
0 & 0 & 0 & 0 & 0 & 1
\end{array}\right)
\end{array}
\tag{2-100}
$$

$$
\vec{C}_{(1/2\ 3/2)} =
\begin{array}{c}
\\
\langle m_1 m_2| \\[4pt]
\langle \tfrac{1}{2}\ \tfrac{3}{2}| \\[6pt]
\langle \tfrac{1}{2}\ \tfrac{1}{2}| \\[6pt]
\langle \tfrac{1}{2}\ -\tfrac{1}{2}| \\[6pt]
\langle \tfrac{1}{2}\ -\tfrac{3}{2}| \\[6pt]
\langle -\tfrac{1}{2}\ \tfrac{3}{2}| \\[6pt]
\langle -\tfrac{1}{2}\ \tfrac{1}{2}| \\[6pt]
\langle -\tfrac{1}{2}\ -\tfrac{1}{2}| \\[6pt]
\langle -\tfrac{1}{2}\ -\tfrac{3}{2}|
\end{array}
\begin{array}{cccccccc}
|Jm\rangle & & & & & & & \\
|1\,1\rangle & |1\,0\rangle & |1-1\rangle & |2\,2\rangle & |2\,1\rangle & |2\,0\rangle & |2-1\rangle & |2-2\rangle \\[4pt]
\left(\begin{array}{cccccccc}
0 & 0 & 0 & 1 & 0 & 0 & 0 & 0 \\[6pt]
\tfrac{1}{2} & 0 & 0 & 0 & \tfrac{1}{2}\sqrt{3} & 0 & 0 & 0 \\[6pt]
0 & \dfrac{1}{\sqrt{2}} & 0 & 0 & 0 & \dfrac{1}{\sqrt{2}} & 0 & 0 \\[6pt]
0 & 0 & \tfrac{1}{2}\sqrt{3} & 0 & 0 & 0 & \tfrac{1}{2} & 0 \\[6pt]
-\tfrac{1}{2}\sqrt{3} & 0 & 0 & 0 & \tfrac{1}{2} & 0 & 0 & 0 \\[6pt]
0 & -\dfrac{1}{\sqrt{2}} & 0 & 0 & 0 & \dfrac{1}{\sqrt{2}} & 0 & 0 \\[6pt]
0 & 0 & -\tfrac{1}{2} & 0 & 0 & 0 & \tfrac{1}{2}\sqrt{3} & 0 \\[6pt]
0 & 0 & 0 & 0 & 0 & 0 & 0 & 1
\end{array}\right)
\end{array}
\tag{2-101}
$$

$\vec{\vec{C}}_{(1\,1)} =$

$\langle m_1 m_2\|$ \\ $\|Jm\rangle$	$\|0\,0\rangle$	$\|1\,1\rangle$	$\|1\,0\rangle$	$\|1\,-1\rangle$	$\|2\,2\rangle$	$\|2\,1\rangle$	$\|2\,0\rangle$	$\|2\,-1\rangle$	$\|2\,-2\rangle$
$\langle 1\,1\|$	0	0	0	0	1	0	0	0	0
$\langle 1\,0\|$	0	$\frac{1}{\sqrt{2}}$	0	0	0	$\frac{1}{\sqrt{2}}$	0	0	0
$\langle 1\,-1\|$	$\frac{1}{\sqrt{3}}$	0	$\frac{1}{\sqrt{2}}$	0	0	0	$\frac{1}{\sqrt{6}}$	0	0
$\langle 0\,1\|$	0	$-\frac{1}{\sqrt{2}}$	0	0	0	$\frac{1}{\sqrt{2}}$	0	0	0
$\langle 0\,0\|$	$\frac{1}{\sqrt{3}}$	0	0	0	0	0	$\sqrt{\frac{2}{3}}$	0	0
$\langle 0\,-1\|$	0	0	0	$\frac{1}{\sqrt{2}}$	0	0	0	$\frac{1}{\sqrt{2}}$	0
$\langle -1\,1\|$	$\frac{1}{\sqrt{3}}$	0	$-\frac{1}{\sqrt{2}}$	0	0	0	$\frac{1}{\sqrt{6}}$	0	0
$\langle -1\,0\|$	0	0	0	$-\frac{1}{\sqrt{2}}$	0	0	0	$\frac{1}{\sqrt{2}}$	0
$\langle -1\,-1\|$	0	0	0	0	0	0	0	0	1

$$(2\text{-}102)$$

These matrices are real and unitary, but not symmetric. The $\vec{\vec{C}}_{(3/2\,1/2)}$ matrix is easily obtained by the appropriate rearrangement of $\vec{\vec{C}}_{(1/2\,3/2)}$. Various authors have tabulated the Clebsch–Gordon coefficients that are needed to generate higher spin cases.[2]

2-8 PERTURBATION THEORY

In magnetic resonance studies some terms in the Hamiltonian $\mathscr{H}$ are frequently rather small in comparison to others. The leading or dominant terms are denoted by H_0, and the small perturbations are designated as $\mathscr{H}'$, to give

$$\mathscr{H} = \mathscr{H}_0 + \mathscr{H}' \tag{2-103}$$

It is assumed that the energies E_{0i} and eigenkets $|i\rangle = \psi_i$ for $\mathscr{H}_0$ alone are already known. The shifts and splittings of the energy levels and the corrections to the wavefunctions arising from the presence of $\mathscr{H}'$ are desired; these may be easily approximated from perturbation theory.

The general expression for the energy to second order is

$$E_i = E_{0i} + \langle i|\mathscr{H}'|i\rangle + \sum_{j \neq i} \frac{\langle i|\mathscr{H}'|j\rangle\langle j|\mathscr{H}'|i\rangle}{E_{0i} - E_{0j}} \tag{2-104}$$

where E_{0i} is the zero-order energy of the ith quantum state,

$$E_{0i} = \langle i | \mathcal{H}_0 | i \rangle \tag{2-105}$$

One should note that the first-order term is a diagonal matrix element, and the second-order one contains a sum over off-diagonal matrix elements. The first-order wavefunction ψ_i is given by

$$\psi_i = \psi_{0i} + \sum_{j \neq i} \frac{\langle i | \mathcal{H}' | j \rangle}{E_{0i} - E_{0j}} \psi_{0j} \tag{2-106}$$

It is possible to write expressions for higher-order terms in the energy and wavefunctions, but these are not ordinarily employed in practice. Standard books on quantum mechanics[2-5] may be consulted for further details.

REFERENCES

1. E. U. Condon and G. H. Shortley, *The Theory of Atomic Spectra*, Cambridge University Press, Cambridge, 1953.
2. M. Rotenberg, R. Bivins, N. Metropolis, and J. K. Wooten, *The 3-j and 6-j Symbols*, Technology Press, Cambridge, MA, 1959.
3. A. Messiah, *Quantum Mechanics*, Wiley, New York, 1962, Vol. II, Appendix C.
4. M. E. Rose, *Elementary Theory of Angular Momentum*, Wiley, New York, 1957.
5. L. I. Schiff, *Quantum Mechanics*, McGraw-Hill, New York, 1949, p. 144.
6. M. Tinkham, *Group Theory and Quantum Mechanics*, McGraw-Hill, New York, 1964.

GENERAL TWO-SPIN $(\frac{1}{2}, \frac{1}{2})$ SYSTEM

3-1 PRELIMINARY REMARKS

By far the most important spin system in magnetic resonance is the two-spin $(\frac{1}{2}, \frac{1}{2})$ pair, the members of which interact isotropically with each other and with an external magnetic field. This is a prototype for high-resolution NMR since it takes into account both the chemical shift and spin–spin interactions, and most high-resolution NMR research deals exclusively with spin-$\frac{1}{2}$ nuclei. This spin system is also a prototype for ESR, in which one ordinarily measures an unpaired spin $S = \frac{1}{2}$, and the most common hyperfine interaction is with a spin-$\frac{1}{2}$ proton. The NMR and ESR cases represent widely differing values for the Hamiltonian parameters, and the specialization of the general formalism to these cases emphasizes both the basic unity of magnetic resonance and the versatility of the formalism presented here.

3-2 THE HAMILTONIAN

The interaction of two spins $\vec{I}_1$ and $\vec{I}_2$, with an external magnetic field and with each other may be described by the spin Hamiltonian with the Zeeman ($g\beta H I_z$) and spin–spin ($T\vec{I}_1 \cdot \vec{I}_2$) terms:

$$\mathcal{H} = \beta H(g_1 I_{1z} + g_2 I_{2z}) + T\vec{I}_1 \cdot \vec{I}_2 \quad \text{General} \qquad (3\text{-}1a)$$

$$= H(g_e \beta S_z - g_N \beta_N I_z) + T\vec{S} \cdot \vec{I} \quad \text{(ESR)} \qquad (3\text{-}1b)$$

$$= \hbar H(\gamma_1 I_{1z} + \gamma_2 I_{2z}) + J\vec{I}_1 \cdot \vec{I}_2 \quad \text{(NMR)} \qquad (3\text{-}1c)$$

The first equation is the one employed in this text with the assumptions $T > 0$, $g_1 > g_2 > 0$. The results for negative g-factors and spin–spin coupling constants are easily deduced by the appropriate sign changes in the final answers. We express all g-factors relative to the Bohr magneton, β, which exceeds the nuclear magneton, β_N, by the ratio of the proton mass, M, to the nuclear mass, m:

$$\beta = \frac{e\hbar}{2mc}, \qquad \beta_N = \frac{e\hbar}{2Mc}, \qquad \frac{\beta}{\beta_N} = \frac{M}{m} = 1836 \qquad (3\text{-}2)$$

It is more conventional to treat nuclear g-factors in terms of the nuclear magneton, as indicated in the specialized ESR Hamilton of Eq. 3-1b corresponding to an electronic spin S and a nuclear spin I with a hyperfine coupling constant T. In NMR it is customary to express the Zeeman interactions in terms of the gyromagnetic ratio γ:

$$\gamma\hbar = g\beta = g_N\beta_N \qquad (3\text{-}3)$$

in accordance with Eq. 3-1c, and to employ the symbol J for the spin–spin coupling constant. Our general development employs the notation of Eq. 3-1a, and later we specialize to the ESR and NMR cases.

3-3 THE SECULAR EQUATION AND ENERGY LEVELS

The spin Hamiltonian has a dimensionality $(2I_1 + 1)(2I_2 + 1) = 4$ for the case $I_1 = I_2 = \frac{1}{2}$, and there are a corresponding number of basis functions or ket vectors, $|I_1 I_2 m_1 m_2\rangle$, which will be written in the shortened notation $|m_1 m_2\rangle$, where $m_1 = \pm\frac{1}{2}$, $m_2 = \pm\frac{1}{2}$. The eigenfunctions, ψ_j, are linear combinations of these basis functions.

Each operator in the spin Hamiltonian is a 2×2 matrix, and the unit 2×2 matrix is inserted where needed, as explained in Section 2-7. As a result Eq. 3-1a becomes

$$\mathcal{H} = \frac{\beta H}{2}\left[g_1\begin{pmatrix} 1 & 0 \\ 0 & -1 \end{pmatrix} \times \begin{pmatrix} 1 & 0 \\ 0 & 1 \end{pmatrix} + g_2\begin{pmatrix} 1 & 0 \\ 0 & 1 \end{pmatrix} \times \begin{pmatrix} 1 & 0 \\ 0 & -1 \end{pmatrix} \right]$$

$$+ \frac{T}{4}\left[\begin{pmatrix} 0 & 1 \\ 1 & 0 \end{pmatrix} \times \begin{pmatrix} 0 & 1 \\ 1 & 0 \end{pmatrix} + \begin{pmatrix} 0 & -i \\ i & 0 \end{pmatrix} \times \begin{pmatrix} 0 & -i \\ i & 0 \end{pmatrix} + \begin{pmatrix} 1 & 0 \\ 0 & -1 \end{pmatrix} \times \begin{pmatrix} 1 & 0 \\ 0 & -1 \end{pmatrix} \right]$$

$$(3\text{-}4)$$

The direct product expansion (Eq. 2-44) and the matrix addition operation (Eq. 2-42) provide the following 4×4 Hamiltonian matrix:

	$\lvert++\rangle$	$\lvert+-\rangle$	$\lvert-+\rangle$	$\lvert--\rangle$
$\langle++\rvert$	$(g_1+g_2)\dfrac{\beta H}{2}+\dfrac{T}{4}$	0	0	0
$\langle+-\rvert$	0	$(g_1-g_2)\dfrac{\beta H}{2}-\dfrac{T}{4}$	$\dfrac{T}{2}$	0
$\langle-+\rvert$	0	$\dfrac{T}{2}$	$-(g_1-g_2)\dfrac{\beta H}{2}-\dfrac{T}{4}$	0
$\langle--\rvert$	0	0	0	$-(g_1+g_2)\dfrac{\beta H}{2}+\dfrac{T}{4}$

$$(3\text{-}5)$$

This is partially diagonal, since it is in the form

$$
\begin{array}{|c|cc|c|}
\hline
1\times1 & 0 & 0 & 0 \\
\hline
0 & & & 0 \\
0 & \multicolumn{2}{c}{2\times2} & 0 \\
\hline
0 & 0 & 0 & 1\times1 \\
\hline
\end{array}
$$

$$(3\text{-}6)$$

and the eigenvalues in the upper left and lower right corners are

$$E_1 = \tfrac{1}{2}(g_1+g_2)\beta H + \tfrac{1}{4}T$$
$$E_4 = -\tfrac{1}{2}(g_1+g_2)\beta H + \tfrac{1}{4}T$$

$$(3\text{-}7)$$

The remaining two eigenvalues result from the quadratic equation

$$
\begin{vmatrix}
(g_1-g_2)\dfrac{\beta H}{2}-\dfrac{1}{4}T-E & \tfrac{1}{2}T \\[3mm]
\tfrac{1}{2}T & -(g_1-g_2)\dfrac{\beta H}{2}-\dfrac{1}{4}T-E
\end{vmatrix} = 0
$$

$$(3\text{-}8)$$

with the solutions E_2 and E_3, which have the explicit forms

$$E_2 = -\tfrac{1}{4}T + \tfrac{1}{2}[T^2 + (g_1-g_2)^2\beta^2 H^2]^{1/2}$$
$$E_3 = -\tfrac{1}{4}T - \tfrac{1}{2}[T^2 + (g_1-g_2)^2\beta^2 H^2]^{1/2}$$

$$(3\text{-}9)$$

These are valid for all values of g_1, g_2, and T.

Experimental work is often carried out under conditions in which one of the two terms under the square root is much less than the other. Accordingly, it is appropriate to write down the expansion formulas for these conditions. When $T^2 \gg (g_1-g_2)^2\beta^2 H^2$, one has

$$E_2 = \frac{1}{4}T + \frac{(g_1 - g_2)^2\beta^2H^2}{4T} - \cdots$$

$$E_3 = -\frac{3}{4}T - \frac{(g_1 - g_2)^2\beta^2H^2}{4T} + \cdots \tag{3-10}$$

and in the opposite limit, $(g_1 - g_2)^2\beta^2H^2 \gg T^2$, the expansion gives

$$E_2 = \frac{1}{2}(g_1 - g_2)\beta H - \frac{1}{4}T + \frac{T^2}{4(g_1 - g_2)\beta H}$$

$$E_3 = -\frac{1}{2}(g_1 - g_2)\beta H - \frac{1}{4}T - \frac{T^2}{4(g_1 - g_2)\beta H} \tag{3-11}$$

The appendix of Chapter 3 of the first edition gives alternate forms for Eqs. 3-9 to 3-11.

3-4 EIGENFUNCTIONS

The two energy levels, E_1 and E_4, are exact when calculated with the basis set $|m_1m_2\rangle$, and hence their two respective eigenfunctions or eigenkets, $|+ +\rangle$ and $|- -\rangle$, are exact. The remaining two eigenkets, ψ_2 and ψ_3, for the energies E_2 and E_3 have the forms

$$\psi_2 = \alpha|+ -\rangle + \gamma|- +\rangle$$

$$\psi_3 = -\gamma^*|+ -\rangle + \alpha^*|- +\rangle \tag{3-12}$$

with the normalization condition

$$\alpha\alpha^* + \gamma\gamma^* = 1 \tag{3-13}$$

The coefficients α and γ may be evaluated from the expression

$$\begin{pmatrix} (g_1 - g_2)\dfrac{\beta H}{2} - \dfrac{T}{4} & \dfrac{T}{2} \\ \dfrac{T}{2} & -(g_1 - g_2)\dfrac{\beta H}{2} - \dfrac{T}{4} \end{pmatrix} \begin{pmatrix} \alpha \\ \gamma \end{pmatrix} = E_2 \begin{pmatrix} \alpha \\ \gamma \end{pmatrix} \tag{3-14}$$

A similar equation may be written for the column vector $\begin{pmatrix} -\gamma^* \\ \alpha^* \end{pmatrix}$ and E_3, but it is redundant.

Matrix multiplication of Eq. 3-14 gives the two expressions

$$\left[(g_1 - g_2)\frac{\beta H}{2} - \frac{T}{4} - E_2\right]\alpha + \frac{T}{2}\gamma = 0$$

$$\frac{T}{2}\alpha + \left[-(g_1 - g_2)\frac{\beta H}{2} - \frac{T}{4} - E_2\right]\gamma = 0 \tag{3-15}$$

Either one may be employed to determine α and γ by use of normalization condition 3-13 with the result

$$\alpha = \left\{1 + \left[\frac{(g_1 - g_2)\beta H}{T} - \sqrt{1 + \frac{(g_1 - g_2)^2\beta^2 H^2}{T^2}}\right]^2\right\}^{-1/2}$$

$$\gamma = \left\{1 + \left[\frac{(g_1 - g_2)\beta H}{T} - \sqrt{1 + \frac{(g_1 - g_2)^2\beta^2 H^2}{T^2}}\right]^{-2}\right\}^{-1/2}$$

(3-16)

The appendix of Chapter 3 of the first edition gives an alternate form of these expressions and Section 4-3 of this edition explains the conventions employed in NMR. Figure 5-6 shows how α and γ depend on the ratio $g\beta H/T$.

The four eigenvalues ψ_i have the forms

$$\psi_1 = |+ +\rangle$$
$$\psi_2 = \alpha|+ -\rangle + \gamma|- +\rangle$$
$$\psi_3 = -\gamma|+ -\rangle + \alpha|- +\rangle$$
$$\psi_4 = |- -\rangle$$

(3-17)

with the real coefficients α and γ given by Eqs. 3-16. Thus the original basis functions $|m_1 m_2\rangle$ used to form secular matrix 3-5 approach true eigenfunctions in the high-field limit, $(g_1 - g_2)\beta H \gg T$. The ket vectors $|+ +\rangle$ and $|- -\rangle$ are always exact eigenfunctions.

At low fields, $(g_1 - g_2)\beta H \ll T$, the eigenfunctions ψ_2 and ψ_3 approach the limit corresponding to $\alpha = \gamma = \sqrt{\frac{1}{2}}$. For $H = 0$ one obtains the singlet- and triplet-state eigenfunctions, which may be labeled by the F, M_F quantum numbers; these assume the values $F = 0$, 1 and $-F \le M_F \le F$. These functions with their corresponding $|FM_F\rangle$ ket vectors are as follows:

$$\psi_1 = |1\ 1\rangle = |\tfrac{1}{2}\ \tfrac{1}{2}\rangle \qquad\qquad M_F = 1$$

$$\psi_2 = |1\ 0\rangle = \frac{1}{\sqrt{2}}[|\tfrac{1}{2} - \tfrac{1}{2}\rangle + |- \tfrac{1}{2}\ \tfrac{1}{2}\rangle] \qquad M_F = 0$$

$$\psi_4 = |1\ -1\rangle = |- \tfrac{1}{2} - \tfrac{1}{2}\rangle \qquad\qquad M_F = -1$$

Triplet state $F = 1$ (3-18a)

$$\psi_3 = |0\ 0\rangle = \frac{1}{\sqrt{2}}[-|\tfrac{1}{2} - \tfrac{1}{2}\rangle + |- \tfrac{1}{2}\ \tfrac{1}{2}\rangle] \quad M_F = 0$$

Singlet state $F = 0$ (3-18b)

These two states correspond to the total angular momentum, $\vec{F} = \vec{I}_1 + \vec{I}_2$, where the quantum number F can take on the values from $|I_1 - I_2|$ to $(I_1 + I_2)$. The kets $|+ -\rangle$ and $|- +\rangle$ are true eigenfunctions for $T = 0$, and wavefunctions 3-18 are true eigenfunctions at zero field.

3-5 TRANSITION PROBABILITIES

The induced transition probability, $P_{if} = P_{fi}$, between the initial state, E_i, and the final state, E_f, is given by the expression

$$P_{if} = \frac{2\pi}{\hbar^2} |\langle f|V_{(t)}|i\rangle|^2 \delta_{(\omega_{if} - \omega)} \tag{3-19}$$

where $V_{(t)}$ is the perturbation which induces the transitions. The delta function $\delta_{(\omega_{if} - \omega)}$ assumes infinitely narrow lines, corresponding to the case of being exactly "on resonance." Lineshapes resulting from $\omega \neq \omega_{if}$ are treated in Section 11-7.

Magnetic resonance transitions are induced by the interaction between the magnetic moment operator $\beta\vec{M}$ and the time-varying radiofrequency field $\vec{H}_1 \cos \omega t$ to give

$$V = -\beta\vec{M} \cdot \vec{H}_1 \tag{3-20}$$

$$= -\beta(g_1\vec{I}_1 + g_2\vec{I}_2) \cdot \vec{H}_1 \tag{3-21}$$

The operator $\vec{M}$ may be written in either Cartesian or raising and lowering operator form:

$$\vec{M} = g_1(I_{1x}\hat{i} + I_{1y}\hat{j} + I_{1z}\hat{k}) + g_2(I_{2x}\hat{i} + I_{2y}\hat{j} + I_{2z}\hat{k}) \tag{3-22}$$

$$= \tfrac{1}{2} g_1[I_1^+(\hat{i} - i\hat{j}) + I_1^-(\hat{i} + i\hat{j}) + 2I_{1z}\hat{k}]$$

$$+ \tfrac{1}{2} g_2[I_2^+(\hat{i} - i\hat{j}) + I_2^-(\hat{i} + i\hat{j}) + 2I_{2z}\hat{k}] \tag{3-23}$$

The magnetic moment operator matrix $\langle f|\vec{M}|i\rangle$, which provides the transition probability through the expression

$$P_{if} = \frac{2\pi\beta^2}{\hbar^2} |\langle f|\vec{M}|i\rangle \cdot \vec{H}_1|^2 \delta_{(\omega_{if} - \omega)} \tag{3-24}$$

must be written in terms of the true wavefunctions (3-17). It has the form

	$\psi_1 = \lvert + + \rangle$	ψ_2	ψ_3	$\psi_4 = \lvert - - \rangle$
$\psi_1^* = \langle + +\rvert$	—	$\frac{1}{\sqrt{2}}(\alpha g_2 + \gamma g_1)(\hat{i} - i\hat{j})$	$\frac{1}{\sqrt{2}}(g_1\alpha - g_2\gamma)(\hat{i} - i\hat{j})$	0
ψ_2^*	$\frac{1}{\sqrt{2}}(g_2\alpha + g_1\gamma)(\hat{i} + i\hat{j})$	—	$\alpha\gamma(g_1 - g_2)\hat{k}$	$\frac{1}{\sqrt{2}}(\alpha g_1 + \gamma g_2)(\hat{i} - i\hat{j})$
ψ_3^*	$\frac{1}{\sqrt{2}}(g_1\alpha - g_2\gamma)(\hat{i} + i\hat{j})$	$\alpha\gamma(g_1 - g_2)\hat{k}$	—	$\frac{1}{\sqrt{2}}(\alpha g_2 - \gamma g_1)(\hat{i} - i\hat{j})$
$\psi_4^* = \langle - -\rvert$	0	$\frac{1}{\sqrt{2}}(\alpha g_1 + \gamma g_2)(\hat{i} + i\hat{j})$	$\frac{1}{\sqrt{2}}(\alpha g_2 - \gamma g_1)(\hat{i} + i\hat{j})$	—

$$\tag{3-25}$$

Thus we see that the transitions $\psi_1 \leftrightarrow \psi_4$ are always forbidden, the transitions $\psi_2 \leftrightarrow \psi_3$ may be induced by parallel polarization of the radiofrequency field H_1 along the static field direction, and all other possible transitions may be induced by perpendicular (M_x or M_y) or circular (M^+ or M^-) polarizations.

The usual experimental condition is linear polarization along a particular direction such as the x axis, and for this special case, wherein

$$V = -\beta(g_1 I_{1x} + g_2 I_{2x})H_1 \cos \omega t \tag{3-26}$$

we have the matrix

	$\psi_1 = |{+}\,{+}\rangle$	ψ_2	ψ_3	$\psi_4 = |{-}\,{-}\rangle$
$\psi_1^* = \langle {+}\,{+}|$	—	$\alpha g_2 + \gamma g_1$	$\alpha g_1 - \gamma g_2$	0
ψ_2^*	$\alpha g_2 + \gamma g_1$	—	0	$\alpha g_1 + \gamma g_2$
ψ_3^*	$\alpha g_1 - \gamma g_2$	0	—	$\alpha g_2 - \gamma g_1$
$\psi_4^* = \langle {-}\,{-}|$	0	$\alpha g_1 + \gamma g_2$	$\alpha g_2 - \gamma g_1$	—

$$\tag{3-27}$$

where the unit vector $\hat{i}$ is suppressed for convenience. Special cases of these transition matrices are treated in later chapters.

Before concluding it may be mentioned that the matrix $\langle f|\vec{\mathsf{M}}|i\rangle$ can be obtained by using the operator properties of I_1 and I_2 on eigenvfunctions 3-17, or it can be obtained directly by transforming the matrix form of $\vec{\mathsf{M}}$ to the representation in which the Hamiltonian (1) is diagonal:

$$\vec{\mathsf{M}}\big|_{H\,\text{diag}} = \vec{\mathsf{U}}^{-1}\vec{\mathsf{M}}\vec{\mathsf{U}} \tag{3-28}$$

where $\vec{\mathsf{U}}$ is given explicitly by 3-31 of the first edition. The latter method seems preferable. For the special case of 3-26 $\vec{\mathsf{M}}$ has the initial matrix representation

$$\vec{\mathsf{M}} = g_1 \begin{pmatrix} 0 & 1 \\ 1 & 0 \end{pmatrix} \times \begin{pmatrix} 1 & 0 \\ 0 & 1 \end{pmatrix} + g_2 \begin{pmatrix} 1 & 0 \\ 0 & 1 \end{pmatrix} \times \begin{pmatrix} 0 & 1 \\ 1 & 0 \end{pmatrix} \tag{3-29}$$

and operation 3-28 provides the final matrix (3-27).

3-6 CONCLUSIONS

In this chapter the generalized matrices for the Hamiltonian and transition probability of the two-spin $I_1 = I_2 = \frac{1}{2}$ case were derived and written down explicitly. All the desired information is contained in these general solutions. However, the appearances of the energy level diagram and of the observed spectra depend quite critically on the relative magnitudes of the three Hamiltonian terms of Eqs. 3-1. In Chapters 4 and 5 the two important particular cases of NMR and ESR are discussed in detail, using the results of this chapter. This procedure serves to emphasize the essential unity of the general field of magnetic resonance.

NMR TWO-SPIN
($\frac{1}{2}$, $\frac{1}{2}$) SYSTEM

4-1 PRELIMINARY REMARKS

In this chapter the general solution for the two-spin ($\frac{1}{2}$, $\frac{1}{2}$) system is specialized for the case wherein the two g-factors are very close to each other. In a typical case such as two protons the g-factors differ from each other by about 1 part per million:

$$\left| \frac{g_1 - g_2}{g_1 + g_2} \right| = \left| \frac{\gamma_1 - \gamma_2}{\gamma_1 + \gamma_2} \right| \sim 10^{-6} \qquad (4\text{-}1)$$

and so a perturbation approach is certainly valid. A complication that often arises in nuclear magnetic resonance is due to a spin–spin coupling constant that has the same magnitude as the difference in g-factor energy. This produces a mixing of the wavefunctions and the appearance of a symmetric four-line spectrum whose intensities depend on the magnitude of the coupling constant. The experimental technique of spin decoupling may be employed to remove the effect of such a spin–spin interaction.

The approach followed in this book consists of the calculation of the energy levels and wavefunctions, which are then employed to obtain the separations and the relative intensities, respectively, of the spectral lines. The calculation of chemical shifts and spin–spin coupling constants has been discussed by various authors.[1-14]

4-2 ENERGY LEVELS AND LINE SPACINGS

The energies (Eqs. 3-7 and 3-9) of the general two-spin system for the NMR case are most conveniently written in the notation of the following spin Hamiltonian (3-1c):

$$H = \hbar[H(\gamma_1 I_{1z} + \gamma_2 I_{2z}) + J \vec{I}_1 \cdot \vec{I}_2] \tag{4-2}$$

where J is the spin–spin coupling constant, and the $\gamma_j = \omega_0/H_j$ are the gyromagnetic ratios.[1,15] These energies are

$$\frac{E_1}{\hbar} = \omega_0 + \tfrac{1}{4}J$$

$$\frac{E_2}{\hbar} = C - \tfrac{1}{4}J$$

$$\frac{E_3}{\hbar} = -C - \tfrac{1}{4}J \tag{4-3}$$

$$\frac{E_4}{\hbar} = -\omega_0 + \tfrac{1}{4}J$$

where the quantity C is defined to agree with the convention of Emsley, Feeney, and Sutcliffe:[1]

$$C = \tfrac{1}{2}(\omega_0^2 \delta^2 + J^2)^{1/2} \tag{4-4}$$

The average Zeeman energy, $\hbar\omega$, is related to the mean applied field strength, H_0, through

$$E_0 = \hbar\omega_0 = \frac{\hbar}{2}(\gamma_1 + \gamma_2)H$$

and the dimensionless chemical shift, δ, has the usual definition:

$$\delta = \frac{\gamma_1 - \gamma_2}{\tfrac{1}{2}(\gamma_1 + \gamma_2)} \tag{4-5}$$

where we assume that both gyromagnetic ratios are positive and $\gamma_1 > \gamma_2$. Since δ is in parts per million due to the closeness of γ_1 and γ_2, the relative chemical shift is very closely approximated by the expressions

$$\delta = \frac{\gamma_1 - \gamma_2}{\gamma_1}$$

$$= \frac{\gamma_1 - \gamma_2}{\gamma_2} \tag{4-6}$$

These two definitions of δ agree to within 1 part in 10^6, a negligible discrepancy.

The energy levels normalized in relation to the parameter $(\omega_0^2 \delta^2 + J^2)$ depend on the field $H_0 = \omega_0/\gamma$ in the manner shown in Fig. 4-1. The range in which NMR measurements are made is far off the figure on the right, where $\omega_0 \delta/J \sim 1$. At this point the four levels are effectively parallel and close to

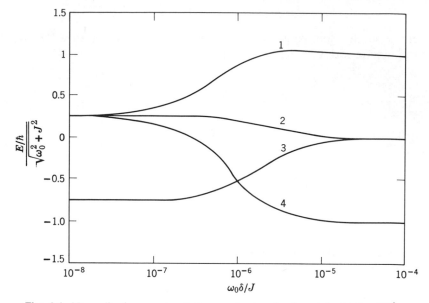

Fig. 4-1. Normalized energy level diagram for two $I=\frac{1}{2}$ nuclei with $\delta=10^{-6}$.

their asymptotic values. As a result it is appropriate to make a plot which expands the region near the ordinate values $(0,\pm\omega_0)$, and is shown on Fig. 4-2. The abscissa $\omega_0\delta/J$ is chosen to encompass the field region near the point where the chemical shift, $\omega_0\delta$, is comparable in magnitude to the coupling constant, J. The upper and lower levels are shown in relation to the basic energy, $\hbar\omega_0$, and the two center levels are shown centered about the zero energy point. For the upper and lower levels of Fig. 4-1 the energy $E=E_0+J\hbar/4$, while for the two center levels $E=\hbar(\pm C-J/4)$.

The four transitions with the energies

$$\omega_2'=\frac{E_1-E_3}{\hbar}=\omega_0+(C+\tfrac{1}{2}J)$$

$$\omega_2=\frac{E_2-E_4}{\hbar}=\omega_0+(C-\tfrac{1}{2}J)$$

$$\omega_1=\frac{E_1-E_2}{\hbar}=\omega_0-(C-\tfrac{1}{2}J)$$

$$\omega_1'=\frac{E_3-E_4}{\hbar}=\omega_0-(C+\tfrac{1}{2}J)$$

(4-7)

provide a symmetric spectrum centered about the frequency ω_0 with the line spacings shown in Fig. 4-3. The behavior of the two doublets as a function of the ratio $\omega_0\delta/J$ is given in Fig. 4-4; each case is as follows:

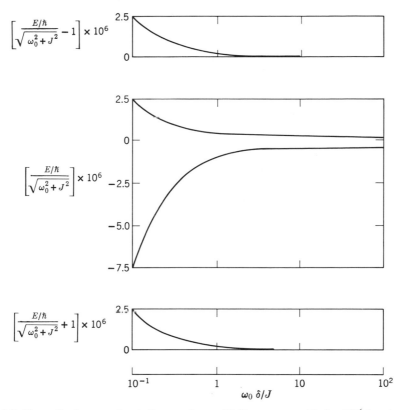

Fig. 4-2. Normalized energy level diagram for an NMR spectrum with $\delta = 10^{-6}$ for the case $I_1 = I_2 = \frac{1}{2}$. This figure shows details in the neighborhood of each level and suppresses the magnitudes of the main transitions.

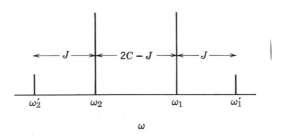

Fig. 4-3. Symmetric quartet spectrum obtained in the intermediate coupling region where $\omega_0 \delta / J \sim 1$.

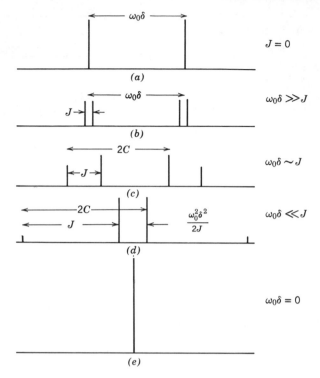

Fig. 4-4. NMR spectra for various ratios of $\omega_0\delta/J$.

(a) In the limit $J = 0$ one obtains two lines with the separation $\omega_0\delta$. This is the case for two nuclei with no spin–spin coupling, since each produces a singlet at the positions $\omega_0(1 \pm \delta/2)$, where it would be in the absence of the other nucleus, as shown in the upper spectrum of Fig. 4-4.

(b) For a small spin–spin coupling ($J \ll \omega_0\delta$) each singlet at the positions $\omega_0(1 \pm \delta/2)$ splits into a doublet centered about these same positions with the spacing J, as shown in the second from the top spectrum of Fig. 4-4. Such a spectrum is easy to interpret experimentally, since the splitting within each doublet is J and the separation between the doublets is $\omega_0\delta$.

(c) In the intermediate case in which $J \sim \omega_0\delta$, shown in the center of Fig. 4-4, the spin–spin coupling constant, J, may be read directly off the spectrum and the chemical shift $\omega_0\delta$ is given by the exact equation

$$\omega_0\delta = \sqrt{(\omega_2 - \omega_1)(\omega_2' - \omega_1')} \tag{4-8}$$

using the notation of Fig. 4-3.

(d) For the case $\omega_0\delta \ll J$ there are two strong lines centered at ω_0, with the approximate spacing $\omega_0^2\delta^2/2J$, and two weak lines in the wings, each at a distance J from its strong counterpart.

(e) The spectrum is a singlet in the limit $\delta = 0$. This occurs with the merger of the two strong lines at the position ω_0 and the passage of the two weak lines to $\pm\infty$ (since $\delta = 0$ corresponds to $J = \infty$ in our mathematical formalism).

Equation 4-8 applies to all five cases. The spacing J between lines ω_2' and ω_2 and between ω_1 and ω_1' is exact. The expressions for the spacing $\omega_0\delta$ of (b) and $\omega_0^2\delta^2/2J$ of (d) are approximations valid for limiting cases.

Before proceeding it is helpful to say a few words about the nomenclature that is customary for the above cases. Nonequivalent chemical shift differences similar in magnitude to the coupling constants are denoted by the letters $A, B, C, \ldots$, and the corresponding nuclei constitute a basic group. Additional nonequivalent nuclei separated from the $A, B, \ldots$ sets by large chemical shifts, but separated from each other by chemical shifts of magnitudes similar to the coupling constants involved, are denoted by $X, Y, Z, \ldots$. A further set of such nuclei, separated from these two sets by large chemical shifts, is labeled $P, Q, R, \ldots$. For example, monofluoroethene ($FHC{=}CH_2$) is an $ABCX$ system, since the three protons have chemical shifts comparable to their coupling constants, and the fluorine nucleus has a very large chemical shift in relation to protons. Case (e) with equivalent nuclei is referred to as an A_2 spectrum, (c) and (d) are AB spectra, and (a) and (b) are AX spectra. The sign of a coupling constant can not be determined from an AB spectrum alone.

4-3 INTENSITIES

The relative intensity of a transition $E_i \rightarrow E_j$ is obtained from the square of the corresponding matrix element, $|\langle \psi_i|\hbar(\gamma_1 I_{1x} + \gamma_2 I_{2x})|\psi_j\rangle|^2$. Matrix 3-27 displays these matrix elements, so it is necessary only to square them to obtain the relative intensities. Since $\gamma_1 \cong \gamma_2$ to within 1 part in 10^6, these gyromagnetic ratios may be omitted, to give the intensities $(\alpha + \gamma)^2$ for the inner lines and $(\alpha - \gamma)^2$ for the outer lines.

In the high-resolution NMR literature the relative intensities are often presented in a different form. The parameters α and γ of Eqs. 3-12 may be employed to define the angle θ:

$$\alpha = \frac{1}{\sqrt{1 + Q^2}} = \cos\theta$$

$$\gamma = \frac{Q}{\sqrt{1 + Q^2}} = \sin\theta$$

(4-9)

where

$$Q = \frac{1}{J}\left(\sqrt{J^2 + \omega_0^2\delta^2} - \omega_0\delta\right)$$

(4-10)

as may be proved by direct substitution. This is equivalent to

$$Q = \frac{J}{\omega_0 \delta + 2C} \tag{4-11}$$

These substitutions lead to the result

$$\begin{aligned} \omega_0 \delta &= 2C \cos 2\theta \\ J &= 2C \sin 2\theta \end{aligned} \tag{4-12}$$

where θ is given by

$$\cos 2\theta = \frac{1 - Q^2}{1 + Q^2} \tag{4-13}$$

When these expressions are used, the relative intensities of the inner and outer lines are given by

$$|\langle \psi_i | I_{1x} + I_{2x} | \psi_j \rangle|^2 + \begin{cases} \alpha^2 + \gamma^2 \sim 1 + \sin 2\theta & \text{(Inner lines)} \\ \alpha^2 - \gamma^2 \sim 1 - \sin 2\theta & \text{(Outer lines)} \end{cases} \tag{4-14}$$

The central spectrum of Fig. 4-4 is drawn for the condition $\omega_0 \delta = J$, which corresponds to $\tan 2\theta = 1$ and an intensity ratio $(\sqrt{2}+1)/(\sqrt{2}-1)$ for the inner to outer lines.

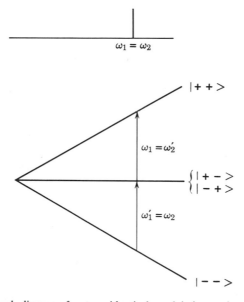

Fig. 4-5. Energy level diagram for two identical nuclei ($\gamma_1 = \gamma_2$) without any spin–spin interaction ($J = 0$). The four allowed transition frequencies, ω_1, ω_1', ω_2, and ω_2', are equal, so that only one line appears in the spectrum shown above.

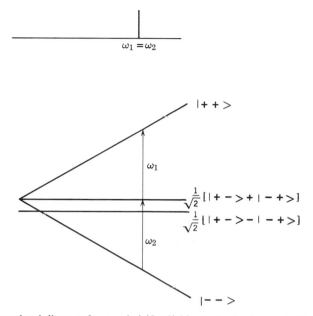

Fig. 4-6. Energy level diagram for coupled $(J \neq 0)$ identical spins $(\gamma_1 = \gamma_2)$. Note that only one transition, $\omega_1 = \omega_2$, is observed.

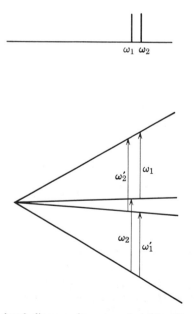

Fig. 4-7 Two-spin energy level diagram for uncoupled $(J = 0)$, nearly equivalent $(\gamma_1 \sim \gamma_2)$ spins. Note that $\omega_1 = \omega_1'$ and $\omega_2 = \omega_2'$, so that only two lines appear in the typical NMR spectrum shown above.

4-4 LIMITING CASES

To gain more insight into the mechanisms behind the energy level diagram of Figs. 4-1 and 4-2 it is appropriate to draw diagrams for special cases. These diagrams plot E versus H without any special expanded scales. Figure 4-5 presents the case of two equivalent nuclei with $J = 0$, for which there is only one transition, as expected. For coupled identical spins $\delta = 0$ and $Q = 1$ in Eq. 4-10. As a result $\alpha = \gamma$, and again only one transition, $\omega_1 = \omega_1'$, is allowed, as shown in Fig. 4-6. This is equivalent to saying that only transitions within the triplet level are allowed. The spin–spin coupling constant J does not affect the spectrum.

For two uncoupled ($J = 0$) inequivalent spins a doublet is observed with components at the positions corresponding to each individual nuclear transition. This means that the transitions $\omega_1 = \omega_1'$ and $\omega_2 = \omega_2'$ in Fig. 4-7, and it corresponds to $Q = 0$, $\alpha = 1$, and $\gamma = 0$ in Eqs. 4-9. In the general case $J \neq 0$, $\delta \neq 0$, the four transitions ω_1, ω_1', ω_2, and ω_2' are unequal, and a four-line spectrum results. Figure 4-1 shows the manner in which the energy levels shift in this general case.

4-5 CONCLUSIONS

Many common spin-$\frac{1}{2}$ nuclei have been widely studied, such as hydrogen[16] carbon-13,[16,17] fluorine-19, and phosphorus-31,[18] and articles on them may be found in annual review type publications.[19-21]

REFERENCES

1. J. W. Emsley, J. Feeney, and L. H. Sutcliffe, *High Resolution Nuclear Magnetic Resonance Spectroscopy*, Pergamon, New York, 1965, Vols. 1 and 2.

2. I. V. Aleksandrov, *Theory of Nuclear Magnetic Resonance*, C. P. Poole, Jr., translation ed., Academic Press, New York, 1966.

3. J. D. Memory, *Quantum Theory of Magnetic Resonance Parameters*, McGraw-Hill, New York, 1968.

4. A. Abragam, *Principles of Nuclear Magnetism*, Oxford University Press, New York, 1961.

5. A. Abragam and M. Goldman, *Nuclear Magnetism: Order and Disorder*, Oxford University Press, New York, 1982.

6. U. Haeberlen, *High Resolution NMR in Solids* (Advances in Magnetic Resonance Series: Supplement 1), Academic Press, New York, 1976.

7. R. Harris and B. Mann, *NMR and the Periodic Table*, Academic Press, New York, 1978.

8. P. Laszlo, ed., *NMR of Newly Accessible Nuclei*: Vol. 1, *Chemical and Biochemical Applications*; Vol. 2, *Chemically and Biochemically Important Elements*, Academic Press, New York, 1983.

9. M. Barfield and D. M. Grant, *Adv. Magn. Reson.*, **1**, 149 (1965).

10. A. A. Bothner-By, *Adv. Magn. Reson.*, **1** 195 (1969).

11. W. N. Lipscomb, *Adv. Magn. Reson.*, **2**, 138 (1966).

12. J. I. Musher, *Adv. Magn. Reson.*, **2**, 177 (1966).

13. R. H. Contreras, M. A. Natiello, and G. E. Scuseria, *Magn. Reson. Rev.*, **9**, 239 (1985).

14. P. Sohar, ed., *Nuclear Magnetic Resonance Spectroscopy*, Vol. I, CRC Press, Boca Raton, FL, 1983.

15. P. L. Corio, *Structures of High Resolution NMR Spectra*, Academic Press, New York, 1966.

16. Q. T. Pham, R. Petiaud, and H. Waton, *Proton and Carbon NMR Spectra of Polymers*, Vol. 2, Wiley-Interscience, New York, 1974.

17. G. C. Levy, *Topics in Carbon-13 NMR Spectroscopy*, Vol. 1, Wiley-Interscience, New York, 1974.

18. D. G. Gorenstein, ed., *Phosphorus-31 NMR: Principles and Applications*, Academic Press, New York, 1984.

19. J. W. Emsley, J. Feeney, and L. M. Sutcliffe, Eds., *Progress in Nuclear Magnetic Resonance Spectroscopy*, Pergamon Press, New York. A continuing series started in 1966.

20. E. F. Mooney and G. A. Webb, eds., *Annual Reports on NMR Spectroscopy*, Academic Press, New York. A continuing series starting in 1968.

21. R. J. Abraham, *Nuclear Magnetic Resonance*, Royal Chemical Society of London. A continuing series starting in 1972.

ESR TWO-SPIN
($\frac{1}{2}$, $\frac{1}{2}$) SYSTEM

5-1 PRELIMINARY REMARKS

In this chapter the general solution for the two-spin ($\frac{1}{2}$, $\frac{1}{2}$) case is special-ized in the limit opposite to that presented in Chapter 4, namely, for the situation wherein

$$\left| \frac{g_1}{g_2} \right| \sim 10^3 \tag{5-1}$$

As a result, the electronic Zeeman term completely dominates the nuclear Zeeman term. The spectrum is ordinarily fairly simple with two equal intensity lines centered about the electron Zeeman field. A complication that sometimes arises is due to an anisotropic hyperfine coupling constant which is comparable in magnitude to the nuclear Zeeman term. This will produce a symmetric four-line spectrum analogous to that discussed in the NMR case. The intensities depend on the magnitude of the hyperfine coupling constant relative to that of the electronic Zeeman term. The present chapter concentrates on the simpler isotropic case, and Chapter 6 treats anisotropies.

Many books[1-8] discuss the case presented here, and experimental data may be obtained, for example, from literature reviews.[9,10]

5-2 ENERGY LEVELS AND LINE SPACINGS

The energies (Eqs. 3-7 and 3-9) of the general two-spin system for the electron spin resonance case[1-8] are most conveniently written in the notation of the following Hamiltonian (3-1b):

$$H = H(g\beta S_z - g_N\beta_N I_z) + T\vec{S} \cdot \vec{I} \tag{5-2}$$

where T is the hyperfine coupling constant, and the signs on the two Zeeman terms arise from the signs of the corresponding charges on the electron and nucleus. Since $\beta/\beta_N = 1836$, the nuclear Zeeman term is negligible relative to the electronic Zeeman term, but it is retained in the Hamiltonian since it is sometimes comparable to the hyperfine interaction.

The two middle energies (Eqs. 3-9) have the exact forms

$$E_2 = -\tfrac{1}{4}T + \tfrac{1}{2}[T^2 + (g\beta + g_N\beta_N)^2 H^2]^{1/2}$$
$$E_3 = -\tfrac{1}{4}T - \tfrac{1}{2}[T^2 + (g\beta + g_N\beta_N)^2 H^2]^{1/2} \tag{5-3}$$

and in the ESR approximation, $g\beta H \gg T$, the four energies are (Eqs. 3-7 and 3-11)

$$E_1 = \tfrac{1}{2}(g\beta - g_N\beta_N)H + \tfrac{1}{4}T$$

$$E_2 = \tfrac{1}{2}(g\beta + g_N\beta_N)H - \tfrac{1}{4}T + \frac{T^2}{4(g\beta + g_N\beta_N)H}$$

$$E_3 = -\tfrac{1}{2}(g\beta + g_N\beta_N)H - \tfrac{1}{4}T - \frac{T^2}{4(g\beta + g_N\beta_N)H} \tag{5-4}$$

$$E_4 = -\tfrac{1}{2}(g\beta - g_N\beta_N)H + \tfrac{1}{4}T$$

where the expressions for E_1 and E_4 are exact, and the other two energies are written down to second order. The same expressions may be obtained from perturbation theory using Eqs. 2-104 and 3-5.

An overall picture of the energy level behavior is obtained by normalizing the energy in relation to $(T^2 + g^2\beta^2 H^2)^{1/2}$ and plotting it in relation to $g\beta H/T$. Such a diagram is given in Fig. 5-1, which shows the ordering of the energy levels and two main allowed transitions. The location of the spectra for hydrogen atoms and a $\cdot$CH fragment at both the x band and k band are indicated, where one has approximately

$$T = 500 \text{ G} \quad (\text{H atom})$$
$$= 25 \text{ G} \quad (\cdot\text{CH fragment}) \tag{5-5}$$

There is a crossing over of the two upper energy levels at the point where $2g_N\beta_N H = T$. Figure 5-1 and the discussion in this section assume positive values of g, g_N, and T. The effect of other choices of sign is discussed in Section 5-4.

The usual way to present an energy level diagram is to plot energy E versus $g\beta H/T$. Such a diagram is shown in Fig. 5-2, in which the allowed transitions are indicated. At magnetic field values larger than those shown in Fig. 5-2 the two upper levels cross. Figure 5-3 illustrates another manner of representing the energy level shifts that arise from each term. One should note that the energy level changes due to the nuclear Zeeman term are not

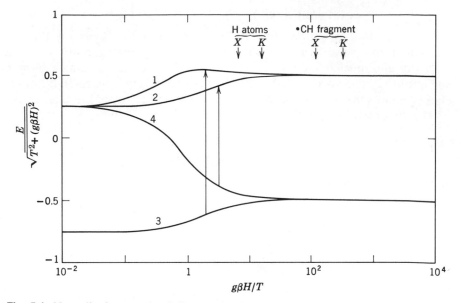

Fig. 5-1. Normalized energy level diagram for the case of positive T, g, and g_N in Eq. 5-2, where $g\beta/g_N\beta_N = 660$. The hydrogen atom and ·CH fragment (with $T > 0$) spectra occur where indicated at x and k bands.

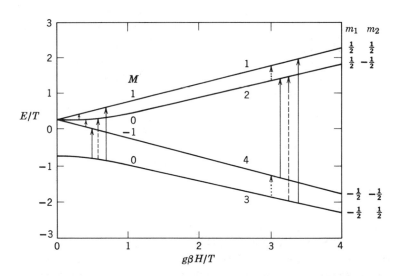

Fig. 5-2. Energy level diagram for two spin-$\frac{1}{2}$ particles with $g\beta/g_N\beta_N = 660$ showing the strong (—) and weak (· ·) transitions with perpendicularly polarized rf fields, and the parallel field (---) transitions. The weak perpendicular transitions correspond to NMR lines.

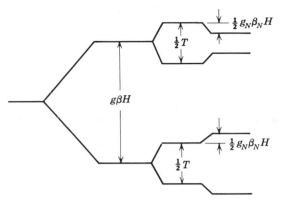

Fig. 5-3. Shift in energy levels of $S = \frac{1}{2}$, $I = \frac{1}{2}$ case due to the hyperfine term $T\vec{S} \cdot \vec{I}$ and the nuclear Zeeman term $g_N \beta_N H I_z$.

observed since the upper and lower levels of each transition shift by the same amount.

5-3 INTENSITIES

The relative intensities for the various transitions are obtained by squaring the matrix elements in expression 3-25 and neglecting terms with the nuclear g-factor g_2. At high magnetic fields where $\gamma \ll \alpha$ we have, for the microwave field perpendicular to the applied field,

$$|\langle \psi_i | S_x | \psi_j \rangle|^2 = \begin{cases} \frac{1}{2}\alpha^2 & (3 \rightarrow 1 \text{ transition}) \\ \frac{1}{2}\alpha^2 & (4 \rightarrow 2 \text{ transition}) \\ \frac{1}{2}\left(\gamma + \dfrac{g_N \beta_N}{g\beta}\alpha\right)^2 & (2 \rightarrow 1 \text{ transition}) \\ \frac{1}{2}\left(\gamma - \dfrac{g_N \beta_N}{g\beta}\alpha\right)^2 & (3 \rightarrow 4 \text{ transition}) \end{cases} \quad (5\text{-}6)$$

At high fields with the microwave magnetic field polarized parallel to the large applied one, there is only one transition with a nonzero transition probability:

$$|\langle \psi_i | S_x | \psi_j \rangle|^2 = \alpha^2 \gamma^2 \quad (3 \rightarrow 2 \text{ transition}) \tag{5-7}$$

If expressions 3-16 or 3-41 and 3-42 of the first edition are expanded in powers of $T/g\beta H$, neglecting $g_N \beta_N H$, then for $T/g\beta H \ll 1$ the coefficients of the wavefunction have the approximate values

$$\alpha = \left[1 - \left(\frac{T}{2g\beta H}\right)^2\right]^{1/2}$$

$$\gamma = \frac{T}{2g\beta H} \tag{5-8}$$

In typical cases $\gamma \sim 10^{-2}$, so that the parallel field transitions will be about 0.01% as strong as the usual $3 \rightarrow 1$, $4 \rightarrow 2$ ones.

At low magnetic fields where $g\beta H < T$, all five transitions become appreciable in magnitude, since both α and γ become appreciable. In the limit $g\beta H \ll T$, one has explicitly

$$\alpha = \left(\frac{1 + g\beta H/T}{2}\right)^{1/2}$$

$$\gamma = \left(\frac{1 - g\beta H/T}{2}\right)^{1/2} \tag{5-9}$$

to render all five low-field transitions of Fig. 5-4 comparable in magnitude.

The appearance of the overall spectra at high field and low field is shown in Fig. 5-4. Most ESR research work detects only the $4 \rightarrow 2$ and $3 \rightarrow 1$ lines

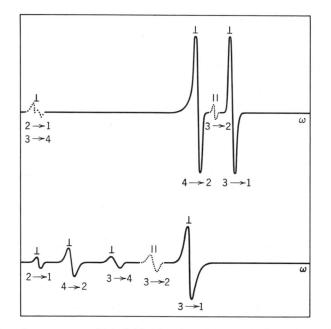

Fig. 5-4. Overall spectrum at high field (above) and low field (below), showing all lines corresponding to the transitions of Fig. 5-2. The amplitudes of the weak transitions at high field are exaggerated. The upper spectrum covers a much wider frequency range and has exaggerated linewidths in relation to the lower spectrum.

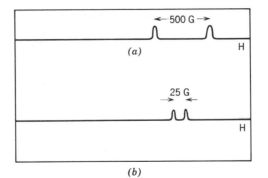

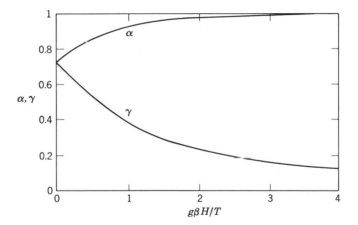

Fig. 5-5. ESR absorption spectrum for (a) hydrogen atoms and (b) the ·CH fragment, with the splitting given in gauss.

Fig. 5-6. Magnitudes of the wavefunction coefficients α and γ as a function of $|g\beta H/T|$.

at high field. The dominant high-field spectra for hydrogen atoms and a ·CH fragment are shown in Fig. 5-5.

The dependence of the wavefunction coefficients α and γ on the ratio $|g\beta H/T|$ is shown in Fig. 5-6.

5-4 SIGNS OF HAMILTONIAN TERMS

The mathematical analysis in this chapter has assumed that the parameters g, g_N, and T are all positive. The treatment is similar for all choices of signs, and the various formulas that were derived are valid for these different possibilities. If T is changed in sign, the singlet and triplet levels interchange at low field. The relative sign of the ratio g_N/T determines whether the high-field crossover point occurs with the upper or the lower pair of levels. The exact labeling of the levels with their m_F and $|m_s m_I\rangle$ symbols depends on the various choices of sign. The precise form of the energy levels for those cases is shown in Fig. 5-7 for the Hamiltonian written as follows:

$$H = g\beta H S_z - g_N \beta_N H I_z + T\vec{S} \cdot \vec{I} \qquad (5\text{-}10)$$

The case discussed above ($g > 0$, $g_N > 0$, $T > 0$) corresponds to free hydrogen atoms, where the negative sign of the nuclear Zeeman term is explicitly inserted into the Hamiltonian of Eq. 5-2. The hyperfine structure constant is 1420 MHz or 510 G.[11-14] For the methyl radical the isotropic part of the hyperfine tensor is negative ($a_H = -23$ G) while the isotropic part of the ^{13}C

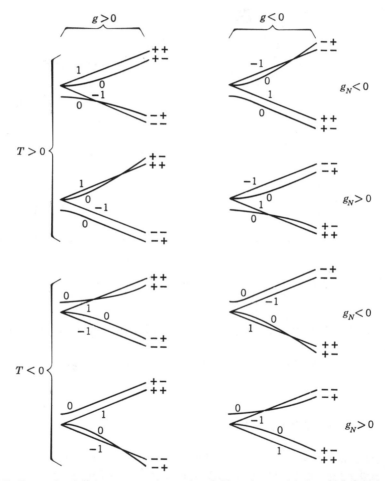

Fig. 5-7. Energy level diagrams, m_F values at low field, and $m_s m_I$ labels at high field for ($\frac{1}{2}$, $\frac{1}{2}$) spin system with positive (left column) and negative (right column) g-factors, positive (upper four) and negative (lower four) hyperfine coupling constants, and positive and negative nuclear g-factors (labeled at right) for the Hamiltonian of Eq. 5-10.

coupling constant is positive ($a_C = +41$ G). The ^{13}C spectrum occurs with a natural abundance of 1.1%. The ·CH fragment indicated in Fig. 5-1 actually has $T < 0$; nevertheless, it is included on the figure for comparison with a hydrogen atom.

The choice of signs in g, g_N, and T that corresponds to a particular spin system may be determined experimentally. The sign of the electronic g-factor is found by the use of polarized microwaves, as described elsewhere,[15,16] and g_N may be deduced from the corresponding NMR experiment.[16,17] Measurements of the ESR spectra at low and high fields will clarify whether the two-triplet levels cross at high field, or whether the crossing occurs with a singlet and a triplet level. A combination of these experiments will serve to identify which of the eight cases of Fig. 5-7 corresponds to the $(\frac{1}{2}, \frac{1}{2})$ spin system under study. Similar techniques are applicable to nuclear spins $I > 0$ and to hyperfine patterns arising from couplings to more than one nucleus.

5-5 PERTURBATION SOLUTIONS

Most ESR studies are carried out at sufficiently high fields so that the second-order term, $T^2/4(g\beta + g_N\beta_N)H$, of Eq. 5-4 is negligible. As a result, this correction is ordinarily not taken into account. For the sake of completeness it is of interest to show the effects of neglecting the correction at somewhat lower fields.

The energy levels E_1 and E_4 are known exactly, as Eq. 5-4 indicates. The levels E_2 and E_3 of Eq. 5-3 may be written in the following manner at low fields:

$$E_2 = -\tfrac{1}{4}T + \tfrac{1}{2}T\left[1 + \frac{(g\beta + g_N\beta_N)H^2}{2T^2}\right]$$

$$E_3 = -\tfrac{1}{4}T - \tfrac{1}{2}T\left[1 + \frac{(g\beta + g_N\beta_N)H^2}{2T^2}\right]$$

(5-11)

where

$$|(g\beta + g_N\beta_N)H| \ll |T| \tag{5-12}$$

At high fields, on the other hand, the appropriate expressions are

$$E_2 = -\tfrac{1}{4}T + \tfrac{1}{2}(g\beta + g_N\beta_N)H\left[1 + \frac{T^2}{2(g\beta + g_N\beta_N)^2H^2}\right]$$

$$E_3 = -\tfrac{1}{4}T - \tfrac{1}{2}(g\beta + g_N\beta_N)H\left[1 + \frac{T^2}{2(g\beta + g_N\beta_N)^2H^2}\right]$$

(5-13)

where

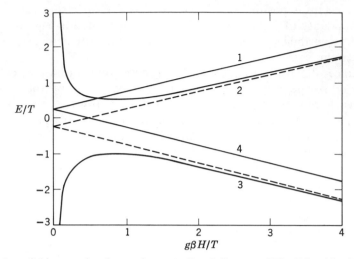

Fig. 5-8. Low-field approximation to the energy level diagram of Fig. 5-2, taking into account the terms up to first order (---) and second order (—) in the factor $(g\beta + g_N\beta_N)H/T$ of Eq. 5-11. The levels E_1 and E_4 are linear in all orders of perturbation.

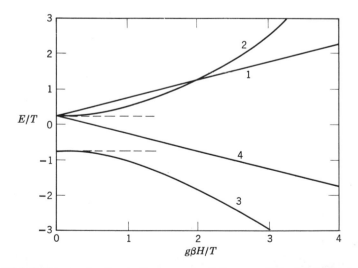

Fig. 5-9. High-field approximation to the energy level diagram of Fig. 5-2, taking into account terms up to first order (---) and second order (—) in the factor $T/(g\beta + g_N\beta_N)H$ of Eq. 5-13. The levels E_1 and E_4 are linear in all orders of perturbation.

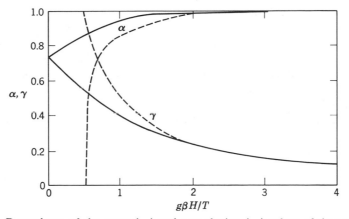

Fig. 5-10. Dependence of the exact (—) and perturbation (---) values of the wavefunction coefficients α and γ on the magnetic field strength. The former correspond to Eq. 3-16, and the latter to Eq. 5-8.

$$|(g\beta + g_N\beta_N)H| \gg |T| \qquad (5\text{-}14)$$

The first- and second-order approximations of Eq. 5-11 are plotted in Fig. 5-8, and the corresponding behavior of Eqs. 5-13 is shown in Fig. 5-9. The field dependence of the wavefunction coefficients (Eqs. 5-8) which are valid in the high-field approximation 5-14 is presented in Fig. 5-10.

REFERENCES

1. G. E. Pake and T. L. Estle, *The Physical Principles of Electron Paramagnetic Resonance*, 2nd ed., Benjamin-Cummings, Menlo Park, CA, 1973.

2. S. A. Al'tshuler and B. M. Kozyrev, *Electron Paramagnetic Resonance*, C. P. Poole, Jr., translation ed., Academic Press, New York, 1964.

3. A. Carrington and A. D. McLachlan, *Introduction to Magnetic Resonance*, Harper, New York, 1967.

4. A. Abragam and B. Bleaney, *Electron Paramagnetic Resonance of Transition Ions*, Clarendon Press, Oxford, 1970.

5. J. W. Orton, *Electron Paramagnetic Resonance*, Iliffe Books, London, 1968.

6. J. E. Harriman, ed., *Theoretical Foundations of Electron Spin Resonance*, Academic Press, New York, 1979.

7. J. E. Wertz and J. R. Bolton, *Electron Spin Resonance*, Chapman & Hall, New York, 1986.

8. W. T. Dixon, *Theory & Interpretation of Magnetic Resonance Spectra*, Plenum, New York, 1972.

9. D. R. Ayscough, ed., *Electron Spin Resonance*, American Chemical Society of New York and Royal Society of Chemistry, London, Vol. 1, 1973; Vol. 2, 1974; Vol. 3, 1976; Vol. 4, 1977; Vol. 5, 1979.

10. *Magnetic Resonance Review*, Gordon & Breach, New York.

11. J. P. Wittke and R. H. Dicke, *Phys. Rev.*, **103**, 620 (1956).

12. R. Livingston, H. Zeldes, and E. H. Taylor, *Discuss. Faraday Soc.*, **19**, 166 (1955).

13. C. K. Jen, S. N. Foner, E. L. Cochran, and V. A. Bowers, *Phys. Rev.*, **104**, 846 (1956).

14. C. K. Jen, S. N. Foner, E. L. Cochran, and V. A. Bowers, *Phys. Rev.*, **112**, 1169 (1958).

15. J. R. Eshbach and M. W. P. Strandberg, *Rev. Sci. Instrum.*, **23**, 623 (1952).

16. C. P. Poole, Jr., *Electron Spin Resonance*, Wiley, New York, 1967; see Sec. 8 M; 2nd ed. 1983, see Sec. 5-O.

17. A. Abragam, *Principles of Nuclear Magnetism*, Oxford University Press, London, 1961.

ANISOTROPIC HAMILTONIANS

6-1 INTRODUCTION

The problem of finding the energy levels of a single spin, $S = \frac{1}{2}$, in a magnetic field is rather simple for an isotropic g-factor, but when g becomes anisotropic the task is not so trivial.[1-4] It is instructive to solve the problem for spins with magnitudes greater than $\frac{1}{2}$ since this provides an insight into the quantum mechanics of anisotropic tensors. After doing this we will treat the case of an anisotropic hyperfine tensor,[4-12] and then the more complex situation when both are anisotropic.[4,12] Many data are available on these systems.[11-16]

6-2 ENERGY LEVELS

The interaction between the spin $\vec{S}$ and a magnetic field $\vec{H}$ is given by the Hamiltonian

$$\mathcal{H} = \beta \vec{S} \cdot \overset{\leftrightarrow}{g} \cdot H \tag{6-1}$$

where $\overset{\leftrightarrow}{g}$ is a symmetric tensor ($g_{ij} = g_{ji}$). It is most convenient to select z as the magnetic field direction, to give

$$\mathcal{H} = \beta H(g_{xz}S_x + g_{yz}S_y + g_{zz}S_z) \tag{6-2a}$$

which may also be expressed in terms of raising and lowering operators:

$$\mathcal{H} = \beta H[\tfrac{1}{2}S^+(g_{xz} + ig_{yz}) + \tfrac{1}{2}S^-(g_{xz} - ig_{yz}) + S_z g_{zz}] \tag{6-2b}$$

Either expression, together with the spin operators of Section 2-6, may be employed to write the Hamiltonian matrix. For the spins $S = \frac{1}{2}$, 1, and $\frac{3}{2}$ these matrices have the explicit forms

$$\begin{pmatrix} \frac{1}{2}g_{zz}\beta H & \frac{1}{2}(g_{xz}-ig_{yz})\beta H \\ \frac{1}{2}(g_{xz}+ig_{yz})\beta H & -\frac{1}{2}g_{zz}\beta H \end{pmatrix} \quad (S=\tfrac{1}{2}) \tag{6-3a}$$

$$\begin{pmatrix} g_{zz}\beta H & \frac{1}{\sqrt{2}}(g_{xz}-ig_{yz})\beta H & 0 \\ \frac{1}{\sqrt{2}}(g_{xz}+ig_{yz})\beta H & 0 & \frac{1}{\sqrt{2}}(g_{xz}-ig_{yz})\beta H \\ 0 & \frac{1}{\sqrt{2}}(g_{xz}+ig_{yz})\beta H & -g_{zz}\beta H \end{pmatrix} \quad (S=1)$$

$$\tag{6-3b}$$

$$\begin{pmatrix} \frac{3}{2}g_{zz}\beta H & \frac{\sqrt{3}}{2}(g_{xz}-ig_{yz})\beta H & 0 & 0 \\ \frac{\sqrt{3}}{2}(g_{xz}+ig_{yz})\beta H & \frac{1}{2}g_{zz}\beta H & (g_{xz}-ig_{yz})\beta H & 0 \\ 0 & (g_{xz}+ig_{yz})\beta H & -\frac{1}{2}g_{zz}\beta H & \frac{\sqrt{3}}{2}(g_{xz}-ig_{yz})\beta H \\ 0 & 0 & \frac{\sqrt{3}}{2}(g_{xz}+ig_{yz})\beta H & -\frac{3}{2}g_{zz}\beta H \end{pmatrix} \quad (S=\tfrac{3}{2})$$

$$\tag{6-3c}$$

These are equivalent to the polynomial equations

$$E^2 - \tfrac{1}{4}g^2\beta^2 H^2 = 0 \qquad\qquad (S=\tfrac{1}{2}) \tag{6-4a}$$

$$E^3 - g^2\beta^2 H^2 E = 0 \qquad\qquad (S=1) \tag{6-4b}$$

$$E^4 - \tfrac{5}{2}g^2\beta^2 H^2 E^2 + \tfrac{9}{16}g^4\beta^4 H^4 = 0 \qquad (S=\tfrac{3}{2}) \tag{6-4c}$$

with the respective roots

$$E = \pm\tfrac{1}{2}g\beta H \qquad\qquad (S=\tfrac{1}{2}) \tag{6-5a}$$

$$E = 0,\ \pm g\beta H \qquad\qquad (S=1) \tag{6-5b}$$

$$E = \pm\tfrac{1}{2}g\beta H,\ \pm\tfrac{3}{2}g\beta H \qquad (S=\tfrac{3}{2}) \tag{6-5c}$$

where in each case

$$g^2 = g_{xz}^2 + g_{yz}^2 + g_{zz}^2 \tag{6-6}$$

One may write for eigenenergies of a general spin, S:

$$E_m = g\beta H m \tag{6-7}$$

where g is given by Eq. 6-6 and $-S \le m \le S$. Thus even with an anisotropic g-factor the energy levels are equally spaced with the effective g-factor of 6-6. The situation is more complex when additional interactions are included in the Hamiltonian.[4,12]

6-3 INTENSITIES

To determine the relative intensities of the transitions one may assume a harmonically varying radiofrequency field H_1 in the x direction. The Hamiltonian

$$\mathcal{H}_1 = \beta H_1 \cos \omega t (g_{xx} S_x + g_{xy} S_y + g_{xz} S_z) \tag{6-8}$$

may be written in matrix form to provide the transition probabilities which are proportional to $|\langle f | \mathcal{H}_1 | i \rangle|^2$. It is easy to write the transition probability matrix corresponding to 6-8 following the procedure whereby matrices 6-3 are formed from the Hamiltonian 6-2. This is to be done, using the eigenfunctions which diagonalize Eqs. 6-3. For the $S = \frac{1}{2}$ case we have the eigenkets $|\psi_1\rangle$ and $|\psi_2\rangle$, corresponding to the energies $E = +g\beta H/2$ and $E = -g\beta H/2$, respectively:

$$|\psi_1\rangle = \alpha|+\rangle + \gamma|-\rangle \tag{6-9a}$$

$$|\psi_2\rangle = -\gamma^*|+\rangle + \alpha^*|-\rangle \tag{6-9b}$$

where

$$\alpha = \frac{g_{xz} - ig_{yz}}{[2g(g - g_{zz})]^{1/2}} \tag{6-9c}$$

$$\gamma = \frac{g - g_{zz}}{[2g(g - g_{zz})]^{1/2}} \tag{6-9d}$$

These expressions were derived by the method of Eq. 2-36. The transition probability is proportional to

$$|\langle \psi_1 | \mathcal{H}_1 | \psi_2 \rangle|^2 = \frac{1}{4}\beta^2 H_1^2 \langle \cos^2 \omega t \rangle |\alpha\alpha^*(g_{xx} - ig_{xy})$$
$$- (\alpha\gamma^* + \gamma\alpha^*)g_{xz} - \gamma\gamma^*(g_{xx} + ig_{xy})|^2 \tag{6-10}$$

which has a strong orientation dependence. The explicit dependence of the intensity on the angle may be obtained by substituting the appropriate quantities from Eqs. 6-19 to 6-21 into Eq. 6-10. The resulting expression is quite complex. There is also a temperature dependence, and at high temperature where $g\beta H \ll kT$ the result for spin $S = \frac{1}{2}$ is

$$\text{Intensity} \sim |\langle f | \mathcal{H}_1 | i \rangle|^2 \left| \frac{g\beta H}{(2S + 1)kT} \right| \tag{6-11}$$

6-4 ANGULAR DEPENDENCE OF THE g-FACTOR

The g-factor tensor for an arbitrary orientation (x, y, z),

$$\overset{\leftrightarrow}{g} = \begin{pmatrix} g_{xx} & g_{xy} & g_{xz} \\ g_{yx} & g_{yy} & g_{yz} \\ g_{zx} & g_{zy} & g_{zz} \end{pmatrix} \tag{6-12}$$

is obtained from the diagonal g-factor tensor in the $x'y'z'$ principal axis system (Fig. 6-1):

$$\overset{\leftrightarrow}{g}_{(\text{diag})} = \begin{pmatrix} g_{x'x'} & 0 & 0 \\ 0 & g_{y'y'} & 0 \\ 0 & 0 & g_{z'z'} \end{pmatrix} = \begin{pmatrix} g_1 & 0 & 0 \\ 0 & g_2 & 0 \\ 0 & 0 & g_3 \end{pmatrix} \tag{6-13}$$

by means of the direction cosine matrix R:

$$\overset{\leftrightarrow}{g} = \overset{\leftrightarrow}{R} \overset{\leftrightarrow}{g}_{(\text{diag})} \overset{\leftrightarrow}{R} \tag{6-14}$$

where

$$\overset{\leftrightarrow}{R} = \begin{pmatrix} \cos\theta_{xx'} & \cos\theta_{xy'} & \cos\theta_{xz'} \\ \cos\theta_{yx'} & \cos\theta_{yy'} & \cos\theta_{yz'} \\ \cos\theta_{zx'} & \cos\theta_{zy'} & \cos\theta_{zz'} \end{pmatrix} \tag{6-15}$$

The matrices $\overset{\leftrightarrow}{R}$ and $\overset{\leftrightarrow}{g}$ are real, but only the latter is symmetric:

$$\cos\theta_{ij'} \neq \cos\theta_{ji'}$$

$$g_{ij} = g_{ji} \tag{6-16}$$

If the rotation is carried out about a principal direction, $\overset{\leftrightarrow}{R}$ reduces to a much simpler form, and (6-15) is considerably simplified. For example, if $\overset{\leftrightarrow}{R}_x^{\theta}$ is a rotation about the $x' = x$ direction, one obtains

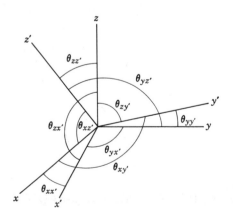

Fig. 6-1. Direction cosines, $\cos\theta_{ij'}$, between the laboratory coordinate system (xyz) and the principal axis system $(x'y'z')$.

$$\vec{\vec{R}}_x^{\theta} = \begin{pmatrix} 1 & 0 & 0 \\ 0 & \cos\theta & \sin\theta \\ 0 & -\sin\theta & \cos\theta \end{pmatrix} \tag{6-17}$$

and $\vec{\vec{g}}_{(\theta)} = \vec{\vec{R}}_x^{\theta}\vec{\vec{g}}_{(\text{diag})}\vec{\vec{R}}_x^{\theta}$ from Eq. 6-14 becomes

$$\vec{\vec{g}} = \begin{pmatrix} g_1 & 0 & 0 \\ 0 & g_2\cos^2\theta + g_3\sin^2\theta & (g_3 - g_2)\sin\theta\cos\theta \\ 0 & (g_3 - g_2)\sin\theta\cos\theta & g_2\sin^2\theta + g_3\cos^2\theta \end{pmatrix} \tag{6-18}$$

with analogous expressions for rotations about the y' and z' principal directions.

The g-factor for an arbitrary orientation of the magnetic field may be determined by writing the Hamiltonian (6-1) in the principal axis system where the g-factor is diagonal (Eq. 6-13):

$$\mathcal{H} = \beta(g_1 H'_x S'_x + g_2 H'_y S'_y + g_3 H'_z S'_z) \tag{6-19}$$

It is convenient to know the g-factor in spherical coordinates (Fig. 6-2), and so we make use of the expressions

$$H'_x = H_0 \sin\theta \cos\phi$$
$$H'_y = H_0 \sin\theta \sin\phi$$
$$H'_z = H_0 \cos\theta$$

to give

$$\mathcal{H} = \beta H_0[(g_1 \sin\theta \cos\phi)S_x + (g_2 \sin\theta \cos\phi)S_y + (g_3 \cos\theta)S_z] \tag{6-20}$$

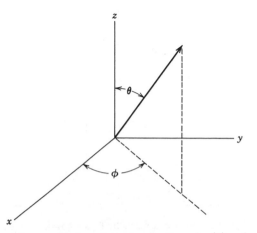

Fig. 6-2. Spherical or polar coordinate system, showing the polar (θ) and azimuthal (ϕ) angles.

This equation has the same form as Eq. 6-2a with different coefficients on the spin operators. As a result solution 6-5 provides the g-factor 6-6, and by comparing coefficients one may write

$$g^2 = (g_1^2 \cos^2 \phi + g_2^2 \sin^2 \phi) \sin^2 \theta + g_3^2 \cos^2 \theta \qquad (6\text{-}21)$$

This expression gives the observed g-factor in polar coordinates relative to its principal axis coordinate system, where $\theta = \pi/2$, $\phi = 0$ and $\pi/2$ for x' and y', respectively, and $\theta = 0$ for z'. Section 6-6 of the first edition gives a geometric interpretation of Eq. 6-21.

6-5 ANGULAR VARIATIONS OF EXPERIMENTAL SPECTRA

In an experimental spectrum the principal directions of the g-factor tensor are now known. Hence the experimental problem is to rotate the g-tensor expressed in the laboratory coordinate system abc shown in Fig. 6-3,

$$\overset{\leftrightarrow}{g}_{(abc)} = \begin{pmatrix} g_{aa} & g_{ab} & g_{ac} \\ g_{ab} & g_{bb} & g_{bc} \\ g_{ac} & g_{bc} & g_{cc} \end{pmatrix} \qquad (6\text{-}22)$$

about the three perpendicular laboratory coordinate axes x, y, and z defined in Fig. 6-3, using the following angular rotation matrices:

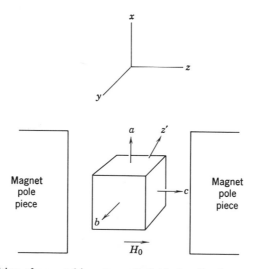

Fig. 6-3. Initial position of a crystal in a magnetic field, showing the crystal axes (a, b, c), the laboratory coordinate system (x, y, z), and one of the principal axes, z'.

$$\vec{\vec{R}}_x = \begin{pmatrix} 1 & 0 & 0 \\ 0 & \cos\theta_1 & \sin\theta_1 \\ 0 & -\sin\theta_1 & \cos\theta_1 \end{pmatrix}, \quad \vec{\vec{R}}_y = \begin{pmatrix} \cos\theta_2 & 0 & -\sin\theta_2 \\ 0 & 1 & 0 \\ \sin\theta_2 & 0 & \cos\theta_2 \end{pmatrix},$$

$$\vec{\vec{R}}_z = \begin{pmatrix} \cos\theta_3 & \sin\theta_3 & 0 \\ -\sin\theta_3 & \cos\theta_3 & 0 \\ 0 & 0 & 1 \end{pmatrix} \tag{6-23}$$

The transformation for $\vec{\vec{R}}_x$ will be illustrated in detail, and then the final equations for the other two planes will be written down.

If the rotation in the yz plane is carried out to give

$$\vec{\vec{g}}(\theta_1) = \vec{\vec{R}}_x \vec{\vec{g}}_{(abc)} \vec{\vec{R}}_x \tag{6-24}$$

then the g-factor $g_{(\theta_1)}$ has the angular variation shown in Fig. 6-4:

$$g^2_{(\theta_1)} = P_1 \cos^2\theta_1 + Q_1 \sin^2\theta_1 - 2R_1 \sin\theta_1 \cos\theta_1 \tag{6-25}$$

where

$$\begin{aligned} P_1 &= g^2_{ac} + g^2_{bc} + g^2_{cc} \\ Q_1 &= g^2_{ab} + g^2_{bb} + g^2_{bc} \\ R_1 &= g_{ab}g_{ac} + g_{bb}g_{bc} + g_{bc}g_{cc} \end{aligned} \tag{6-26}$$

with the average value $\frac{1}{2}(P+Q)$ and the amplitude

$$K_1 = +[(P_1 - Q_1)^2 + 4R_1^2]^{1/2} \tag{6-27}$$

as shown on Fig. 6-4.

To determine the principal values of g^2 a rotation is carried out about the three perpendicular directions x, y, and z, using transformation matrices 6-23 (Fig. 6-5). The data in each plane satisfy the equation

$$g^2_{(\theta_i)} = P_i \cos^2\theta_i + Q_i \sin^2\theta_i - 2R_i \sin\theta_i \cos\theta_i \tag{6-28}$$

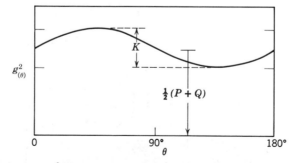

Fig. 6-4. Dependence of $g^2(\theta)$ on the angle. The average value $(P+Q)/2$ and the amplitude K of the sinusoidal angular dependence are indicated.

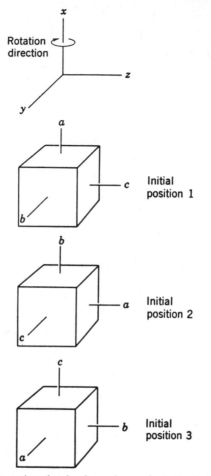

Fig. 6-5. Initial crystal orientations for the three planes of rotation in the laboratory coordinate system.

where

$$P_1 = Q_2, \qquad P_2 = Q_3, \qquad P_3 = Q_1 \qquad (6\text{-}29)$$

Matrix multiplication of the tensor $\vec{\vec{g}}$ by itself gives the tensor $\vec{\vec{g}}^2$. By carrying out this operation one obtains

$$
\vec{\vec{g}}^2 =
\begin{pmatrix}
g_{aa} & g_{ab} & g_{ac} \\
g_{ab} & g_{bb} & g_{bc} \\
g_{ac} & g_{bc} & g_{cc}
\end{pmatrix}
\begin{pmatrix}
g_{aa} & g_{ab} & g_{ac} \\
g_{ab} & g_{bb} & g_{bc} \\
g_{ac} & g_{bc} & g_{cc}
\end{pmatrix}
$$

$$
=
\begin{pmatrix}
[P_2 = Q_3] & R_3 & R_2 \\
R_3 & [P_3 = Q_1] & R_1 \\
R_2 & R_1 & [P_1 = Q_2]
\end{pmatrix}
\qquad (6\text{-}30)
$$

Thus we see that a determination of the nine parameters P_i, Q_i, and R_i provides the matrix of $\vec{\vec{g}}^2$. This matrix may be diagonalized by an orthogonal rotation matrix $\vec{\vec{R}}$ to give the principal values

$$\vec{\vec{R}}\vec{\vec{g}}^2\vec{\vec{R}} = \begin{pmatrix} g_1^2 & 0 & 0 \\ 0 & g_2^2 & 0 \\ 0 & 0 & g_3^2 \end{pmatrix} \tag{6-31}$$

The same transformation diagonalizes $\vec{\vec{g}}$ itself, since

$$\vec{\vec{R}}\vec{\vec{g}}^2\vec{\vec{R}} = \vec{\vec{R}}\vec{\vec{g}}\vec{\vec{R}}\vec{\vec{R}}\vec{\vec{g}}\vec{\vec{R}} \tag{6-32}$$

and

$$\vec{\vec{R}}\vec{\vec{g}}\vec{\vec{R}} = \begin{pmatrix} g_1 & 0 & 0 \\ 0 & g_2 & 0 \\ 0 & 0 & g_3 \end{pmatrix}$$

The matrix $\vec{\vec{R}}$ gives the direction cosines (Eq. 6-15) of the crystallographic coordinate system abc relative to the principal axis system $x'y'z'$ shown on Fig. 6-3. One should note that only the magnitude of each g_i is determined. Experimental papers frequently list the principal g-values and direction cosines of the g-tensor relative to the crystallographic axes.[4]

6-6 HYPERFINE HAMILTONIAN MATRIX

When hyperfine structure is present anisotropies in the spin Hamiltonian produce greater complications. The general interaction Hamiltonian for an electronic spin $S = \frac{1}{2}$ coupled to a nuclear spin $I = \frac{1}{2}$ in a magnetic field $\vec{H}$ is

$$\mathcal{H} = \beta \vec{S}_z \cdot \vec{\vec{g}} \cdot \vec{H} - \beta_N \vec{H} \cdot \vec{\vec{g}}_N \cdot \vec{I} + \vec{S} \cdot \vec{\vec{T}} \cdot \vec{I} \tag{6-33}$$

where the various terms are, respectively, the electronic Zeeman, nuclear Zeeman, and hyperfine interactions. In this and the next section the two g-factors are assumed to be isotropic, whereby the Hamiltonian simplifies to

$$\mathcal{H} = g\beta H S_z - g_N \beta_N H I_z + \vec{S} \cdot \vec{\vec{T}} \cdot \vec{I} \tag{6-34}$$

with the hyperfine tensor $\vec{\vec{T}}$ remaining anisotropic.

The use of the direct product expansion technique allows one to write for the Hamiltonian matrix

$$\begin{pmatrix} \frac{1}{2}[(g\beta - g_N\beta_N)H + \frac{1}{2}T_{zz}] & \frac{1}{4}[T_{xz} - iT_{yz}] & \frac{1}{4}[T_{xz} - iT_{yz}] & \frac{1}{4}[T_{xx} - T_{yy} - 2iT_{xy}] \\ \frac{1}{4}[T_{xz} + iT_{yz}] & \frac{1}{2}[(g\beta + g_N\beta_N)H - \frac{1}{2}T_{zz}] & \frac{1}{4}[T_{xx} + T_{yy}] & -\frac{1}{4}[T_{xz} - iT_{yz}] \\ \frac{1}{4}[T_{xz} + iT_{yz}] & \frac{1}{4}[T_{xx} + T_{yy}] & -\frac{1}{2}[(g\beta + g_N\beta_N)H + \frac{1}{2}T_{zz}] & -\frac{1}{4}[T_{xz} - iT_{yz}] \\ \frac{1}{4}[T_{xx} - T_{yy} + 2iT_{xy}] & -\frac{1}{4}[T_{xz} + iT_{yz}] & -\frac{1}{4}[T_{xz} + iT_{yz}] & -\frac{1}{2}[(g\beta - g_N\beta_N)H - \frac{1}{2}T_{zz}] \end{pmatrix}$$

$$(6\text{-}35)$$

Electron spin resonance studies are ordinarily carried out at high fields where

$$|T_{ij}| \ll |g\beta H| \tag{6-36}$$

and therefore the technique of degenerate first-order perturbation theory may be employed. This allows a neglect of all the off-diagonal matrix elements which connect different S_z states. As a result the Hamiltonian matrix simplifies to two 2×2 submatrices in the following manner:

$$\begin{pmatrix} \frac{1}{2}[(g\beta - g_N\beta_N)H + \frac{1}{2}T_{zz}] & \frac{1}{4}[T_{xz} - iT_{yz}] & 0 & 0 \\ \frac{1}{4}[T_{xz} + iT_{yz}] & \frac{1}{2}[(g\beta + g_N\beta_N)H - \frac{1}{2}T_{zz}] & 0 & 0 \\ 0 & 0 & -\frac{1}{2}[(g\beta + g_N\beta_N)H + \frac{1}{2}T_{zz}] & -\frac{1}{4}[T_{xz} - iT_{yz}] \\ 0 & 0 & -\frac{1}{4}[T_{xz} + iT_{yz}] & -\frac{1}{2}[(g\beta - g_N\beta_N)H - \frac{1}{2}T_{zz}] \end{pmatrix}$$

$$(6\text{-}37)$$

The two submatrices would be negatives of each other if it were not for the presence of the nuclear Zeeman term, $g_N\beta_N H$.

6-7 HYPERFINE PATTERNS

The submatrices of Eq. 6-37 correspond to quadratic equations which are easily solved. The upper left submatrix gives the energies E_1 and E_2; the lower right submatrix, E_3 and E_4:

$$\begin{aligned} E_1 &= \tfrac{1}{2}g\beta H + \tfrac{1}{4}T\sqrt{1 + A_-} \\ E_2 &= \tfrac{1}{2}g\beta H - \tfrac{1}{4}T\sqrt{1 + A_-} \end{aligned} \Bigg\} \quad (m_S = \tfrac{1}{2})$$

$$\begin{aligned} E_3 &= -\tfrac{1}{2}g\beta H + \tfrac{1}{4}T\sqrt{1 + A_+} \\ E_4 &= -\tfrac{1}{2}g\beta H - \tfrac{1}{4}T\sqrt{1 + A_+} \end{aligned} \Bigg\} \quad (m_S = -\tfrac{1}{2})$$

$$(6\text{-}38)$$

where the levels are listed in the order of decreasing energy, and the following notation has been adopted:

$$T = \pm\sqrt{T_{xz}^2 + T_{yz}^2 + T_{zz}^2} \tag{6-39}$$

$$A_\pm = \frac{2N}{T^2}\left(\tfrac{1}{2}N \pm T_{zz}\right) \tag{6-40}$$

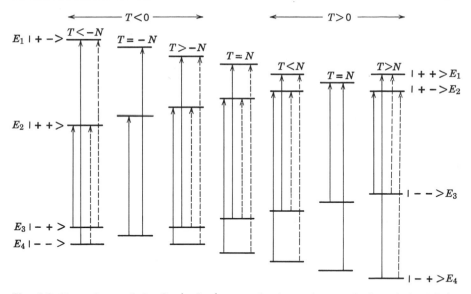

Fig. 6-6. Dependence of the $S = \frac{1}{2}$, $I = \frac{1}{2}$ energy levels on the magnitude and sign of the hyperfine coupling constant and nuclear Zeeman term for $T_{zz} = T$. The strong (—) and weak (---) transitions are shown.

$$N = 2g_N\beta_N H \qquad (6\text{-}41)$$

The dependence of the energy levels on N and T is shown in Fig. 6-6, and the effect of T_{zz} is indicated in Fig. 6-7. Note how the two levels cross for $N = \pm T$.

Along a principal direction, $T_{xz} = T_{yz} = 0$ and $T = T_{zz}$. In this case the four energies (6-37) reduce to the isotropic high-field counterparts of the form $E_i = \pm \frac{1}{2} g\beta H \pm \frac{1}{4}T \pm \frac{1}{4}N$.

The experimentally observed spectrum consists of four lines[17,18] arising from the four possible transitions with the selection rule $\Delta m_S = \pm 1$, $\Delta m_I = 0$. The outside lines arise from the $4 \rightarrow 1$ and $3 \rightarrow 2$ transitions when $|T| > |N|$

$$E_1 - E_4 = g\beta H + \tfrac{1}{4}T(\sqrt{1 + A_+} + \sqrt{1 + A_-})$$
$$E_2 - E_3 = g\beta H - \tfrac{1}{4}T(\sqrt{1 + A_+} + \sqrt{1 + A_-}) \qquad (6\text{-}42)$$

and the inside pair are derived from the $3 \rightarrow 1$ and $4 \rightarrow 2$ transitions:

$$E_1 - E_3 = g\beta H + \tfrac{1}{4}T(\sqrt{1 + A_-} - \sqrt{1 + A_+})$$
$$E_2 - E_4 = g\beta H - \tfrac{1}{4}T(\sqrt{1 + A_-} - \sqrt{1 + A_+}) \qquad (6\text{-}43)$$

as shown in Fig. 6-8 where the line spacings have the explicit magnitudes[18]

$$T_i = \tfrac{1}{2}|T||\sqrt{1 + A_+} - \sqrt{1 + A_-}| \qquad (6\text{-}44a)$$
$$T_o = \tfrac{1}{2}|T|(\sqrt{1 + A_+} + \sqrt{1 + A_-}) \qquad (6\text{-}44b)$$

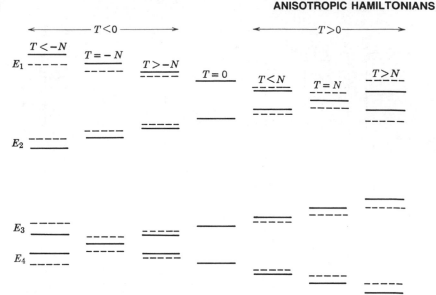

Fig. 6-7. Effect of the ratio T_{zz}/T on the energy levels of Fig. 6-6. Levels are shown for $T_{zz} = T$ (—) and for $|T_{zz}| < |T|$ (---).

One should note that in these expressions T_i and T_o are defined as positive and real, since only their magnitudes are determined experimentally. The quantity $g_N \beta_N H$ is known from nuclear magnetic resonance, T_i and T_o are measured experimentally, and T is the quantity which is to be determined from the spectra. With the use of Eq. 6-40 the sums and differences of Eqs. 6-44a and 6-44b provide the magnitudes of T and T_{zz}, as follows:

$$T^2 = T_i^2 + T_o^2 - N^2 \tag{6-45}$$

$$|T_{zz}| = \frac{T_i T_o}{N} \tag{6-46}$$

where it is always true that

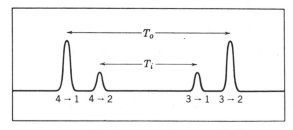

Fig. 6-8. The spectrum for $S = \frac{1}{2}$, $I = \frac{1}{2}$, showing the two strong $(4 \rightarrow 1, 3 \rightarrow 2)$ and two weak $(3 \rightarrow 1, 4 \rightarrow 2)$ transitions for $0 < N < T$.

$$T_i \leq N \leq T_o$$
$$T_i \leq T \leq T_o \qquad (6\text{-}47)$$

The inverse transformation is

$$T_i^2 = \tfrac{1}{2}(T^2 + N^2) - \tfrac{1}{2}\sqrt{(T^2 + N^2)^2 - 4T_{zz}^2 N^2}$$
$$T_o^2 = \tfrac{1}{2}(T^2 + N^2) + \tfrac{1}{2}\sqrt{(T^2 + N^2)^2 - 4T_{zz}^2 N^2} \qquad (6\text{-}48)$$

For convenience T_i is defined in Eq. 6-44 as positive definite, so only $|T_{zz}|$ appears in Eq. 6-46. The quantity T_{zz} can of course change sign in accordance with the expression

$$T_{zz} = \frac{T^2}{4N}(A_+ - A_-) \qquad (6\text{-}49)$$

and the relative sign of T_{zz} and those of the principal values T_a, T_b, T_c of T may be deduced from angular rotation spectra if all four hyperfine lines are observed. Values of T^2 calculated from Eq. 6-45 may be employed in the usual formula, which was derived in Section 6-5 for anisotropic g factors:

$$T^2 = P \cos^2 \theta + Q \sin^2 \theta - 2R \sin \theta \cos \theta \qquad (6\text{-}50)$$
$$= \tfrac{1}{2}(P + Q) + \tfrac{1}{2}(P - Q) \cos 2\theta - R \sin 2\theta \qquad (6\text{-}51)$$

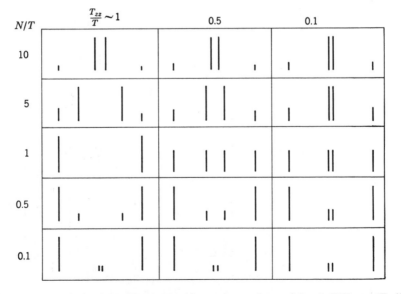

Fig. 6-9. Stick plot of the hyperfine quartet for various values of $2g_N\beta_N H/T$ and T_{zz}/T, keeping T_o constant. Close splittings of T_i are exaggerated. For $T = T_{zz} = N$ only two lines occur.

to evaluate the principal values and direction cosines of T^2. The principal experimental difficulty is that ordinarily T is appreciably greater than $2g_N\beta_N H$, and so only the outer doublet, T_o, is observed. The use of Eq. 6-50 with the assumption $T^2 = T_o^2$ leads to errors in some reported principal values of T.

We show in Section 7-6 of the first edition that the intensities I_i and I_o of the inner and outer lines, respectively, are given by [18]

$$I_i = \frac{N^2 - T_i^2}{T_o^2 - T_i^2} \qquad (6\text{-}52)$$

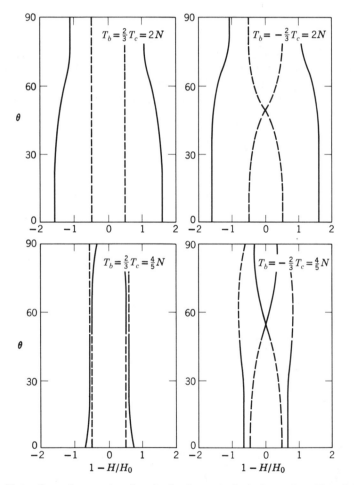

Fig. 6-10. Plots of angular rotation data in the b, c principal plane when T_b and T_c are the same (left) and opposite (right) in sign. The upper curves are for $T_b = 2N$ and the lower two for $T_b = 0.8N$, where $T_b = \pm 2T_c/3$. The allowed (—) and forbidden (---) transitions are indicated.

$$I_o = \frac{T_o^2 - N^2}{T_o^2 - T_i^2} \tag{6-53}$$

Stick plots of those intensities for several values of the ratio N/T and three values of the ratio T_{zz}/T are shown in Fig. 6-9.

The relative signs of two coupling constants are most easily determined by rotating in a plane containing them. For example, if the pair of principal values T_c and T_b in the c and b principal directions, respectively, have the ratio $|T_c/T_b| = 3$, then from Eq. 7-36 of the first edition T_{zz}/T will reach a minimum value of 0.87 if both members have the same sign and 0 if they differ in sign. Figure 6-10 shows the angular dependence curves when the crystal is rotated between two principal directions for the cases $T_b = 2T_c/3$, $T_b = -2T_c/3$. One should note that the two inner lines cross when the sign of the coupling constant changes during the rotation. This is illustrated in the right-hand plots of Fig. 6-10. No such crossing occurs when the coupling constant remains the same in sign, in accordance with the curves on the left-hand side of Fig. 6-10.

6-8 COMBINED *g*-FACTOR AND HYPERFINE ANISOTROPIES

The most obvious way of extending the theory is to consider the simultaneous presence of the *g*-factor and hyperfine anisotropies. This is done now for the two-spin, $S = I = \frac{1}{2}$ system.[4,12,19]

It is most convenient to employ the basis $|\psi_i m_I\rangle$, where ψ_1 and ψ_2 are the eigenfunctions for $T = 0$, as defined by Eq. 6-9, and $m_I = \pm\frac{1}{2}$. The Hamiltonian

$$\mathcal{H} = \beta \vec{S} \cdot \vec{\vec{g}} \cdot \vec{H} + \vec{S} \cdot \vec{\vec{T}} \cdot \vec{I} \tag{6-54}$$

is a 4×4 matrix of the form

	$\lvert\psi_1 +\rangle$	$\lvert\psi_1 -\rangle$	$\lvert\psi_2 +\rangle$	$\lvert\psi_2 -\rangle$
$\langle\psi_1 +\rvert$	To be		Neglect	
$\langle\psi_1 -\rvert$	diagonalized			
$\langle\psi_2 +\rvert$	Neglect		Negative	
$\langle\psi_2 -\rvert$			of upper left	

$$\tag{6-55}$$

It is constructed by writing out the hyperfine matrix with its nine components $S_i T_{ij} I_j$:

$$\vec{S} \cdot \vec{\vec{T}} \cdot \vec{I} = S_x T_{xx} I_x + S_x T_{xy} I_y + \cdots, \tag{6-56}$$

According to standard stationary-state degenerate perturbation theory, it is necessary to diagonalize the upper left submatrix of Eq. 6-55 exactly. When the above prescription is followed, this submatrix is of the form

$$
\begin{pmatrix}
\frac{1}{2} g\beta H + \frac{1}{4} U & \dfrac{V - iW}{4} \\[2ex]
\dfrac{V + iW}{4} & \frac{1}{2} g\beta H - \frac{1}{4} U
\end{pmatrix}
\tag{6-57}
$$

and the energies are

$$
\begin{aligned}
E &= \tfrac{1}{2} g\beta H \pm \tfrac{1}{4} (U^2 + V^2 + W^2)^{1/2} \\
&= \tfrac{1}{2} g\beta H \pm \tfrac{1}{4} T_{\exp}
\end{aligned}
\tag{6-58}
$$

which corresponds to two lines separated by the energy $T_{\exp} = (U^2 + V^2 + W^2)^{1/2}$, as shown in Fig. 6-11. The quantities U, V, and W are given by

$$
U^2 = \frac{1}{g^2} (XT_{xz} + YT_{yz} + ZT_{zz})
$$

$$
V^2 = \frac{1}{g^2} (XT_{xx} + YT_{yx} + ZT_{zx})
\tag{6-59}
$$

$$
W^2 = \frac{1}{g^2} (XT_{xy} + YT_{yy} + ZT_{zy})
$$

$$
\begin{aligned}
X &= l_1 g_{xx} + l_2 g_{xy} + l_3 g_{xz} \\
Y &= l_1 g_{yx} + l_2 g_{yy} + l_3 g_{yz} \\
Z &= l_1 g_{zx} + l_2 g_{zy} + l_3 g_{zz} \\
l_1 &= \sin\theta\cos\phi, \qquad l_2 = \sin\theta\sin\phi, \qquad l_3 = \cos\theta
\end{aligned}
\tag{6-60}
$$

According to our notation, the T_{ij} are components of the hyperfine tensor that appears in the Hamiltonian of Eq. 6-54. The quantity $T_{\exp}$, on the other hand, is the experimentally measured hyperfine splitting, as shown in Fig. 6-11. It may be expressed in matrix notation as

$$
T_{\exp}^2 = \frac{1}{g^2} \vec{l} \cdot \vec{\vec{g}} \vec{\vec{T}}^2 \vec{\vec{g}} \cdot \vec{l}
\tag{6-61}
$$

as is easily verified by multiplying out the right-hand side where $\vec{l}$ is a unit vector in the magnetic field direction, in accordance with Eqs. 6-19 and 6-60. Note that the angular dependence of $T_{\exp}^2$ arises from both the vector $\vec{l}$ and the scalar g^2, defined by Eq. 6-6.

To experimentally evaluate the principal values of T^2 we follow the procedure of rotating about three perpendicular axes, as described in

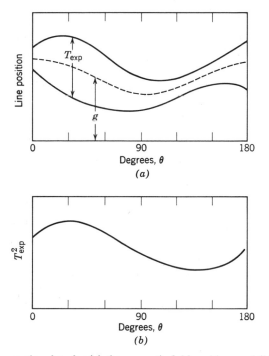

Fig. 6-11. Angular rotation data for (a) the magnetic field positions and (b) hyperfine splitting squared due to a proton interacting with an unpaired electron.

Section 6-10. It is convenient to simplify the notation by defining a new matrix $\vec{\mathbf{K}}^2$:

$$\vec{\mathbf{K}}^2 = \vec{\mathbf{g}}\vec{\mathbf{T}}^2\vec{\mathbf{g}} \tag{6-62}$$

In analogy with Eq. 6-28 of the g-factor case a rotation about x axis gives

$$g^2 T_{exp}^2 = P_1' \cos^2 \theta + Q_1' \sin^2 \theta - 2R_1' \sin \theta \cos \theta \tag{6-63}$$

where the parameters P_1', Q_1', and R_1' are independent of the angle θ:

$$\begin{aligned} P_1' &= K_{xz}^2 + K_{yz}^2 + K_{zz}^2 \\ Q_1' &= K_{xy}^2 + K_{yy}^2 + K_{yz}^2 \\ R_1' &= K_{xy}K_{xz} + K_{yy}K_{yz} + K_{yz}K_{zz} \end{aligned} \tag{6-64}$$

Rotations about the y and z axes give P_2', Q_2', R_2', and P_3', Q_3', and R_3', respectively. The quantities K_{ij} are complicated functions of their counterparts, T_{ij}, and the elements, g_{ij}, of the g-factor tensor. The experimentally measured P'-Q'-R' matrix K^2 is easily written from Eq. 6-63 and its counterparts in the other two planes:

$$\begin{pmatrix} P_2' & R_3' & R_2' \\ R_3' & P_3' & R_1' \\ R_2' & R_1' & P_1' \end{pmatrix} = \begin{pmatrix} (K^2)_{xx} & (K^2)_{xy} & (K^2)_{xz} \\ (K^2)_{xy} & (K^2)_{yy} & (K^2)_{yz} \\ (K^2)_{xz} & (K^2)_{yz} & (K^2)_{zz} \end{pmatrix} \qquad (6\text{-}65)$$

The $\vec{\vec{K}}^2$ matrix (6-65) is converted to the physically meaningful matrix $\vec{\vec{T}}^2$ by an inversion of Eq. 6-62,

$$\vec{\vec{T}}^2 = \vec{\vec{g}}^{-1} \vec{\vec{K}}^2 \vec{\vec{g}}^{-1} \qquad (6\text{-}66)$$

to give the square of the hyperfine coupling matrix, which appears in the Hamiltonian of Eq. 6-54.

The principal values of $\vec{\vec{T}}^2$ may be obtained in the usual manner by calculating the direction cosine matrix R on a computer. These direction cosines relate the principal axes of the hyperfine tensor to the coordinate system of the laboratory where the measurements are carried out. Usually each T_i is the same sign, and the square root of each T_i^2 gives the hyperfine coupling constant to within a sign.

The manner in which both the g-factor and the observed hyperfine splitting T_{exp} vary with the angle is shown in Fig. 6-11. Spectra at three typical angles are presented in Fig. 6-12.

The first edition discusses anisotropies involving several nuclear spins (Section 8-3) and proton spin flip satellite lines[17-24] (Section 8-4).

6-9 ALPHA AND BETA PROTONS

In an organic free radical the unpaired electron is ordinarily in a p_z atomic orbital localized on a carbon atom called C_α. Most hyperfine structure arises

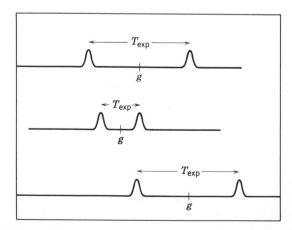

Fig. 6-12. Hyperfine patterns at three orientations for a spin-$\frac{1}{2}$ nucleus with an associated anisotropic g-factor.

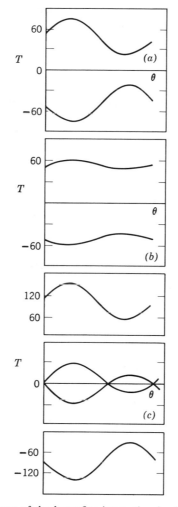

Fig. 6-13. Angular dependence of the hyperfine interaction for (*a*) α-protons, (*b*) β-protons, and (*c*) both present simultaneously. The ordinate T is in megahertz.

from α-protons which are bonded to this α-carbon, and β-protons which are bonded to an adjacent carbon atom denoted by C_β. A typical radical is

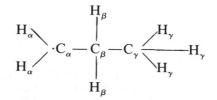

where the coupling to γ and more distant protons is negligible.

A β-proton exhibits a small hyperfine anisotropy as shown in Fig. 6-13b whereas an α-proton is more anisotropic, as shown in Fig. 6-13a. If both protons are present simultaneously, then the composite pattern illustrated in Fig. 6-13c is obtained.

6-10 EXPERIMENTAL DETERMINATION OF THE g- AND T-TENSORS

To determine the principal values and direction cosines of the g-factor and hyperfine tensors of a particular solid it is best to record the spectrum at regular angular intervals over a range of 180° in three perpendicular planes. For this purpose the three coordinate systems illustrated in Fig. 6-3 are used:

(a) The laboratory system (x, y, z), with the applied magnetic field along the z direction and the axis of rotation x directed vertically.

(b) The (a, b, c) frame of reference, which is fixed in the crystal and chosen in relation to the crystallographic planes.

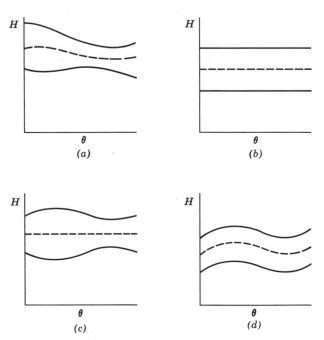

Fig. 6-14. Hyperfine doublet with (a) anisotropic g and T, (b) isotropic g and T, (c) isotropic g and anisotropic T, and (d) anisotropic g and isotropic T. The spectral line positions (——) and their average value (– – –) corresponding to the g-factor angular dependence from Eq. 6-58 are shown.

(c) The principal axes (x', y', z') system in which a given g- or T-tensor is diagonal.

The initial crystal orientations for the three rotations are shown in Fig. 6-5. These provide the relations P_i and Q_j for $i, j = 1, 2, 3$ in Eq. 6-29 for use in Eq. 6-50.

It is of interest to summarize the procedure for experimentally obtaining the principal values of $\vec{\vec{g}}$ and $\vec{\vec{T}}$.

1. The crystal is rotated successively about three perpendicular axes in the crystal, and spectra are recorded every few degrees. Three typical spectra are shown in Fig. 6-12 and angular rotation data are plotted in Figs. 6-11 and 6-14.

2. The g-factor curve (---) of Fig. 6-11 is fitted to Eq. 6-28 for each axis.

3. The matrix (P, Q, R) is formed and diagonalized according to Eqs. 6-30 to 6-32.

4. The T_{exp} curves for the three axes (Fig. 6-5) are fitted to Eq. 6-50.

5. The inverse of the g-factor matrix, $\vec{\vec{g}}^{-1}$, is calculated from

$$
\begin{pmatrix} C_{11} & C_{21} & C_{31} \\ C_{12} & C_{22} & C_{32} \\ C_{13} & C_{23} & C_{33} \end{pmatrix}
\begin{pmatrix} \dfrac{1}{g_1} & 0 & 0 \\ 0 & \dfrac{1}{g_2} & 0 \\ 0 & 0 & \dfrac{1}{g_3} \end{pmatrix}
\begin{pmatrix} C_{11} & C_{12} & C_{13} \\ C_{21} & C_{22} & C_{23} \\ C_{31} & C_{32} & C_{33} \end{pmatrix}
$$

$$
= \begin{pmatrix} \left(\dfrac{1}{g}\right)_{xx} & \left(\dfrac{1}{g}\right)_{xy} & \left(\dfrac{1}{g}\right)_{xz} \\ \left(\dfrac{1}{g}\right)_{yx} & \left(\dfrac{1}{g}\right)_{yy} & \left(\dfrac{1}{g}\right)_{yz} \\ \left(\dfrac{1}{g}\right)_{zx} & \left(\dfrac{1}{g}\right)_{zy} & \left(\dfrac{1}{g}\right)_{zz} \end{pmatrix}
\tag{6-67}
$$

6. The matrix $\vec{\vec{T}}^2$ is computed from Eq. 6-66.

7. The matrix $\vec{\vec{T}}^2$ is diagonalized to find its principal values and their direction cosines.

A number of experimental data have been published without taking into account the influence of the nuclear Zeeman term.

8. The crystal is reoriented so that the principal directions x', y', and z' replace those of a, b, and c in Fig. 6-3.

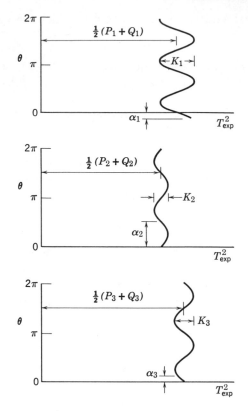

Fig. 6-15. Angular rotation results in three planes, illustrating a typical small-anisotropy case.

Steps 1 to 7 are now repeated in the principal planes, and more accurate principal values and direction cosines are obtained. This more refined measurement largely compensates for errors arising from the influence of the nuclear Zeeman term.

For the two-spin $S = \frac{1}{2}$, $I = \frac{1}{2}$ system there are four possible sets of anisotropies, as illustrated in Fig. 6-14. Typical angular rotation curves for combined anisotropies are shown in Fig. 6-15. In the initial inspection of experimental data one should try to classify them in one of these four categories.

REFERENCES

1. D. S. Schonland, *Proc. Phys. Soc.*, **73**, 788 (1959).
2. J. A. Weil and J. H. Anderson, *J. Chem. Phys.*, **28**, 864 (1958).
3. A. Lund and T. Vänngärd, *J. Chem. Phys.*, **42**, 2979 (1965).
4. C. P. Poole, Jr. and H. A. Farach, *Adv. Magn. Reson.*, **5**, 229 (1971).

5. A. Carrington and A. D. McLachlan, *Introduction to Magnetic Resonance*, Harper & Row, New York, 1967. Chapter 7 covers much of the material presented here but from a different viewpoint.

6. C. P. Slichter, *Principles of Magnetic Resonance*, 2nd ed., Harper & Row, New York, 1980.

7. I. Miyagawa and W. Gordy, *J. Chem. Phys.*, **32**, 255 (1960).

8. H. M. McConnell, C. Heller, T. Cole, and R. W. Fessenden, *J. Am. Chem. Soc.*, **82**, 776 (1960).

9. J. A. Weil and J. H. Anderson, *J. Chem. Phys.*, **35**, 1410 (1961).

10. J. R. Morton, *Chem. Rev.*, **64**, 453 (1964).

11. H. Fischer, *Magnetic Properties of Free Radicals*, Vol. 1, *Group II (Atomic and Molecular Physics) of Landolt-Bornstein*, K. H. Hellwege, ed., Springer-Verlag, Berlin, 1965. Tabulation of data on irradiated organic free radicals.

12. J. E. Wertz and J. R. Bolton, *Electron Spin Resonance*, Chapman & Hall, New York, 1986.

13. L. A. Sorin and M. V. Viasova, *Electron Spin Resonance of Paramagnetic Crystals*, Plenum, New York, 1973.

14. G. Lancaster, *Electron Spin Resonance in Semiconductors*, Plenum, New York, 1967.

15. T. F. Yen, ed., *Electron Spin Resonance of Metal Complexes*, Plenum, New York, 1969.

16. H. A. Buckmaster and B. D. Delay, *Magn. Reson. Rev.*, **3**, 127 (1974); **4**, 63 (1976); **5**, 25, 121 (1979); **6**, 85, 139 (1980); **8**, 285 (1983).

17. G. T. Trammell, H. Zeldes, and R. Livingston, *Phys. Rev.*, **110**, 630 (1958).

18. C. P. Poole, Jr. and H. A. Farach, *J. Magn. Reson.*, **4**, 312 (1971); **5**, 305 (1971).

19. H. A. Farach and C. P. Poole, Jr., *Nuovo Cimento*, **4**, 51 (1971).

20. N. M. Atherton and D. H. Whiffen, *Mol. Phys.*, **3**, 1 (1960).

21. W. C. Lin, C. A. McDowell, and J. R. Rowlands, *J. Chem. Phys.*, **35**, 757 (1961).

22. H. N. Rexroad, Y. H. Hahn, and W. J. Temple, *J. Chem. Phys.*, **42**, 324 (1965).

23. M. Jagannadha and R. S. Anderson, *J. Chem. Phys.*, **42**, 2899 (1965).

24. D. V. G. L. N. Rao and W. Gordy, *J. Chem. Phys.*, **35**, 362 (1966).

MULTISPIN SYSTEMS

7-1 INTRODUCTION

The direct product method facilitates the treatment of Hamiltonians with several nuclear spins, but the calculations become tedious as the number of spins becomes large. For this reason we present a detailed treatment of the three-spin $(\frac{1}{2}, \frac{1}{2}, \frac{1}{2})$ case in general at high field and low field, and also in the NMR and ESR limits. The results obtained for this case can be generalized to larger groups of coupled spins.

7-2 THREE-SPIN ZEEMAN INTERACTIONS

The three-spin $(\frac{1}{2}, \frac{1}{2}, \frac{1}{2})$ case is quite straightforward in the absence of spin–spin coupling. The Hamiltonian assumes the rather simple form

$$\mathcal{H} = (g_1 I_{1z} + g_2 I_{2z} + g_3 I_{3z})\beta H \tag{7-1}$$

where only isotropic Zeeman terms are included, and we assume positive g-factors with

$$g_3 < g_2 < g_1 \tag{7-2}$$

The eight energies are

$$E = (g_1 m_1 + g_2 m_2 + g_3 m_3)\beta H \tag{7-3}$$

$$= \frac{(\pm g_1 \pm g_2 \pm g_3)\beta H}{2} \tag{7-4}$$

where each possible combination of plus and minus signs corresponds to a particular energy level. The energy level diagram for $g_3 < g_2 \ll g_1$ is shown in Fig. 7-1, and that for $g_1 \sim g_2 \sim g_3$, subject to the condition of Eq. 7-2, is

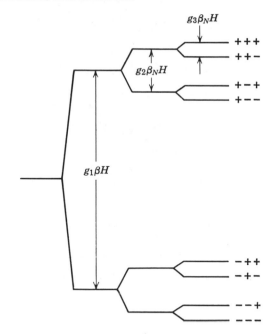

Fig. 7-1. Energy level diagram for three spin-$\frac{1}{2}$ magnetic moments in a magnetic field where $g_3 < g_2 \ll g_1$ and no spin–spin (or hyperfine) coupling exists.

given in Fig. 7-2. The eight eigenfunctions $|m_1 m_2 m_3\rangle$ for the Hamiltonian of Eq. 7-1 in several common notations are as follows:

$$
\begin{aligned}
\psi_1 &= |{\textstyle\frac{1}{2}}\ {\textstyle\frac{1}{2}}\ {\textstyle\frac{1}{2}}\rangle & &= |+++\rangle = \alpha\alpha\alpha \\
\psi_2 &= |{\textstyle\frac{1}{2}}\ {\textstyle\frac{1}{2}}\ -{\textstyle\frac{1}{2}}\rangle & &= |++-\rangle = \alpha\alpha\beta \\
\psi_3 &= |{\textstyle\frac{1}{2}}\ -{\textstyle\frac{1}{2}}\ {\textstyle\frac{1}{2}}\rangle & &= |+-+\rangle = \alpha\beta\alpha \\
\psi_4 &= |-{\textstyle\frac{1}{2}}\ {\textstyle\frac{1}{2}}\ {\textstyle\frac{1}{2}}\rangle & &= |-++\rangle = \beta\alpha\alpha \\
\psi_5 &= |{\textstyle\frac{1}{2}}\ -{\textstyle\frac{1}{2}}\ -{\textstyle\frac{1}{2}}\rangle & &= |+--\rangle = \alpha\beta\beta \\
\psi_6 &= |-{\textstyle\frac{1}{2}}\ {\textstyle\frac{1}{2}}\ -{\textstyle\frac{1}{2}}\rangle & &= |-+-\rangle = \beta\alpha\beta \\
\psi_7 &= |-{\textstyle\frac{1}{2}}\ -{\textstyle\frac{1}{2}}\ {\textstyle\frac{1}{2}}\rangle & &= |--+\rangle = \beta\beta\alpha \\
\psi_8 &= |-{\textstyle\frac{1}{2}}\ -{\textstyle\frac{1}{2}}\ -{\textstyle\frac{1}{2}}\rangle & &= |---\rangle = \beta\beta\beta
\end{aligned}
\tag{7-5}
$$

One should note that

$$
\mathscr{H}|m_1 m_2 m_3\rangle = (g_1 m_1 + g_2 m_2 + g_3 m_3)\beta H|m_1 m_2 m_3\rangle
\tag{7-6}
$$

since

$$
\begin{aligned}
I_{1z}|m_1 m_2 m_3\rangle &= m_1|m_1 m_2 m_3\rangle \\
I_{2z}|m_1 m_2 m_3\rangle &= m_2|m_1 m_2 m_3\rangle \\
I_{3z}|m_1 m_2 m_3\rangle &= m_3|m_1 m_2 m_3\rangle
\end{aligned}
\tag{7-7}
$$

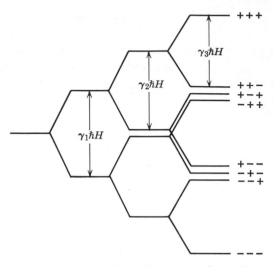

Fig. 7-2. Energy level diagram for three almost identical spin-$\frac{1}{2}$ nuclei in a magnetic field where $\gamma_3 < \gamma_2 < \gamma_1$ ($g_3 < g_2 < g_1$) and no spin–spin coupling exists.

where

$$|m_1 m_2 m_3\rangle = |m_1\rangle|m_2\rangle|m_3\rangle \tag{7-8}$$

In other words, the kets $|m_1 m_2 m_3\rangle$ are eigenfunctions of the three z components of spin, and the absence of a spin–spin coupling term automatically makes them eigenfunctions of the Hamiltonian.

If spin–spin coupling is present then linear combinations of the eigenfunctions ψ_i are more appropriate. For example, when a spin of one type is coupled to two equivalent spins of another type the following eigenfunctions are convenient to use:

$$\phi_1 = |\tfrac{1}{2}\ \tfrac{1}{2}\ \tfrac{1}{2}\rangle = \alpha\alpha\alpha$$

$$\phi_2 = \frac{1}{\sqrt{2}}\left[|\tfrac{1}{2}\ \tfrac{1}{2}\ -\tfrac{1}{2}\rangle + |\tfrac{1}{2}\ -\tfrac{1}{2}\ \tfrac{1}{2}\rangle\right] = \frac{1}{\sqrt{2}}\ \alpha(\alpha\beta + \beta\alpha)$$

$$\phi_3 = |-\tfrac{1}{2}\ \tfrac{1}{2}\ \tfrac{1}{2}\rangle = \beta\alpha\alpha$$

$$\phi_4 = \frac{1}{\sqrt{2}}\left[|\tfrac{1}{2}\ \tfrac{1}{2}\ -\tfrac{1}{2}\rangle - |\tfrac{1}{2}\ -\tfrac{1}{2}\ \tfrac{1}{2}\rangle\right] = \frac{1}{\sqrt{2}}\ \alpha(\alpha\beta - \beta\alpha) \tag{7-9}$$

$$\phi_5 = \frac{1}{\sqrt{2}}\left[|-\tfrac{1}{2}\ \tfrac{1}{2}\ -\tfrac{1}{2}\rangle - |-\tfrac{1}{2}\ -\tfrac{1}{2}\ \tfrac{1}{2}\rangle\right] = \frac{1}{\sqrt{2}}\ \beta(\alpha\beta - \beta\alpha)$$

$$\phi_6 = |\tfrac{1}{2}\ -\tfrac{1}{2}\ -\tfrac{1}{2}\rangle = \alpha\beta\beta$$

$$\phi_7 = \frac{1}{\sqrt{2}}\left[|-\tfrac{1}{2}\ \tfrac{1}{2}\ -\tfrac{1}{2}\rangle + |-\tfrac{1}{2}\ -\tfrac{1}{2}\ \tfrac{1}{2}\rangle\right] = \frac{1}{\sqrt{2}}\ \beta(\alpha\beta + \beta\alpha)$$

$$\phi_8 = |-\tfrac{1}{2}\ -\tfrac{1}{2}\ -\tfrac{1}{2}\rangle = \beta\beta\beta$$

This set leaves unchanged the four ket vectors, ψ_1, ψ_4, ψ_5, and ψ_8, and takes linear combinations of the other vectors in pairs. These wavefunctions are convenient for use with the AB_2 group in NMR and the CH_2 radical in ESR, since they single out one spin and couple the other two spins in an equivalent manner.

7-3 GENERAL THREE-SPIN ($\frac{1}{2}$, $\frac{1}{2}$, $\frac{1}{2}$) CASE

The spin Hamiltonian for three coupled spins, I_1, I_2, and I_3, in a magnetic field H is

$$\mathcal{H} = g_1\beta\vec{H}\cdot\vec{I}_1 + g_2\beta\vec{H}\cdot\vec{I}_2 + g_3\beta\vec{H}\cdot\vec{I}_3 + \vec{I}_1\cdot\vec{\vec{T}}_{12}\cdot\vec{I}_2 + \vec{I}_2\cdot\vec{\vec{T}}_{23}\cdot\vec{I}_3$$

$$+ \vec{I}_3\cdot\vec{\vec{T}}_{31}\cdot\vec{I}_1 \tag{7-10}$$

For isotropic interactions this simplifies to

$$\mathcal{H} = \beta H(g_1 I_{1z} + g_2 I_{2z} + g_3 I_{3z}) + T_{12}\vec{I}_1\cdot\vec{I}_2 + T_{23}\vec{I}_2\cdot\vec{I}_3 + T_{31}\vec{I}_3\cdot\vec{I}_1 \tag{7-11}$$

where the external magnetic field H is selected in the z direction. With the eight fundamental wavefunctions $|m_1 m_2 m_3\rangle$ of Eq. 7-5 used as a basis set, the direct product expansions are made in the following manner, exemplified by one component of the Zeeman interaction:

$$\beta H g_2 I_{2z} = \tfrac{1}{2}g_2\beta H \begin{pmatrix} 1 & 0 \\ 0 & 1 \end{pmatrix} \times \begin{pmatrix} 1 & 0 \\ 0 & -1 \end{pmatrix} \times \begin{pmatrix} 1 & 0 \\ 0 & 1 \end{pmatrix} \tag{7-12}$$

$$= \tfrac{1}{2}g_2\beta H \left(\begin{array}{cccc|cccc} 1 & 0 & 0 & 0 & 0 & 0 & 0 & 0 \\ 0 & 1 & 0 & 0 & 0 & 0 & 0 & 0 \\ 0 & 0 & -1 & 0 & 0 & 0 & 0 & 0 \\ 0 & 0 & 0 & -1 & 0 & 0 & 0 & 0 \\ \hline 0 & 0 & 0 & 0 & 1 & 0 & 0 & 0 \\ 0 & 0 & 0 & 0 & 0 & 1 & 0 & 0 \\ 0 & 0 & 0 & 0 & 0 & 0 & -1 & 0 \\ 0 & 0 & 0 & 0 & 0 & 0 & 0 & -1 \end{array}\right)$$

$$\tag{7-13}$$

For the spin–spin term between spins I_1 and I_2,

$$T_{12}\vec{I}_1 \cdot \vec{I}_2 = T_{12}(I_{1x}I_{2x} + I_{1y}I_{2y} + I_{1z}I_{2z}) \tag{7-14}$$

each part expands in the manner exemplified by $I_{1y}I_{2y}$,

$$T_{12}I_{1y}I_{2y} = \tfrac{1}{4}T_{12}\begin{pmatrix} 0 & -i \\ i & 0 \end{pmatrix} \times \begin{pmatrix} 0 & -i \\ i & 0 \end{pmatrix} \times \begin{pmatrix} 1 & 0 \\ 0 & 1 \end{pmatrix} \tag{7-15}$$

$$= \tfrac{1}{4}\left(\begin{array}{cccc|cccc}
0 & 0 & 0 & 0 & 0 & 0 & -1 & 0 \\
0 & 0 & 0 & 0 & 0 & 0 & 0 & -1 \\
0 & 0 & 0 & 0 & 1 & 0 & 0 & 0 \\
0 & 0 & 0 & 0 & 0 & 1 & 0 & 0 \\
\hline
0 & 0 & 1 & 0 & 0 & 0 & 0 & 0 \\
0 & 0 & 0 & 1 & 0 & 0 & 0 & 0 \\
-1 & 0 & 0 & 0 & 0 & 0 & 0 & 0 \\
0 & -1 & 0 & 0 & 0 & 0 & 0 & 0
\end{array} \right)$$

$$\tag{7-16}$$

The other parts of the Hamiltonian are expanded in like manner, keeping the spins in the sequence I_1, I_2, I_3 and inserting a unit matrix wherever the corresponding spin is missing.

After each Hamiltonian term has been expanded by direct products and the resulting 8×8 matrices have been added together, one obtains the following overall Hamiltonian matrix:

	$\lvert + + + \rangle$	$\lvert + + - \rangle$	$\lvert + - + \rangle$
$\langle + + + \rvert$	$\left[\begin{array}{c} \frac{1}{2}(g_1 + g_2 + g_3)\beta H \\ + \frac{1}{4}(T_{12} + T_{23} + T_{31}) \end{array} \right]$	0	0
$\langle + + - \rvert$	0	$\left[\begin{array}{c} \frac{1}{2}(g_1 + g_2 - g_3)\beta H \\ + \frac{1}{4}(T_{12} - T_{23} - T_{31}) \end{array} \right]$	$\frac{1}{2}T_{23}$
$\langle + - + \rvert$	0	$\frac{1}{2}T_{23}$	$\left[\begin{array}{c} \frac{1}{2}(g_1 - g_2 + g_3)\beta H \\ + \frac{1}{4}(-T_{12} - T_{23} + T_{31}) \end{array} \right]$
$\langle - + + \rvert$	0	$\frac{1}{2}T_{13}$	$\frac{1}{2}T_{12}$
$\langle + - - \rvert$	0	0	0
$\langle - + - \rvert$	0	0	0
$\langle - - + \rvert$	0	0	0
$\langle - - - \rvert$	0	0	0

This matrix has all the Zeeman terms on the diagonal and all the spin–spin terms either on the diagonal or at locations where $\Delta m = (\Delta m_1 + \Delta m_2 + \Delta m_3) = 0$ in the corresponding matrix element. One should note that the rows and columns of the Hamiltonian matrix of Eq. 7-17 have been rearranged to group them by their m values.

The two exact energies from the upper left- and lower right-hand corners of Eq. 7-17 are

$$E = \pm \tfrac{1}{2}(g_1 + g_2 + g_3) + \tfrac{1}{4}(T_{12} + T_{23} + T_{31}) \tag{7-18}$$

and the remaining six energies arise from the solution of the cubic equations

$$(E - K_1^{\pm})(E - K_2^{\pm})(E - K_3^{\pm})$$
$$- \tfrac{1}{4}[T_{23}^2(E - K_1^{\pm}) + T_{31}^2(E - K_2^{\pm}) + T_{12}^2(E - K_3^{\pm})] - \tfrac{1}{4}T_{12}T_{23}T_{31} = 0 \tag{7-19}$$

where

$$K_i^{\pm} = \tfrac{1}{4}(T_{jk} - T_{ij} - T_{ik}) \pm \tfrac{1}{2}\beta H(g_i - g_j - g_k) \tag{7-20}$$

for $i, j, k = 1, 2, 3$ in cyclic order, and $T_{ij} = T_{ji}$.

Three energies arise from the use of K_i with a plus sign, and three originate from the minus sign case. Since this cubic equation is quite complex, the general closed-form solution will not be written down. It should be pointed out that at high fields, where the Zeeman terms dominate, the spectrum becomes symmetric to first order, corresponding to the plus and minus sign cases of Eqs. 7-18 to 7-20. At low fields such symmetry does not exist, and at zero field one obtains from Eq. 7-19

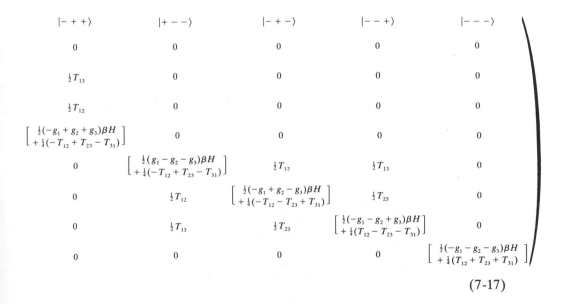

$$(7\text{-}17)$$

$$E^3 + \tfrac{1}{4}E^2(T_{12} + T_{23} + T_{31}) - \tfrac{1}{16}E[5(T_{12}^2 + T_{23}^2 + T_{31}^2)$$
$$-2(T_{12}T_{23} + T_{23}T_{31} + T_{31}T_{12})]$$
$$- \tfrac{3}{64}[T_{12}^3 + T_{23}^3 + T_{31}^3 - T_{12}^2(T_{23} + T_{31}) - T_{23}^2(T_{31} + T_{12})$$
$$- T_{31}^2(T_{12} + T_{23}) - 6T_{12}T_{23}T_{31}] \tag{7-21}$$

which has three different roots in general. Hence at zero field there is a doublet at $E = \tfrac{1}{4}(T_{12} + T_{23} + T_{31})$ from the two exact energy levels and three more doublets given by Eq. 7-21 since this equation occurs twice in the matrix of Eq. 7-17. For three equally coupled spins Eq. 7-21 gives $E = 3T/4$ once and $-3T/4$ twice, as expected

This three-spin $(\tfrac{1}{2}, \tfrac{1}{2}, \tfrac{1}{2})$ case is quite complex, and hence the particular subcases which occur frequently in ESR and NMR are discussed separately in the next three sections.

7-4 THREE-SPIN ESR CASE

In electron spin resonance studies it is quite common to encounter experimental situations in which two nuclear spins, I_1 and I_2, interact with an electron spin S in a strong magnetic field H where the electronic Zeeman energy $g\beta H$ far exceeds all other energies. The Hamiltonian has the form

$$\mathcal{H} = g\beta H S_z - \beta_N H(g_1 I_{1z} + g_2 I_{2z}) + T_1 \vec{S} \cdot \vec{I}_1 + T_2 \vec{S} \cdot \vec{I}_2 \tag{7-22}$$

where the terms of Eq. 7-11 have been relabeled in an obvious manner, and the spin–spin coupling between the two nuclear spins has been neglected. This Hamiltonian is valid for all field strengths. The 3×3 submatrix from 7-17 has the form

$$\begin{pmatrix} \pm\tfrac{1}{2}[g\beta H - (g_1 - g_2)\beta_N H] \\ -\tfrac{1}{4}(-T_1 + T_2) \end{pmatrix} \qquad 0 \qquad \tfrac{1}{2}T_2 \\[2mm] \quad 0 \qquad \begin{matrix} \pm\tfrac{1}{2}[g\beta H + (g_1 - g_2)\beta_N H] \\ -\tfrac{1}{4}(T_1 - T_2) \end{matrix} \qquad \tfrac{1}{2}T_1 \\[2mm] \quad \tfrac{1}{2}T_2 \qquad \tfrac{1}{2}T_1 \qquad \begin{matrix} \pm\tfrac{1}{2}[-g\beta H - (g_1 + g_2) \\ \beta_N H] - \tfrac{1}{4}(T_1 + T_2) \end{matrix} \end{pmatrix}$$

$$\tag{7-23}$$

We now discuss two limiting cases.

At zero field $(H = 0)$ the energies are solutions of the cubic equation

$$E^3 + \tfrac{1}{4}E^2(T_1 + T_2) - \tfrac{1}{16}E[5(T_1^2 + T_2^2) - 2T_1 T_2]$$
$$- \tfrac{3}{16}[T_1^3 + T_2^3 - T_1 T_2(T_1 + T_2)] = 0 \tag{7-24}$$

which contains only the two hyperfine coupling constants T_1 and T_2. This gives three doublets in addition to the doublet at $E = \frac{1}{4}(T_1 + T_2)$. For equal hyperfine coupling $T_1 = T_2$ the zero field energies are

$$E = \begin{cases} \frac{1}{2}T & \text{(twice)} \\ 0 \\ -T \end{cases} \tag{7-25}$$

When one coupling constant can be neglected relative to the other (i.e., $T_2 = 0$) we have at zero field

$$E = \begin{cases} \pm \frac{1}{4}T \\ \pm \dfrac{1}{\sqrt{2}}\,T \end{cases} \tag{7-26}$$

These zero field limits are not usually studied, but they give an insight into the overall energy level diagram of Figs. 7-3 and 7-4.

At high fields where $|g\beta H| \gg |T_i|$ it is most convenient to use perturbation theory to obtain the energy levels. From Eqs. 2-104 and 7-21 we have the energies shown below:

$$E_1 = \tfrac{1}{2}g\beta H - \tfrac{1}{2}(g_1 + g_2)\beta_N H + \tfrac{1}{4}(T_1 + T_2) \qquad (+\,+\,+)$$

$$E_2 = \tfrac{1}{2}g\beta H - \tfrac{1}{2}(g_1 - g_2)\beta_N H + \tfrac{1}{4}(T_1 - T_2) + \frac{T_2^2}{4g\beta H} \qquad (+\,+\,-)$$

$$E_3 = \tfrac{1}{2}g\beta H + \tfrac{1}{2}(g_1 - g_2)\beta_N H - \tfrac{1}{4}(T_1 - T_2) + \frac{T_1^2}{4g\beta H} \qquad (+\,-\,+)$$

$$E_4 = \tfrac{1}{2}g\beta H + \tfrac{1}{2}(g_1 + g_2)\beta_N H - \tfrac{1}{4}(T_1 + T_2) + \frac{T_1^2 + T_2^2}{4g\beta H} \qquad (+\,-\,-)$$

$$E_5 = -\tfrac{1}{2}g\beta H + \tfrac{1}{2}(g_1 + g_2)\beta_N H + \tfrac{1}{4}(T_1 + T_2) \qquad (-\,-\,-)$$

$$E_6 = -\tfrac{1}{2}g\beta H + \tfrac{1}{2}(g_1 - g_2)\beta_N H + \tfrac{1}{4}(T_1 - T_2) - \frac{T_2^2}{4g\beta H} \qquad (-\,-\,+)$$

$$E_7 = -\tfrac{1}{2}g\beta H - \tfrac{1}{2}(g_1 - g_2)\beta_N H - \tfrac{1}{4}(T_1 - T_2) - \frac{T_1^2}{4g\beta H} \qquad (-\,+\,-)$$

$$E_8 = -\tfrac{1}{2}g\beta H - \tfrac{1}{2}(g_1 + g_2)\beta_N H - \tfrac{1}{4}(T_1 + T_2) - \frac{T_1^2 + T_2^2}{4g\beta H} \qquad (-\,+\,+)$$

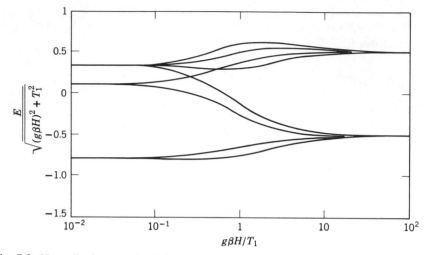

Fig. 7-3. Normalized energy level diagram for one electronic spin interacting with two spin-$\frac{1}{2}$ nuclei for the case $T_1 = 3T_2$.

The zero-order wavefunctions $|m_s m_1 m_2\rangle$ are given for each energy level. The levels shown in Fig. 7-4 do not extend to a field strength high enough for the nuclear Zeeman term to produce level crossings of the type shown in Fig. 7-3.

The four allowed transitions, which occur in accordance with the selection rules $\Delta m_S = \pm 1$, $\Delta m_1 = \Delta m_2 = 0$, have the following energy splittings up to second order:

$$E_1 - E_8 = g\beta H + \tfrac{1}{2}(T_1 + T_2) + \frac{T_1^2 + T_2^2}{4g\beta H}$$

$$E_2 - E_7 = g\beta H + \tfrac{1}{2}(T_1 - T_2) + \frac{T_1^2 + T_2^2}{4g\beta H}$$

$$E_3 - E_6 = g\beta H - \tfrac{1}{2}(T_1 - T_2) + \frac{T_1^2 + T_2^2}{4g\beta H}$$

$$E_4 - E_5 = g\beta H - \tfrac{1}{2}(T_1 + T_2) + \frac{T_1^2 + T_2^2}{4g\beta H}$$

$$(7\text{-}27)$$

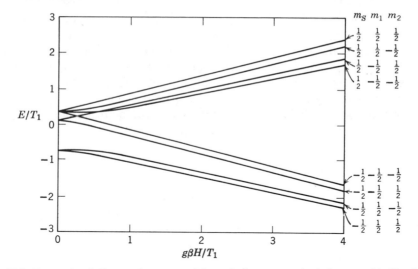

Fig. 7-4. Energy level diagram for two nuclei coupled to an unpaired electron with $T_1 = 3T_2$.

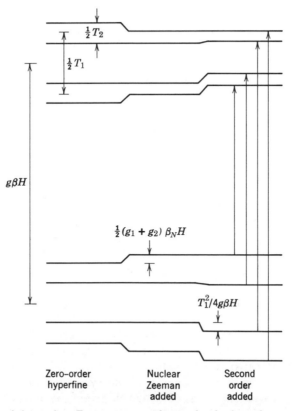

Fig. 7-5. Effect of the nuclear Zeeman term and second-order hyperfine term on the energy levels of two nuclei coupled to an unpaired electron with $T_1 = 3T_2$ and $g_1 = g_2$.

Figure 7-5 shows explicitly the effect of the second-order shifts which add for level pairs exhibiting allowed transitions. In contrast to this the nuclear Zeeman term shifts levels equally for allowed transitions, and so it is absent from the level splittings of Eq. 7-27. Thus it is not observed experimentally. The nuclear Zeeman term does manifest itself experimentally, however, when T is anisotropic, as discussed in Chapter 6.

The transition probability is proportional to the matrix element

$$\frac{1}{g^2\beta^2} |\langle i| g\beta S_x - g_1\beta_N I_{1x} - g_2\beta_N I_{2x} |j\rangle|^2 \qquad (7\text{-}28)$$

and the matrix of Eq. 9-46 of the first edition gives the proportionality factor for each transition. The forbidden transitions are weaker by six orders of magnitude than the allowed ones.

7-5 THREE-SPIN AB_2 NMR CASE

The AB_2 system of NMR[1-4] consists of two magnetically equivalent nuclei $(g_2 = g_3 = g_B)$ coupled to a third nonequivalent species $(g_1 = g_A)$ such that the chemical shift difference $(g_A - g_B)\beta_N H$ is comparable in magnitude to the coupling constant J. An example of such a system is a symmetrically substituted 1,2,3-benzene derivative such as

where R and R' are noninteracting atoms or groups.

As will be explained in Section 7-7, the coupling constant between two equivalent nuclei has no effect on the spectrum observed. Therefore only the coupling constants $J_{AB} = J$ between the nonequivalent A and B nuclei need be considered. Hence the Hamiltonian may be written in the form

$$\mathcal{H} = \hbar H[\gamma_A I_{Az} + \gamma_B(I_{2z} + I_{3z})] + \hbar J I_A(I_2 + I_3) \qquad (7\text{-}29)$$

For convenience the $\hbar$ is suppressed in the subsequent discussion. If these Hamiltonian terms are used to form matrix 7-17, one obtains the usual two exact energies, and two cubic equations similar to 7-24 to be solved.

It is somewhat easier to make use of the symmetry of the AB_2 system, and to select the wavefunctions (7-9) which group the B nuclei into singlet and triplet states. Using this set, one obtains the following Hamiltonian matrix $\mathcal{H}/\hbar$:

$$
\begin{bmatrix}
\left[\begin{array}{c}\frac{1}{2}(\gamma_A + 2\gamma_B)H \\ + \frac{1}{2}J\end{array}\right] & 0 & 0 & 0 & 0 & 0 & 0 & 0 \\
0 & \frac{1}{2}\gamma_A H & \frac{1}{\sqrt{2}}J & 0 & 0 & 0 & 0 & 0 \\
0 & \frac{1}{\sqrt{2}}J & \left[\begin{array}{c}\frac{1}{2}(-\gamma_A + 2\gamma_B)H \\ -\frac{1}{2}J\end{array}\right] & 0 & 0 & 0 & 0 & 0 \\
0 & 0 & 0 & \frac{1}{2}\gamma_A H & 0 & 0 & 0 & 0 \\
0 & 0 & 0 & 0 & -\frac{1}{2}\gamma_A H & 0 & 0 & 0 \\
0 & 0 & 0 & 0 & 0 & \left[\begin{array}{c}-\frac{1}{2}(-\gamma_A + 2\gamma_B)H \\ -\frac{1}{2}J\end{array}\right] & \frac{1}{\sqrt{2}}J & 0 \\
0 & 0 & 0 & 0 & 0 & \frac{1}{\sqrt{2}}J & -\frac{1}{2}\gamma_A H & 0 \\
0 & 0 & 0 & 0 & 0 & 0 & 0 & \left[\begin{array}{c}-\frac{1}{2}(\gamma_A + 2\gamma_B)H \\ +\frac{1}{2}J\end{array}\right]
\end{bmatrix}
\tag{7-30}
$$

In this basis the secular equation contains four exact energies and two quadratic equations to solve.

To simplify the final results it is conventional to adopt the notation

$$\omega_0 \delta = H_0 (\gamma_B - \gamma_A) \tag{7-31}$$

where δ is the dimensionless chemical shift, and ω_0 and H_0 are the angular frequency and the magnetic field, respectively, for a resonance at the mean position of that corresponding to A and B nuclei:

$$\omega_0 = \tfrac{1}{2}(\gamma_A + \gamma_B) H_0 \tag{7-32}$$

where for simplicity both gyromagnetic ratios are assumed positive. The solutions to the quadratic equations are expressed most conveniently in terms of the quantities

$$C_{\pm} = \tfrac{1}{2}\sqrt{(\omega_0 \delta)^2 \pm \omega_0 \delta J + \tfrac{9}{4} J^2} \tag{7-33}$$

and the angles $\theta_{\pm}$, which range from 0 to π defined by

$$\sin 2\theta_{\pm} = \frac{J}{\sqrt{2} C_{\pm}}$$

$$\cos 2\theta_{\pm} = \frac{\omega_0 \delta \pm \tfrac{1}{2} J}{2 C_{\pm}} \tag{7-34}$$

These same angles are used to express the linear combinations of the ϕ_i of Eqs. 7-9 which form the true eigenfunctions ψ_i. The eigenfunctions and energies are given in Table 7-1.

The intensities are easily calculated and may be presented in the form of a matrix such as 7-30. Instead, the transition energies and intensities are presented in Table 7-2, which exhibits several interesting features. The strongest line, line 3, gives the true chemical shift for the A nucleus, and the true chemical shift of the B nucleus is given by the mean position of bands 5 and 7. From an experimental spectrum one may determine which nucleus is

TABLE 7-1   Wavefunctions and Energies in Units of $\hbar$ for the AB_2 Spin System

Wavefunction	Energy/$\hbar$
$\psi_1 \equiv \phi_1$	$E_1 = \tfrac{1}{2}(\gamma_A + 2\gamma_B)H + \tfrac{1}{2}J$
$\psi_2 = \phi_2 \cos \theta_+ + \phi_3 \sin \theta_+$	$E_2 = \tfrac{1}{2}\gamma_B H - \tfrac{1}{4}J + C_+$
$\psi_3 = -\phi_2 \sin \theta_+ + \phi_3 \cos \theta_+$	$E_3 = \tfrac{1}{2}\gamma_B H - \tfrac{1}{4}J - C_+$
$\psi_4 = \phi_4$	$E_4 = \tfrac{1}{2}\gamma_A H$
$\psi_5 = \phi_5$	$E_5 = -\tfrac{1}{2}\gamma_A H$
$\psi_6 = \phi_6 \cos \theta_- + \phi_7 \sin \theta_-$	$E_6 = -\tfrac{1}{2}\gamma_B H - \tfrac{1}{4}J + C_-$
$\psi_7 = -\phi_6 \sin \theta_- + \phi_7 \cos \theta_-$	$E_7 = -\tfrac{1}{2}\gamma_B H - \tfrac{1}{4}J - C_-$
$\psi_8 = \phi_8$	$E_8 = -\tfrac{1}{2}(\gamma_A + 2\gamma_B)H + \tfrac{1}{2}J$

TABLE 7-2 Transition Energies in Units of $\hbar$ and Relative Intensities for the AB_2 Spin System

Band Number	Type	Transition	Energy Difference	Relative Intensity
1	A	$\psi_3 \rightarrow \psi_1$	$\frac{1}{2}(\gamma_A + \gamma_B)H + \frac{3}{4}J + C_+$	$(\sqrt{2}\sin\theta_+ - \cos\theta_+)^2$
2	A	$\psi_7 \rightarrow \psi_2$	$\gamma_B H + C_+ + C_-$	$[\sqrt{2}\sin(\theta_+ - \theta_-) + \cos\theta_+\cos\theta_-]^2$
3	A	$\psi_5 \rightarrow \psi_4$	$\gamma_A H$	1
4	A	$\psi_8 \rightarrow \psi_6$	$\frac{1}{2}(\gamma_A + \gamma_B)H - \frac{3}{4}J + C_-$	$(\sqrt{2}\sin\theta_- + \cos\theta_-)^2$
5	B	$\psi_6 \rightarrow \psi_2$	$\gamma_B H + C_+ - C_-$	$[\sqrt{2}\cos(\theta_+ - \theta_-) + \cos\theta_+\sin\theta_-]^2$
6	B	$\psi_2 \rightarrow \psi_1$	$\frac{1}{2}(\gamma_A + \gamma_B)H + \frac{3}{4}J - C_+$	$(\sqrt{2}\cos\theta_+ + \sin\theta_+)^2$
7	B	$\psi_7 \rightarrow \psi_3$	$\gamma_B H - C_+ + C_-$	$[\sqrt{2}\cos(\theta_+ - \theta_-) - \sin\theta_+\cos\theta_-]^2$
8	B	$\psi_8 \rightarrow \psi_7$	$\frac{1}{2}(\gamma_A + \gamma_B)H - \frac{3}{4}J - C_-$	$(\sqrt{2}\cos\theta_- - \sin\theta_-)^2$
9	Combination	$\psi_6 \rightarrow \psi_3$	$\gamma_B H - C_+ - C_-$	$[\sqrt{2}\sin(\theta_+ - \theta_-) + \sin\theta_+\sin\theta_-]^2$

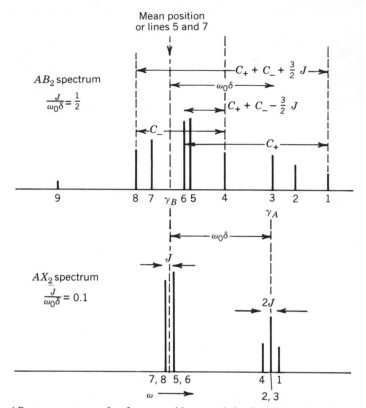

Fig. 7-6. AB_2 type spectrum for J comparable to $\omega_0\delta$ (top) and for J much less than $\omega_0\delta$ (bottom). The latter is ordinarily referred to as an AX_2 spectrum. The values of several line spacings are shown. Note that line 3 remains at the γ_A position and the mean position of lines 5 and 7 does not change with the ratio $J/\omega_0\delta$, thereby permitting the direct evaluation of the chemical shift from the spectrum. In the AX_2 spectrum the lines of the γ_A triplet have the approximate relative intensity ratio 1:2:1 and those of the γ_B doublet have the approximate ratio 4:4. Some of the spectral lines are unresolved in the AX_2 spectrum, and the combination band 9 is too weak to be detectable in this limit.

less shielded, but the sign of the coupling constant cannot be determined from an examination of the spectrum.

The theoretical spectra for two ratios of $J/\omega_0\delta$ which are presented in Fig. 7-6 show how the various Hamiltonian parameters can be evaluated directly from the positions of the lines.

In the limit $J \to 0$ the first four transitions involve the flipping of nucleus A only, while the next four transitions (5–8) involve a flip of only spins B. Such transitions are referred to as fundamental bands. The last transition, $\psi_6 \to \psi_3$, involves a flip of both types of spin and hence is called a combination band. This renders it forbidden in the two limits $J \to 0$ and $\delta \to 0$ and always weak at intermediate values of $\omega_0\delta/J$.

7-6 THREE-SPIN *ABC* AND *ABX* NMR SYSTEMS

The ABC system of NMR corresponds to three nonequivalent nuclei where coupling constants J_{AB}, J_{BC}, and J_{CA} are comparable to the magnitude of the chemical shift differences, $\omega_0\delta_{AB}$, $\omega_0\delta_{BC}$, and $\omega_0\delta_{CA}$. Therefore from Eq. 7-17 we see that two 3×3 determinates must be evaluated. These have the explicit form

$$\begin{pmatrix} \begin{bmatrix} \pm\frac{1}{2}(\gamma_A + \gamma_B - \gamma_C)H \\ +\frac{1}{4}(J_{AB} - J_{BC} - J_{CA}) \end{bmatrix} & \frac{1}{2}J_{BC} & \frac{1}{2}J_{CA} \\ \\ \frac{1}{2}J_{BC} & \begin{bmatrix} \pm\frac{1}{2}(\gamma_A - \gamma_B + \gamma_C)H \\ +\frac{1}{4}(-J_{AB} - J_{BC} + J_{CA}) \end{bmatrix} & \frac{1}{2}J_{AB} \\ \\ \frac{1}{2}J_{CA} & \frac{1}{2}J_{AB} & \begin{bmatrix} \pm\frac{1}{2}(-\gamma_A + \gamma_B + \gamma_C)H \\ +\frac{1}{4}(-J_{AB} + J_{BC} - J_{CA}) \end{bmatrix} \end{pmatrix}$$

$$(7\text{-}35)$$

There are 15 allowed transitions, 12 fundamental frequencies, and 3 combination bands. The analysis of the ABC spectrum is quite complex, and for a discussion of the intricacies involved the reader is referred to the references at the end of the chapter.[1-4]

An ABX system differs from the ABC system in that the magnitude of the chemical shift of the X nucleus is much larger than the various coupling constants. The corresponding 3×3 matrix is identical with 7-30, but now the off-diagonal matrix elements $\frac{1}{2}J_{BX}$ and $\frac{1}{2}J_{CX}$ can be neglected since they connect diagonal elements where γ_X changes sign. The off-diagonal element $\frac{1}{2}J_{AB}$ must be retained, however, since it connects states with $+\gamma_X$ on the diagonal in accordance with standard perturbation theory for nearly degenerate levels. Accordingly matrix 7-35 assumes the simplified form

$$\begin{pmatrix} \begin{bmatrix} \pm\frac{1}{2}(\gamma_A + \gamma_B - \gamma_X)H \\ +\frac{1}{4}(J_{AB} - J_{AX} - J_{BX}) \end{bmatrix} & 0 & 0 \\ \\ 0 & \begin{bmatrix} \pm\frac{1}{2}(\gamma_A - \gamma_B + \gamma_X)H \\ +\frac{1}{4}(-J_{AB} + J_{AX} - J_{BX}) \end{bmatrix} & \frac{1}{2}J_{AB} \\ \\ 0 & \frac{1}{2}J_{AB} & \begin{bmatrix} \pm\frac{1}{2}(-\gamma_A + \gamma_B + \gamma_X)H \\ +\frac{1}{4}(-J_{AB} - J_{AX} + J_{BX}) \end{bmatrix} \end{pmatrix}$$

$$(7\text{-}36)$$

and the energies E_2 and E_7 are given directly.

Following the notation of the previous section, we may define the positive quantities D_+ and D_-,

$$D_\pm = \{[\tfrac{1}{2}\omega_0\delta_{AB} \pm \tfrac{1}{4}(J_{AX} - J_{BX})]^2 + \tfrac{1}{4}J_{AB}^2\}^{1/2} \qquad (7\text{-}37)$$

and the angles θ_+ and θ_-, which have the range from 0 to π:

$$\sin 2\theta_\pm = \frac{J_{AB}}{2D_\pm}$$

$$\cos 2\theta_\pm = \frac{\omega_0 \delta_{AB} \pm \frac{1}{2}(J_{AX} - J_{BX})}{2D_\pm}$$

(7-38)

The energy differences and relative intensities of the 15 possible transitions are given in Table 7-3 in terms of the quantities $D_\pm$ and $\theta_\pm$, and a typical spectrum is shown in Fig. 7-7. One of the transitions has zero intensity, and two are always very weak. The spectrum separates into an AB and an X group of lines, as the figure indicates. This facilitates the analysis of an ABX spectrum. In contrast to this, the major difficulty with the interpretation of an ABC spectrum is that the resonance lines do not separate into two main groups, and the analysis becomes much more complex. The zero intensity transition (number 13) acquires a finite intensity in the ABC spectrum.

As Fig. 7-7 indicates, the AB part of an ABX spectrum is characterized by the presence of two symmetric quartets. The first quartet (ψ_1, ψ_3, ψ_5, ψ_7) is centered at the point $\frac{1}{2}(\gamma_A + \gamma_B)H - \frac{1}{4}(J_{AX} + J_{BX})$, and the second quartet (ψ_2, ψ_4, ψ_6, ψ_8) is centered at $\frac{1}{2}(\gamma_A + \gamma_B)H + \frac{1}{4}(J_{AX} + J_{BX})$. Thus the midpoint between these centers is $\frac{1}{2}(\gamma_A + \gamma_B)H$. The X part of the spectrum is a symmetric sextet centered at the angular frequency $\gamma_X H$. If we let ΔE_i be the energy difference of the ith transition ψ_i, then the following subtraction rules can be written:

$$\Delta E_3 - \Delta E_1 = \Delta E_4 - \Delta E_2 = \Delta E_7 - \Delta E_5 = \Delta E_8 - \Delta E_6$$
$$\Delta E_2 - \Delta E_1 = \Delta E_4 - \Delta E_3 = \Delta E_{11} - \Delta E_9 = \Delta E_{12} - \Delta E_{10} \quad (7\text{-}39)$$
$$\Delta E_{10} - \Delta E_9 = \Delta E_{12} - \Delta E_{11} = \Delta E_6 - \Delta E_5 = \Delta E_8 - \Delta E_7$$

The spin–spin coupling constants and the quantities $D_\pm$ may be evaluated-directly from the spectrum:

Fig. 7-7. Theoretical ABX spectrum for the condition $0 < J_{BX} < J_{AX}$ and $\gamma_B < \gamma_A$. The A, B, and X type groups of lines are labeled, and various difference frequencies are indicated.

TABLE 7-3 Transition Energies Normalized Relative to $\hbar$ and Relative Intensities for the ABX Spectrum

Number	Type	Transition	Energy	Relative Intensity
1	B	$\psi_8 \to \psi_6$	$\frac{1}{2}(\gamma_A + \gamma_B)H - \frac{1}{4}(2J_{AB} + J_{AX} + J_{BX}) - D_-$	$(\cos\theta_- - \sin\theta_-)^2$
2	B	$\psi_7 \to \psi_4$	$\frac{1}{2}(\gamma_A + \gamma_B)H - \frac{1}{4}(2J_{AB} - J_{AX} - J_{BX}) - D_+$	$(\cos\theta_+ - \sin\theta_+)^2$
3	B	$\psi_5 \to \psi_2$	$\frac{1}{2}(\gamma_A + \gamma_B)H + \frac{1}{4}(2J_{AB} - J_{AX} - J_{BX}) - D_-$	$(\cos\theta_- + \sin\theta_-)^2$
4	B	$\psi_3 \to \psi_1$	$\frac{1}{2}(\gamma_A + \gamma_B)H + \frac{1}{4}(2J_{AB} - J_{AX} + J_{BX}) - D_+$	$(\cos\theta_+ + \sin\theta_+)^2$
5	A	$\psi_8 \to \psi_5$	$\frac{1}{2}(\gamma_A + \gamma_B)H - \frac{1}{4}(2J_{AB} + J_{AX} + J_{BX}) + D_-$	$(\cos\theta_+ + \sin\theta_-)^2$
6	A	$\psi_7 \to \psi_3$	$\frac{1}{2}(\gamma_A + \gamma_B)H - \frac{1}{4}(2J_{AB} - J_{AX} - J_{BX}) + D_+$	$(\cos\theta_+ + \sin\theta_+)^2$
7	A	$\psi_6 \to \psi_2$	$\frac{1}{2}(\gamma_A + \gamma_B)H + \frac{1}{4}(2J_{AB} - J_{AX} - J_{BX}) + D_-$	$(\cos\theta_- + \sin\theta_+)^2$
8	A	$\psi_4 \to \psi_1$	$\frac{1}{2}(\gamma_A + \gamma_B)H + \frac{1}{4}(2J_{AB} + J_{AX} + J_{BX}) + D_+$	$(\cos\theta_+ - \sin\theta_-)^2$
9	X	$\psi_8 \to \psi_7$	$\gamma_X H - \frac{1}{2}(J_{AX} + J_{BX})$	1
10	X	$\psi_5 \to \psi_3$	$\gamma_X H + D_+ - D_-$	$\cos^2(\theta_+ - \theta_-)$
11	X	$\psi_6 \to \psi_4$	$\gamma_X H - D_+ + D_-$	$\cos^2(\theta_+ - \theta_-)$
12	X	$\psi_2 \to \psi_1$	$\gamma_X H + \frac{1}{2}(J_{AX} + J_{BX})$	1
13	Combination	$\psi_7 \to \psi_2$	$(\gamma_A + \gamma_B - \gamma_X)H$	0
14	Combination	$\psi_5 \to \psi_4$	$\gamma_X H - D_+ - D_-$	$\sin^2(\theta_+ - \theta_-)$
15	Combination	$\psi_6 \to \psi_3$	$\gamma_X H + D_+ + D_-$	$\sin^2(\theta_+ - \theta_-)$

$$J_{AB} = \Delta E_3 - \Delta E_1 = \Delta E_4 - \Delta E_2 = \Delta E_7 - \Delta E_5 = \Delta E_8 - \Delta E_6$$
$$J_{AX} + J_{BX} = \Delta E_{10} - \Delta E_9$$
$$D_+ = \tfrac{1}{2}(\Delta E_6 - \Delta E_2) = \tfrac{1}{2}(\Delta E_8 - \Delta E_4) \tag{7-40}$$
$$D_- = \tfrac{1}{2}(\Delta E_5 - \Delta E_1) = \tfrac{1}{2}(\Delta E_7 - \Delta E_3)$$

The absolute signs of the coupling constants can not be determined without recourse to additional experiments such as spin decoupling.

7-7 EQUIVALENT SPINS

Magnetically equivalent spins have the same g-factor (gyromagnetic ratio) and couple to the same extent to all other spins in a molecule. In NMR they are denoted by the same letter, such as A_4 for methane and A_3X_3 for CH_3CF_3. For magnetically equivalent nuclei both the transition probabilities and the transition energies are independent of the spin–spin coupling between the equivalent spins, and therefore may be omitted. For example, in an A_2B_2 group the J_{AA} and J_{BB} couplings can not be measured because they do not affect the observed spectrum. Only the gyromagnetic ratios γ_A and γ_B and the spin–spin coupling J_{AB} are determined experimentally. Hence, when nuclei are magnetically equivalent, the internal spin–spin term is omitted from the Hamiltonian, thus simplifying the calculations.

REFERENCES

1. J. W. Emsley, J. Feeney, and L. H. Sutcliffe, *High Resolution Nuclear Magnetic Resonance Spectroscopy*, Pergamon, New York, 1965, Vol. 1, Chap. 8.
2. J. A. Pople, W. G. Schneider, and H. J. Bernstein, *High Resolution Nuclear Magnetic Resonance*, McGraw-Hill, New York, 1959.
3. R. Harris and B. Mann, *NMR and the Periodic Table*, Academic Press, New York, 1978.
4. P. Laszlo, ed., *NMR of Newly Accessible Nuclei*, Vol. 1, *Chemical and Biochemical Applications*, Academic Press, New York, 1983.

HIGH-SPIN SYSTEMS

8-1 INTRODUCTION

The preceding chapters, with the exception of Chapter 7, have dealt mainly with the two-spin $(\frac{1}{2}, \frac{1}{2})$ system, taking into account the Zeeman and spin–spin interactions. It is now appropriate to generalize the treatment to include higher valued spins $(S > \frac{1}{2})$. The direct product mathematical apparatus that is appropriate for this generalization was developed in Sections 2-4 to 2-7.

8-2 SPIN SYSTEMS $S_1 = \frac{1}{2}$, $S_2 = 1$

The two-spin system $(\frac{1}{2}, 1)$ is more common in ESR than in NMR; hence this treatment employs the notation that is appropriate for the former case. Then at the end of the discussion the NMR approximations are treated. Only isotropic Zeeman and spin–spin (hyperfine) interactions are taken into account.

The Hamiltonian

$$\mathcal{H} = \beta \vec{S}_1 \cdot \vec{g}_1 \cdot \vec{H} + \beta \vec{H} \cdot \vec{g}_2 \cdot \vec{S}_2 + \vec{S}_1 \cdot \vec{T} \cdot \vec{S}_2 \tag{8-1}$$

which may be expanded indirect products

$$\mathcal{H} = \frac{g_1 \beta H}{2} \begin{pmatrix} 1 & 0 \\ 0 & -1 \end{pmatrix} \times \begin{pmatrix} 1 & 0 & 0 \\ 0 & 1 & 0 \\ 0 & 0 & 1 \end{pmatrix} + \frac{g_2 \beta_N H}{2} \begin{pmatrix} 1 & 0 \\ 0 & 1 \end{pmatrix} \times \begin{pmatrix} 1 & 0 & 0 \\ 0 & 0 & 0 \\ 0 & 0 & -1 \end{pmatrix}$$

$$+ \frac{T}{2\sqrt{2}} \left[\begin{pmatrix} 0 & 1 \\ 1 & 0 \end{pmatrix} \times \begin{pmatrix} 0 & 1 & 0 \\ 1 & 0 & 1 \\ 0 & 1 & 0 \end{pmatrix} + \begin{pmatrix} 0 & -i \\ i & 0 \end{pmatrix} \times \begin{pmatrix} 0 & -i & 0 \\ i & 0 & -i \\ 0 & i & 0 \end{pmatrix} \right.$$

$$+ \sqrt{2} \begin{pmatrix} 1 & 0 \\ 0 & -1 \end{pmatrix} \times \begin{pmatrix} 1 & 0 & 0 \\ 0 & 0 & 0 \\ 0 & 0 & -1 \end{pmatrix} \Bigg] \tag{8-2}$$

has the explicit matrix form

	$\lvert \tfrac{1}{2}\,1 \rangle$	$\lvert \tfrac{1}{2}\,0 \rangle$	$\lvert \tfrac{1}{2}\,-1 \rangle$	$\lvert -\tfrac{1}{2}\,1 \rangle$	$\lvert -\tfrac{1}{2}\,0 \rangle$	$\lvert -\tfrac{1}{2}\,-1 \rangle$
$\langle \tfrac{1}{2}\,1 \rvert$	$[\tfrac{1}{2}G_1 + G_2 + \tfrac{1}{2}T]$	0	0	0	0	0
$\langle \tfrac{1}{2}\,0 \rvert$	0	$\tfrac{1}{2}G_1$	0	$\dfrac{T}{\sqrt{2}}$	0	0
$\langle \tfrac{1}{2}\,-1 \rvert$	0	0	$[\tfrac{1}{2}G_1 - G_2 - \tfrac{1}{2}T]$	0	$\dfrac{T}{\sqrt{2}}$	0
$\langle -\tfrac{1}{2}\,1 \rvert$	0	$\dfrac{T}{\sqrt{2}}$	0	$[-\tfrac{1}{2}G_1 + G_2 - \tfrac{1}{2}T]$	0	0
$\langle -\tfrac{1}{2}\,0 \rvert$	0	0	$\dfrac{T}{\sqrt{2}}$	0	$-\tfrac{1}{2}G_1$	0
$\langle -\tfrac{1}{2}\,-1 \rvert$	0	0	0	0	0	$[-\tfrac{1}{2}G_1 - G_2 + \tfrac{1}{2}T]$

$$\tag{8-3}$$

where $G_1 = g_1\beta H$ and $G_2 = g_2\beta_N H$. This may be converted to a more convenient form by arranging the bra and ket vectors in the order of decreasing $M = M_1 + M_2$, to give

$$
\begin{array}{cccccc}
[\tfrac{1}{2}G_1 + G_2 + \tfrac{1}{2}T] & 0 & 0 & 0 & 0 & \\
0 & \tfrac{1}{2}G_1 & \dfrac{T}{\sqrt{2}} & 0 & 0 & 0 \\
0 & \dfrac{T}{\sqrt{2}} & [-\tfrac{1}{2}G_1 + G_2 - \tfrac{1}{2}T] & 0 & 0 & 0 \\
0 & 0 & 0 & [\tfrac{1}{2}G_1 - G_2 - \tfrac{1}{2}T] & \dfrac{T}{\sqrt{2}} & 0 \\
0 & 0 & 0 & \dfrac{T}{\sqrt{2}} & -\tfrac{1}{2}G_1 & 0 \\
0 & 0 & 0 & 0 & 0 & -[\tfrac{1}{2}G_1 + G_2 - \tfrac{1}{2}T]
\end{array}
$$

$$\tag{8-4}$$

The $[\tfrac{1}{2}G_1 + G_2 + \tfrac{1}{2}T]$ and $-[\tfrac{1}{2}G_1 + G_2 - \tfrac{1}{2}T]$ energy levels are exact, and the remaining levels are easily obtained from a solution of two 2×2 determinants:

$$
\begin{vmatrix}
\pm \frac{1}{2}G_1 - E & \dfrac{T}{\sqrt{2}} \\[2mm]
\dfrac{T}{\sqrt{2}} & \pm(G_2 - \frac{1}{2}G_1) - \frac{1}{2}T - E
\end{vmatrix} = 0
\qquad (8\text{-}5)
$$

The six energies are as follows:

$$
\begin{aligned}
E_1 &= \tfrac{1}{2}G_1 + G_2 + \tfrac{1}{2}T \\
E_2 &= \tfrac{1}{2}G_2 - \tfrac{1}{4}T + \tfrac{1}{2}[(G_1 - G_2)(G_1 - G_2 + T) + \tfrac{9}{4}T^2]^{1/2} \\
E_3 &= -\tfrac{1}{2}G_2 - \tfrac{1}{4}T + \tfrac{1}{2}[(G_2 - G_1)(G_2 - G_1 + T) + \tfrac{9}{4}T^2]^{1/2} \\
E_4 &= -\tfrac{1}{2}G_1 - G_2 + \tfrac{1}{2}T \\
E_5 &= \tfrac{1}{2}G_2 - \tfrac{1}{4}T - \tfrac{1}{2}[(G_2 - G_1)(G_2 - G_1 + T) + \tfrac{9}{4}T^2]^{1/2} \\
E_6 &= -\tfrac{1}{2}G_2 - \tfrac{1}{4}T - \tfrac{1}{2}[(G_1 - G_2)(G_1 - G_2 + T) + \tfrac{9}{4}T^2]^{1/2}
\end{aligned}
\qquad (8\text{-}6)
$$

These energies are plotted in normalized form in Fig. 8-1 for the case $|G_1| \gg |G_2|$, and in Fig. 8-2 for the case $|G_1| > |G_2|$. The latter corresponds to a spin $S = 1$ interacting with a nucleus with spin $I = \frac{1}{2}$. The former case, $S = \frac{1}{2}$, $I = 1$, is quite common in ESR (e.g., free nitrogen or deuterium atoms). Both cases occur in high-resolution NMR.

For the approximation at high field, where $|g_1\beta H| \gg |T|$ these six energy levels agree with the usual expression:

$$
E = g_1\beta Hm_1 + g_2\beta Hm_2 + Tm_1m_2
\qquad (8\text{-}7)
$$

where terms of order $T^2/(g_1\beta H - g_2\beta_N H)$ and less have been neglected. The high-field region of Fig. 8-3 approximates this expression. This is the usual ESR case for an $I = 1$ nucleus coupled to an unpaired electron.

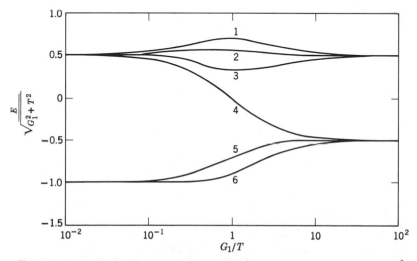

Fig. 8-1. Normalized energy levels for the $S = \frac{1}{2}$, $I = 1$ ESR case $(G_1/G_2 = -10^3)$.

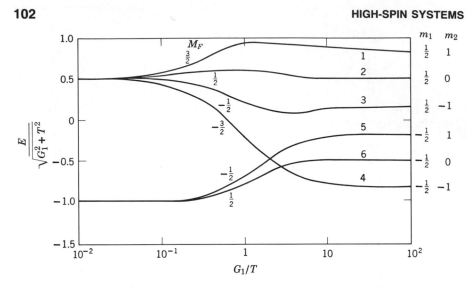

Fig. 8-2. Normalized energy levels for $I_1 = \frac{1}{2}$, $I_2 = 1$ and the ratio $G_1/G_2 = 3$.

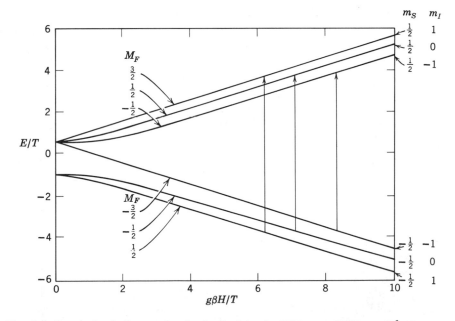

Fig. 8-3. Energy level diagram for $S = \frac{1}{2}$, $I = 1$ in the ESR case $G_1/G_2 = -10^3$. The three high-field transitions are shown.

At low fields the opposite approximation, $|g_1\beta H|, |g_2\beta_N H| \ll |T|$, holds and the energies are given by the following expression, which is derived in Section 8-15:

$$H = g_F\beta HM_F + \tfrac{1}{2}T[F(F+1) - S_1(S_1+1) - S_2(S_2+1)] \qquad (8\text{-}8)$$

where $F = S_1 + S_2 = \tfrac{3}{2}, \tfrac{1}{2}$, and $-F \le M_F \le F$. The g-factor g_F is (cf. Section 8-15)

$$g_F = \tfrac{1}{2}g_1\left[1 + \frac{S_1(S_1+1) - S_2(S_2+1)}{F(F+1)}\right] + \tfrac{1}{2}g_2\left[1 - \frac{S_1(S_1+1) - S_2(S_2+1)}{F(F+1)}\right]$$

$$(8\text{-}9)$$

In the present case one has explicitly

$$E_i = \tfrac{1}{3}(g_i\beta + 2g_2\beta_N)HM_F - T \qquad (F = \tfrac{3}{2}, i = 1, 2, 3, 4)$$

$$E_i = \tfrac{1}{3}(-g_1\beta + 4g_2\beta_N)HM_F + \tfrac{1}{2}T \qquad (F = \tfrac{1}{2}, i = 5, 6) \qquad (8\text{-}10)$$

corresponding to the low-field levels of Fig. 8-3. The M_F labels on the low-field doublet $(F = \tfrac{1}{2})$ remain inverted until very high fields, where the nuclear Zeeman term becomes larger than the hyperfine term. The curves of Figs. 8-1 to 8-3 were not carried to a field high enough to display the crossover of levels.

8-3 QUADRUPOLE INTERACTIONS

Nuclei with spin I greater than $\tfrac{1}{2}$ have quadrupole moments, and we treat them in terms of the high-spin formalism. We begin with a few fundamental remarks about quadrupole moments and their energies. Then the direct measurement of transitions between quadrupole levels is treated. This constitutes a separate field of research referred to as either nuclear quadrupole resonance (NQR) or pure quadrupole resonance (PQR). Finally, the effects of quadrupole moments in electron spin resonance and nuclear magnetic resonance are elucidated.

In pure quadrupole spectroscopy single crystals are used to determine the dependence of the intensity of the transitions on the radiofrequency field orientation. This permits a rough assignment of the principal axes of the field gradient. Single crystals are required for Zeeman studies since otherwise the dependence of the energy levels on the orientation of H_0 will broaden the lines. For axial symmetry the Zeeman pattern simplifies when H_0 is perpendicular to the symmetry axis.

Quadrupole spectroscopy[1-6] is useful for determining the electrostatic potential in solids and molecules. It may be employed for studying electron distributions, intermolecular and intramolecular binding, molecular motions, phase transitions, and other properties of solids and molecules.

8-4 QUADRUPOLE HAMILTONIAN

The quadrupole interaction energy operator

$$\mathcal{H}_Q = \vec{I} \cdot \vec{\vec{Q}} \cdot \vec{I} \tag{8-11}$$

arises from the interaction between the nuclear spin I and the quadrupole energy tensor $\vec{\vec{Q}}$

$$\vec{\vec{Q}}_{(XYZ)} = \begin{pmatrix} Q_{xx} & Q_{xy} & Q_{xz} \\ Q_{yx} & Q_{yy} & Q_{yz} \\ Q_{zx} & Q_{zy} & Q_{zz} \end{pmatrix} \tag{8-12}$$

This tensor is symmetric and traceless,

$$Q_{ij} = Q_{ji}$$
$$Q_{xx} + Q_{yy} + Q_{zz} = 0 \tag{8-13}$$

and so there are five independent components. The quadrupolar Hamiltonian may be expanded to

$$\mathcal{H}_Q = Q_{xx}I_x^2 + Q_{yy}I_y^2 + Q_{zz}I_z^2 + Q_{xy}(I_xI_y + I_yI_x) + Q_{yz}(I_yI_z + I_zI_y)$$
$$+ Q_{zx}(I_zI_x + I_xI_z) \tag{8-14}$$

where $I_jI_k \neq I_kI_j$ since the spin functions obey the commutation relation

$$I_jI_k - I_kI_j = [I_j, I_k] = iI_i \tag{8-15}$$

for i, j, $k = x$, y, z in cyclic order. By using the traceless property 8-13 and the usual identity

$$I_x^2 + I_y^2 + I_z^2 = I(I+1) \tag{8-16}$$

where $I(I+1)$ is the eigenvalue of the operator I^2, 8-14 can be put into the form

$$\mathcal{H}_Q = \tfrac{1}{2}(Q_{xx} - Q_{yy})(I_x^2 - I_y^2) + \tfrac{1}{2}Q_{zz}[3I_z^2 - I(I+1)]$$
$$+ Q_{xy}(I_xI_y + I_yI_x) + Q_{yz}(I_yI_z + I_zI_y) + Q_{zx}(I_zI_x + I_xI_z) \tag{8-17}$$

which is valid for any coordinate system xyz.

The quadrupole energy tensor has a principal coordinate system $x'y'z'$ in which it is diagonal. There is an orthogonal direction cosine transformation R, similar to Eq. 6-15, which converts the quadrupole tensor (8-12) to the principal axis system:

$$\vec{\vec{R}}^{-1}\vec{\vec{Q}}_{(xyz)}\vec{\vec{R}} = \vec{\vec{Q}}_{\text{diag}(x'y'z')} \tag{8-18}$$

where

$$\vec{\vec{Q}}_{\text{diag}} = \begin{bmatrix} Q_{x'x'} & Q & 0 \\ 0 & Q_{y'y'} & 0 \\ 0 & 0 & Q_{z'z'} \end{bmatrix} \tag{8-19}$$

subject to condition 8-13

$$Q_{x'x'} + Q_{y'y'} + Q_{z'z'} = 0 \tag{8-20}$$

since the trace of a matrix is invariant under a unitary transformation. The Hamiltonian now assumes the simplified form

$$\mathcal{H}_Q = \tfrac{1}{2}A[3I_{z'}^2 - I(I+1) + \eta(I_{x'}^2 - I_{y'}^2)] \tag{8-21}$$

where A is proportional to the quadrupole coupling constant, and η is the asymmetry parameter with $|Q_{x'x'}| \le |Q_{y'y'}| \le |Q_{z'z'}|$,

$$\eta = \frac{Q_{x'x'} - Q_{y'y'}}{Q_{z'z'}} \tag{8-22}$$

which vanishes in an axially symmetric field gradient. This allows us to express the $Q_{i'i'}$ in terms of A and η:

$$Q_{x'x'} = -\tfrac{1}{2}A(1-\eta)\,, \qquad Q_{y'y'} = -\tfrac{1}{2}A(1+\eta)\,, \qquad Q_{z'z'} = A$$

$$\tag{8-23}$$

The parameters A and η will be redefined in the next section. (Our definition of A differs by a factor of 2 from that adopted by some other authors.)

For nuclear spin $I = 0$ the spin matrices I_x, I_y, and I_z are all zero, and so H_Q of Eq. 8-21 is zero. For spin $I = \tfrac{1}{2}$ we have, from the well-known properties (2-69) of the Pauli spin matrices $\vec{\sigma}_i = 2\vec{I}_i$,

$$I_i I_j + I_j I_i = 0$$

$$I_x^2 = I_y^2 = I_z^2 = \tfrac{1}{3}I(I+1) \tag{8-24}$$

and so $\mathcal{H}_Q$ is again zero. These expressions 8-24 are not true for $I > \tfrac{1}{2}$, and so all higher spins can exhibit a quadrupole splitting.

When external charges distort the spherical electronic shell of the ion, they enhance the quadrupole coupling by what is called the antishielding factor γ to give[1,7,8]

$$\mathcal{H}_Q = (1 + \gamma)\vec{I} \cdot \vec{\vec{Q}} \cdot \vec{I} \tag{8-25}$$

The antishielding effect is particularly pronounced for heavy ions and in ionic crystals.

8-5 QUADRUPOLE MOMENT

The quadrupole energy tensor matrix elements (8-12) are proportional to the product of the scalar quadrupole moment Q and the gradient of the electric field V_{ij} in the following manner:

$$Q_{ij} = \frac{eQ}{2I(2I-1)} V_{ij} \qquad (8\text{-}26)$$

where e is the electronic charge. The scalar quadrupole moment Q is defined by

$$eQ = \int \rho(r)(3z^2 - r^2)d\tau = \int \rho(r)r^2(3\cos^2\theta - 1)d\tau \qquad (8\text{-}27)$$

where the integration is carried out over the nuclear charge density $\rho(r)$, and θ is the angle that the radius vector $\vec{r}$ makes with the internuclear axis.

The scalar quadrupole moment is a measure of the deviation of the nuclear charge density from spherical symmetry. It can exist only for $I > \frac{1}{2}$. The field gradient $V_{ij} = V_{ji}$ is defined by

$$V_{ij} = -\frac{\partial E_i}{\partial x_j} = \frac{\partial^2 V}{\partial x_i \partial x_j} \qquad (8\text{-}28)$$

where V is the electrostatic potential at the nucleus due to the surrounding charges, and x_i, $x_j = x$, y, z. The field gradient obeys the Laplace equation

$$\nabla^2 V = V_{xx} + V_{yy} + V_{zz} = 0 \qquad (8\text{-}29)$$

which is the reason why the quadrupole energy tensor is traceless (8-13). The zz component of the field gradient is referred to as eq:

$$V_{zz} = eq \qquad (8\text{-}30)$$

The asymmetry parameter, which was introduced in Eqs. 8-22 and 8-23,

$$\eta = \frac{V_{xx} - V_{yy}}{V_{zz}} \qquad 0 \le \eta \le 1 \qquad (8\text{-}31)$$

measures the deviation of the electric field gradient from axial symmetry. If we assume that

$$|V_{xx}| \le |V_{yy}| \le |V_{zz}| \qquad (8\text{-}32)$$

then η will vary from 0 to 1, being zero for axial symmetry,

$$V_{xx} = V_{yy} = -\tfrac{1}{2}eq , \qquad V_{zz} = eq \qquad (8\text{-}33)$$

and unity for the condition

$$V_{xx} = 0 , \qquad V_{yy} = -V_{zz} \qquad (8\text{-}34)$$

a limit which is rather rare.

The proportionality constant of Eq. 8-21 is

$$A = \frac{e^2 qQ}{2I(2I-1)} \qquad (8\text{-}35)$$

and the quadrupole tensor matrix elements are

$$Q_{ij} = A\left(\frac{V_{ij}}{V_{zz}}\right) \tag{8-36}$$

where $A = Q_{z'z'}$ is the quantity that is evaluated experimentally.

The field gradient tensor in a crystallographic coordinate system may be specified completely in terms of the parameters q and η and the three Eulerian angles which describe the relative orientations of the crystallographic and principal coordinate systems. A more convenient method of specifying the orientation of the principal axes entails the use of a direction cosine matrix (6-15).

The object of a pure quadrupole experiment is to determine the magnitudes of A and η and to find the direction cosines of the principal axes of the quadrupole tensor relative to the crystallographic axes. This gives the product of the field gradient q and the quadrupole moment Q.

8-6 SPIN *I* = 1 ENERGY LEVELS AND EIGENFUNCTIONS

The lowest spin nucleus that exhibits a quadrupole moment is $I = 1$, and common examples of this nucleus are deuterium and ^{14}N. To write the quadrupole Hamiltonian for such a nucleus one may use the spin matrices of Eqs. 2-57 to 2-62 and multiply them out in the manner illustrated here for I_y^2:

$$I_y^2 = \begin{pmatrix} 0 & -\dfrac{i}{\sqrt{2}} & 0 \\ \dfrac{i}{\sqrt{2}} & 0 & -\dfrac{i}{\sqrt{2}} \\ 0 & \dfrac{i}{\sqrt{2}} & 0 \end{pmatrix}\begin{pmatrix} 0 & -\dfrac{i}{\sqrt{2}} & 0 \\ \dfrac{i}{\sqrt{2}} & 0 & -\dfrac{i}{\sqrt{2}} \\ 0 & \dfrac{i}{\sqrt{2}} & 0 \end{pmatrix} = \begin{pmatrix} \frac{1}{2} & 0 & -\frac{1}{2} \\ 0 & 0 & 0 \\ -\frac{1}{2} & 0 & \frac{1}{2} \end{pmatrix}$$

$$\tag{8-37}$$

The quadrupole Hamiltonian (8-21) for $I = 1$ then assumes the form

$$\mathcal{H}_Q = \begin{pmatrix} \frac{1}{2}A & 0 & \frac{1}{2}\eta A \\ 0 & -A & 0 \\ \frac{1}{2}\eta A & 0 & \frac{1}{2}A \end{pmatrix} \tag{8-38}$$

where from Eq. 8-35 the parameter A has the explicit value

$$A = \tfrac{1}{2}e^2 qQ \tag{8-39}$$

The secular equation corresponding to $\mathcal{H}_Q$ is easily evaluated to give for the energy levels E_m:

$$E_0 = -A$$
$$E_{\pm 1} = \tfrac{1}{2}A(1 \pm \eta)$$

$$(8\text{-}40)$$

as shown in Fig. 8-4 for $\eta \ll 1$. For axial symmetry only one transition, $0 \rightarrow \pm 1$, exists, while for lower symmetries and $\eta \ll 1$ there will be a high-frequency doublet at $\Delta E = \tfrac{3}{2}A(1 \pm \tfrac{1}{3}\eta)$ and a low-frequency singlet at $\Delta E = A\eta$. The Hamiltonian matrix $\mathcal{H}_Q$ in its diagonal form is

$$\mathcal{H}_Q^{\text{diag}} = \begin{pmatrix} \tfrac{1}{2}A(1+\eta) & 0 & 0 \\ 0 & -A & 0 \\ 0 & 0 & \tfrac{1}{2}A(1-\eta) \end{pmatrix}$$

$$(8\text{-}41)$$

This will be found useful in the next section.

The $m = \pm 1$ levels were mixed in the original Hamiltonian $\mathcal{H}_Q$, and hence its eigenvectors are

$$|\psi_1\rangle = \alpha|1\rangle + \gamma|-1\rangle$$
$$|\psi_0\rangle = |0\rangle$$
$$|\psi_{-1}\rangle = -\gamma^*|1\rangle + \alpha^*|-1\rangle$$

$$(8\text{-}42)$$

subject to the normalization condition

$$\alpha\alpha^* + \gamma\gamma^* = 1$$

$$(8\text{-}43)$$

The coefficients α and γ form a unitary matrix $\vec{U}$ for the eigenvalue equation of the Hamiltonian matrix

$$\mathcal{H}_Q \vec{U} = \vec{U} \mathcal{H}_Q^{\text{diag}}$$

$$(8\text{-}44)$$

which has the explicit form

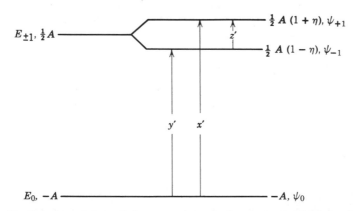

Fig. 8-4. Quadrupole splitting of the $m = \pm 1$ levels for $I = 1$, showing the energies and wavefunctions for each level.

$$\begin{pmatrix} \frac{1}{2}A & 0 & \frac{1}{2}\eta A \\ 0 & -A & 0 \\ \frac{1}{2}\eta A & 0 & \frac{1}{2}A \end{pmatrix} \begin{pmatrix} \alpha & 0 & -\gamma^* \\ 0 & 1 & 0 \\ \gamma & 0 & \alpha^* \end{pmatrix}$$

$$= \begin{pmatrix} \alpha & 0 & -\gamma^* \\ 0 & 1 & 0 \\ \gamma & 0 & \alpha^* \end{pmatrix} \begin{pmatrix} \frac{1}{2}A(1+\eta) & 0 & 0 \\ 0 & -A & 0 \\ 0 & 0 & \frac{1}{2}A(1-\eta) \end{pmatrix} \qquad (8\text{-}45)$$

This provides the relations

$$\alpha + \eta\gamma = (1+\eta)\alpha$$
$$\alpha^*\eta - \gamma^* = -(1-\eta)\gamma^* \qquad (8\text{-}46)$$

and shows that $\alpha = \gamma = 1/\sqrt{2}$. As a result

$$|\psi_1\rangle = \frac{1}{\sqrt{2}}\,[|1\rangle + |-1\rangle]$$

$$|\psi_{-1}\rangle = \frac{1}{\sqrt{2}}\,[-|1\rangle + |-1\rangle] \qquad (8\text{-}47)$$

where, somewhat unexpectedly, α and γ are independent of the asymmetry parameter η. For higher spins the wavefunction coefficients will in general be functions of the asymmetry parameter.

8-7 SPIN *I* = 1 ZEEMAN EFFECT

The Zeeman effect Hamiltonian

$$\mathscr{H} = \vec{I}\cdot\vec{Q}\cdot\vec{I} - g_N\beta_N\vec{H}_0\cdot\vec{I} \qquad (8\text{-}48)$$

has the following matrix form in the principal basis system, $|\psi_1\rangle$, $|\psi_0\rangle$, $|\psi_{-1}\rangle$:

$$\vec{U}^{-1}\mathscr{H}\vec{U} = \begin{pmatrix} \frac{1}{2}A(1+\eta) & ig_N\beta_N H_{0y'} & -g_N\beta_N H_{0z'} \\ -ig_N\beta_N H_{0y'} & -A & -g_N\beta_N H_{0x'} \\ -g_N\beta_N H_{0z'} & -g_N\beta_N H_{0x'} & \frac{1}{2}A(1-\eta) \end{pmatrix} \qquad (8\text{-}49)$$

Instead of diagonalizing this entire matrix one can consider an externally applied magnetic field H_0 along the z' principal direction:

$$\vec{U}^{-1}\mathscr{H}\vec{U} = \begin{pmatrix} \frac{1}{2}A(1+\eta) & 0 & -g_N\beta_N H_0 \\ 0 & -A & 0 \\ -g_N\beta_N H_0 & 0 & \frac{1}{2}A(1-\eta) \end{pmatrix} \qquad (8\text{-}50)$$

This will shift only the $\psi_\pm$ levels, and the energies for H in the z' direction are

$$E_{+1} = \tfrac{1}{2}A + \sqrt{\tfrac{1}{4}A^2\eta^2 + g_N^2\beta_N^2 H_0^2}$$

$$E_{-1} = \tfrac{1}{2}A - \sqrt{\tfrac{1}{4}A^2\eta^2 + g_N^2\beta_N^2 H_0^2} \qquad\qquad (8\text{-}51)$$

$$E_0 = -A$$

For H_0 in the x' direction one has the ψ_1 level unshifted and the ψ_0 and ψ_{-1} levels shifted by the applied field. The energies become

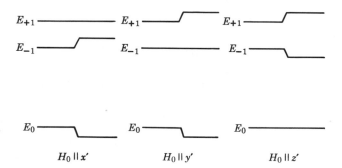

Fig. 8-5. Zeeman effect (right for each case) on the quadrupole energy levels (left for each case) for the applied magnetic field along the x' (left), y' (center), and z' (right) principal axes.

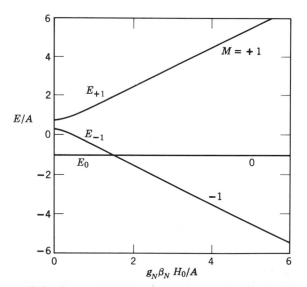

Fig. 8-6. Zeeman splitting for a spin $I = 1$ nucleus with the applied magnetic field along the z' principal axis.

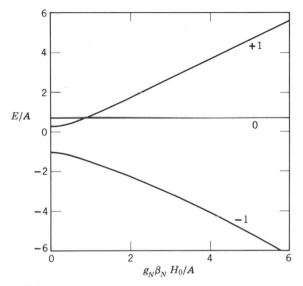

Fig. 8-7. Zeeman splitting for a spin $I = 1$ nucleus with the applied magnetic field H along the x' principal axis. At high field the M values correspond to selecting the quantization axis along H.

$$E_{+1} = \tfrac{1}{2}A(1 + \eta)$$
$$E_{-1} = -\tfrac{1}{4}A(1 + \eta) + \sqrt{[\tfrac{1}{4}A(3 - \eta)]^2 + g_N^2\beta_N^2 H_0^2} \qquad (8\text{-}52)$$
$$E_0 = -\tfrac{1}{4}A(1 + \eta) - \sqrt{[\tfrac{1}{4}A(3 - \eta)]^2 + g_N^2\beta_N^2 H_0^2}$$

For H_0 in the y' direction the energies are

$$E_{+1} = -\tfrac{1}{4}A(1 - \eta) + \sqrt{[\tfrac{1}{4}A(3 + \eta)]^2 + g_N^2\beta_N^2 H_0^2}$$
$$E_{-1} = \tfrac{1}{2}A(1 - \eta) \qquad (8\text{-}53)$$
$$E_0 = -\tfrac{1}{4}A(1 - \eta) - \sqrt{[\tfrac{1}{4}A(3 + \eta)]^2 + g_N^2\beta_N^2 H_0^2}$$

These energy level shifts are shown in Fig. 8-5 for $|g_N\beta_N H_0| \ll |A\eta|$. For low fields an expansion of the square root indicates that each energy level shift depends quadratically on the applied field, whereas for very high fields the shift becomes linear in the field. The dependence of the energy on the magnetic field strength is shown in Fig. 8-6 for the applied field parallel to the z' direction, and in Fig. 8-7 the case for H_0 parallel to x' is plotted. Both figures are drawn to the same scale.

8-8 QUADRUPOLE ENERGIES FOR $I > 1$

Although the methods developed in the preceding few sections may be employed to calculate quadrupole energies for spins greater than 1, the equations become quite complex. A number of authors have obtained general expressions for the energies; but they are rather complicated, and we will not reproduce them here. For axial symmetry the pure quadrupole frequency in the absence of a magnetic field is always of the form

$$E_m = \frac{e^2qQ}{2I(2I-1)}[3m^2 - I(I+1)] \qquad (8\text{-}54)$$

as shown in Fig. 8-8. The splitting between successive levels is

$$E_{m+1} - E_m = \frac{3e^2qQ}{2I(2I-1)}(2m+1) = \tfrac{3}{2}A(2m+1) \qquad (8\text{-}55)$$

which is a fairly simple expression.

For higher spins the various levels begin to depend on the asymmetry parameter to a different extent. For example, with $I = \tfrac{5}{2}$ and $\eta < 0.1$ the two transitions shown in Fig. 8-8 have the approximate differences

$$E_{5/2} - E_{3/2} = \tfrac{1}{2}A(12 - \tfrac{22}{9}\eta^2) \qquad (8\text{-}56)$$

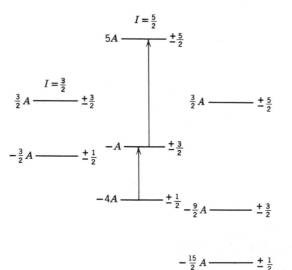

Fig. 8-8. Quadrupole energy levels for $I = \tfrac{3}{2}, \tfrac{5}{2},$ and $\tfrac{7}{2}$, showing the energies and m values. The two allowed transitions are shown for $I = \tfrac{5}{2}$.

$$E_{3/2} - E_{1/2} = \tfrac{1}{2}A(6 + \tfrac{5}{9}\eta^2) \qquad (8\text{-}57)$$

thus permitting a separate determination of A and η.

8-9 ZERO FIELD SPLITTINGS

The next few sections of this chapter deal with zero field effects[9-14] arising from Hamiltonian terms containing multiples of spin operators such as $(S_x S_z S_y S_z)$. The zero field Hamiltonian $\vec{S} \cdot \overset{\leftrightarrow}{D} \cdot \vec{S}$ is identical in form with the quadrupole interaction (8-11), and so its eigenvalues and eigenvectors are already known from the treatment earlier in this chapter. Products of more than two spin operators such as S_x^4 or $S_x^2 S_y^2$ may be computed by matrix multiplication and then formed into a secular equation in the usual manner. Therefore zero field splitting problems may be treated as logical extensions of the spin systems previously considered.

The greatest interest is in the Zeeman effect of spin systems with zero field interactions. These can be treated in the same manner as the Zeeman effect with quadrupole interactions, following the procedure of Section 8-7. Transition elements frequently require both zero field and hyperfine interactions to explain their ESR spectra, and sometimes quadrupole effects are also present.

The two main spin systems to be discussed here are triplet states of organic molecules which have the $\vec{S} \cdot \overset{\leftrightarrow}{D} \cdot \vec{S}$ and electronic Zeeman interactions,[11-14] and transition-metal ions[9-11] which require spin–spin, hyperfine, and perhaps quadrupole interactions to explain their ESR spectra. Specific examples of each type are worked out in detail.

The spin–spin or hyperfine term $\vec{S} \cdot \overset{\leftrightarrow}{T} \cdot \vec{I}$ gives a zero field splitting in the limit of vanishing magnetic field strength. For spin $(\tfrac{1}{2}, \tfrac{1}{2})$ and an isotropic hyperfine tensor $\overset{\leftrightarrow}{T}$ one obtains a singlet at $E = -3T/4$ and a triplet at $E = T/4$. Such zero field splittings are discussed in Sections 8-14 and 8-15.

8-10 ZERO FIELD *D* TERM

The simplest zero field interaction is the D term, and in the presence of a magnetic field the Hamiltonian of a spin S with such an interaction has the form

$$\mathscr{H} = g\beta\vec{H} \cdot \vec{S} + \vec{S} \cdot \overset{\leftrightarrow}{D} \cdot \vec{S} \qquad (8\text{-}58)$$

The D term is symmetric and traceless and it is represented by the matrix

$$\overset{\leftrightarrow}{D}_{(xyz)} = \begin{pmatrix} D_{xx} & D_{xy} & D_{xz} \\ D_{yx} & D_{yy} & D_{yz} \\ D_{zx} & D_{zy} & D_{zz} \end{pmatrix} \qquad (8\text{-}59)$$

subject to the conditions 8-13

$$D_{ij} = D_{ji}$$
$$D_{xx} + D_{yy} + D_{zz} = 0 \tag{8-60}$$

The term $\vec{S} \cdot \vec{D} \cdot \vec{S}$ may be expanded as

$$\vec{S} \cdot \vec{D} \cdot \vec{S} = D_{xx}S_x^2 + D_{yy}S_y^2 + D_{zz}S_z^2 + D_{xy}(S_xS_y + S_yS_x)$$
$$+ D_{yz}(S_yS_z + S_zS_y) + D_{xz}(S_zS_x + S_xS_z) \tag{8-61}$$

This may be put into the form of Eq. 8-17

$$\vec{S} \cdot \vec{D} \cdot \vec{S} = \tfrac{1}{2}(D_{xx} - D_{yy})(S_x^2 - S_y^2) + \tfrac{1}{2}D_{zz}[3S_z^2 - S(S+1)]$$
$$+ D_{xy}(S_xS_y + S_yS_x) + D_{yz}(S_yS_z + S_zS_y) + D_{xz}(S_zS_x + S_xS_z) \tag{8-62}$$

which is valid for any coordinate system xyz.

   This Hamiltonian may be diagonalized by transforming it to the principal axis coordinate system:

$$\vec{R}^{-1}\vec{D}_{(xyz)}\vec{R} = \vec{D}_{\text{diag}(x'y'z')} \tag{8-63}$$

where R is a direction cosine matrix (6-15). In this principal system D has the form

$$D_{\text{diag}} = \begin{pmatrix} D_{x'x'} & 0 & 0 \\ 0 & D_{y'y'} & 0 \\ 0 & 0 & D_{z'z'} \end{pmatrix} \tag{8-64}$$

where it is convenient to define

$$D_{x'x'} = -\tfrac{1}{3}D + E$$
$$D_{y'y'} = -\tfrac{1}{3}D - E$$
$$D_{z'z'} = \tfrac{2}{3}D \tag{8-65}$$

with the inverse transformation

$$D = \frac{3D_{z'z'}}{2} = -\frac{3(D_{x'x'} + D_{y'y'})}{2}$$

$$E = \frac{D_{x'x'} - D_{y'y'}}{2} \tag{8-66}$$

It is customary to select $D_{z'z'}$ as the largest of the three principal values of the D tensor and $D_{x'x'}$ as the smallest in magnitude

$$|D_{z'z'}| \geq |D_{y'y'}| \geq |D_{x'x'}| \tag{8-67}$$

Since the trace $\Sigma_i D_{ii}$ of the $\vec{D}$ tensor vanishes, it follows that $D_{x'x'}$ and $D_{y'y'}$ must have the same sign and be opposite in sign to $D_{z'z'}$:

$$D_{z'z'} = -(D_{x'x'} + D_{y'y'}) \tag{8-68}$$
$$|D_{z'z'}| = |D_{x'x'}| + |D_{y'y'}| \tag{8-69}$$

The analog of the quadrupole asymmetry parameter is

$$\eta_D = \frac{3E}{D} = \frac{D_{x'x'} - D_{y'y'}}{D_{z'z'}} \tag{8-70}$$

and the ratio $3E/D$ is restricted to the range

$$0 \le \frac{3E}{D} \le 1 \tag{8-71}$$

by condition 8-68. The Hamiltonian now assumes the simplified form

$$H = g(H_{x'}S_{x'} + H_{y'}S_{y'} + H_{z'}S_{z'}) + D[S_{z'}^2 - \tfrac{1}{3}S(S+1)] + E(S_{x'}^2 - S_{y'}^2) \tag{8-72}$$

These zero field expressions, written in terms of the quantities D_{ij}, S_i, S_j, are identical with their quadrupolar counterparts of Section 8-4, written in terms of the quantities Q_{ij}, I_i, I_j. Therefore all the results obtained before are valid for the *D*-term case by making the identifications

$$Q_{ij} \to D_{ij}, \qquad I_i \to S_i, \qquad \frac{3A}{2} \to D, \qquad \frac{\eta A}{2} \to E \tag{8-73}$$

with the equivalent expression from Eq. 8-26:

$$\frac{eQV_{ij}}{2I(2I-1)} \to D_{ij} \tag{8-74}$$

As a result of this equivalence the various final expressions of the earlier sections may be written down directly. Equations 8-24 ensure that the zero field term exists only for $S > \tfrac{1}{2}$. The choice of the notation of Eqs. 8-23 for the quadrupolar case and of Eqs. 8-65 for the zero field case is purely a matter of convention. It is necessary to follow conventions to obtain final expressions that are easily compared with the literature.

8-11 PRINCIPAL AXIS MATRICES FOR *D* AND *E*

It is appropriate, for reference purposes, to list the zero field Hamiltonian matrices for several spin values, using the principal axes x, y, z. The unit matrix is assumed to be associated with the term $DS(S+1)/3$ for calculation purposes, where

$$\vec{S} \cdot \vec{D} \cdot \vec{S} = D[S_z^2 - \tfrac{1}{3}S(S+1)] + E(S_x^2 - S_y^2) \tag{8-75}$$

and the S_x^2, S_y^2, and S_z^2 matrices are easily calculated from Eqs. 2-57 to 2-59.

The $\vec{S} \cdot \vec{D} \cdot \vec{S}$ matrices have the explicit forms

$$\vec{S} \cdot \vec{D} \cdot \vec{S} = \begin{pmatrix} 0 & 0 \\ 0 & 0 \end{pmatrix} \qquad\qquad (S = \tfrac{1}{2}) \quad (8\text{-}76)$$

$$= \begin{pmatrix} \tfrac{1}{3}D & 0 & E \\ 0 & -\tfrac{2}{3}D & 0 \\ E & 0 & \tfrac{1}{3}D \end{pmatrix} \qquad\qquad (S = 1) \quad (8\text{-}77)$$

$$= \begin{pmatrix} D & 0 & \sqrt{3}E & 0 \\ 0 & -D & 0 & \sqrt{3}E \\ \sqrt{3}E & 0 & -D & 0 \\ 0 & \sqrt{3}E & 0 & D \end{pmatrix} \qquad (S = \tfrac{3}{2}) \quad (8\text{-}78)$$

$$= \begin{pmatrix} 2D & 0 & \sqrt{6}E & 0 & 0 \\ 0 & -D & 0 & 3E & 0 \\ \sqrt{6}E & 0 & -2D & 0 & \sqrt{6}E \\ 0 & 3E & 0 & -D & 0 \\ 0 & 0 & \sqrt{6}E & 0 & 2D \end{pmatrix} \qquad (S = 2) \quad (8\text{-}79)$$

$$= \begin{pmatrix} \tfrac{10}{3}D & 0 & \sqrt{10}E & 0 & 0 & 0 \\ 0 & -\tfrac{2}{3}D & 0 & 3\sqrt{2}E & 0 & 0 \\ \sqrt{10}E & 0 & -\tfrac{8}{3}D & 0 & 3\sqrt{2}E & 0 \\ 0 & 3\sqrt{2}E & 0 & -\tfrac{8}{3}D & 0 & \sqrt{10}E \\ 0 & 0 & 3\sqrt{2}E & 0 & -\tfrac{2}{3}D & 0 \\ 0 & 0 & 0 & \sqrt{10}E & 0 & \tfrac{10}{3}D \end{pmatrix} \quad (S = \tfrac{5}{2}) \quad (8\text{-}80)$$

These matrices may be employed to extend the examples discussed in this chapter to higher spin values.

8-12 SPIN $S = 1$ ENERGY LEVELS AND EIGENFUNCTIONS

Triplet states of organic molecules have spin $S = 1$, and therefore it is of interest to treat this case in detail. The results of Section 8-6 may be quoted directly, using transformation 8-73. In the absence of a magnetic field the energies shown in Fig. 8-9 have the explicit magnitudes

$$E_0 = -\frac{2D}{3}$$

$$E_{\pm 1} = \frac{D}{3} \pm E \qquad\qquad (8\text{-}81)$$

and the eigenfunctions in terms of $|m_s\rangle$ kets are

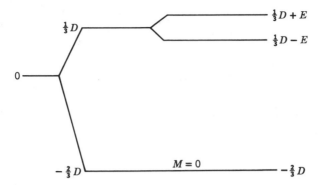

Fig. 8-9. Spin levels for $S = 1$, showing the energies of the levels in the absence of a zero field splitting (left), with an axial splitting (center), and with a completely anisotropic splitting (right).

$$\psi_0 = |0\rangle$$

$$\psi_{\pm 1} = \frac{1}{\sqrt{2}}[|1\rangle \mp |-1\rangle]$$

(8-82)

The condition of Eq. 8-71 requires that the z' axis be chosen so that $|D| > 3|E|$, which is equivalent to the condition that $2E$ be the smallest spacing between two levels on the right-hand side of Fig. 8-10.

The transition probabilities for a radiofrequency magnetic field H_{rf}, oriented in the direction θ, ϕ, arise from the operator

$$H_{rf} = g\beta e^{i\omega t}(H_{1x}S_x + H_{1y}S_y + H_{1z}S_z)$$

(8-83)

and provide the relative intensities

$$|\langle \psi_1 | H_{rf} | \psi_0 \rangle|^2 \sim \sin^2 \theta \sin^2 \phi$$
$$|\langle \psi_0 | H_{rf} | \psi_{-1} \rangle|^2 \sim \sin^2 \theta \cos^2 \phi$$
$$|\langle \psi_1 | H_{rf} | \psi_{-1} \rangle|^2 \sim \cos^2 \theta$$

(8-84)

The overall intensity which is equal to the sum of these three is unity.

As an example we quote the Hamiltonian with an isotropic Zeeman term:

$$\mathcal{H} = g\beta \vec{H}_0 \cdot \vec{S} + \vec{S} \cdot \vec{D} \cdot \vec{S}$$

(8-85)

and write down the Hamiltonian matrix in the principal coordinate system of the D term (vide Eq. 8-49):

$$\begin{pmatrix} \frac{1}{3}D + E & -ig\beta H_{0y'} & g\beta H_{0z'} \\ ig\beta H_{0y'} & -\frac{2}{3}D & g\beta H_{0x'} \\ g\beta H_{0z'} & g\beta H_{0x'} & \frac{1}{3}D - E \end{pmatrix}$$

(8-86)

The energies for the external magnetic field H_0, aligned along the x', y', and z' principal axes, are (compare Eqs. 8-51 to 8-53)

$$\left. \begin{array}{l} E_1 = \frac{1}{3}D + E \\[2mm] E_0 = -\frac{1}{2}(\frac{1}{3}D + E) - \sqrt{[\frac{1}{2}(D - E)]^2 + g^2\beta^2 H_0^2} \\[2mm] E_{-1} = -\frac{1}{2}(\frac{1}{3}D + E) + \sqrt{[\frac{1}{2}(D - E)]^2 + g^2\beta^2 H_0^2} \end{array} \right\} \quad (H_0 \| x') \quad (8\text{-}87)$$

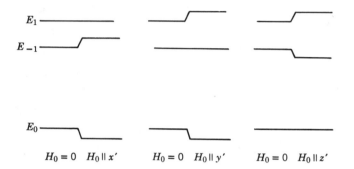

$$E_1 \qquad\qquad E_{-1} \qquad\qquad E_0$$

$$H_0 = 0 \quad H_0 \| x' \qquad H_0 = 0 \quad H_0 \| y' \qquad H_0 = 0 \quad H_0 \| z'$$

Fig. 8-10. Effect of a small Zeeman interaction (right for each case) on a large zero field D-term splitting (left for each case) with the applied magnetic field along the x' (left), y' (center), and z' (right) principal axes.

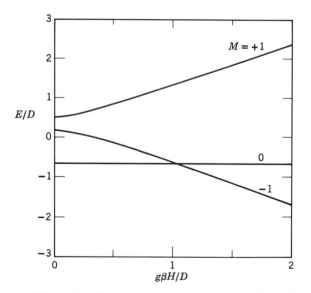

Fig. 8-11. Zeeman splitting of the D-term levels for a spin $S = 1$ electron configuration with an applied magnetic field oriented along the z' principal axis.

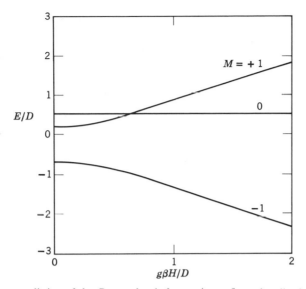

Fig. 8-12. Zeeman splitting of the D-term levels for a spin configuration $S = 1$ with an applied magnetic field H oriented along the x' principal axis. At high field the m values correspond to selecting the quantization axis along H.

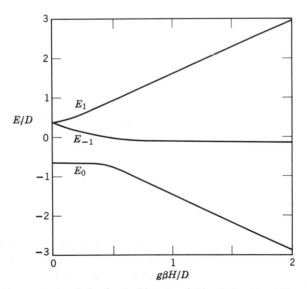

Fig. 8-13. Zeeman energy levels for $S = 1$ with a zero field splitting D and the magnetic field H in the 101 direction.

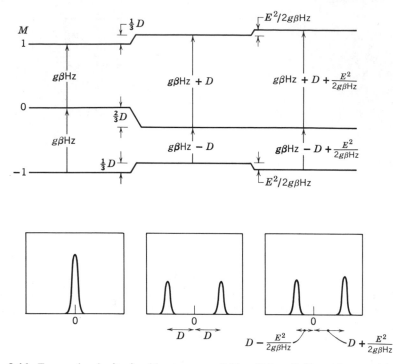

Fig. 8-14. Energy levels (top) without a zero field splitting (left), with an axial splitting (center), and with an added lower symmetry splitting (right). The corresponding spectra are shown at the bottom of the diagram.

$$
\left.
\begin{aligned}
E_1 &= -\tfrac{1}{2}(\tfrac{1}{3}D - E) + \sqrt{[\tfrac{1}{2}(D + E)]^2 + g^2\beta^2 H_0^2} \\
E_0 &= -\tfrac{1}{2}(\tfrac{1}{3}D - E) - \sqrt{[\tfrac{1}{2}(D + E)]^2 + g^2\beta H_0^2} \\
E_{-1} &= \tfrac{1}{3}D - E
\end{aligned}
\right\} \quad (H_0 \| y') \quad (8\text{-}88)
$$

$$
\left.
\begin{aligned}
E_1 &= \tfrac{1}{3}D + \sqrt{E^2 + g^2\beta^2 H_0^2} \\
E_0 &= -\tfrac{2}{3}D \\
E_{-1} &= \tfrac{1}{3}D - \sqrt{E^2 + g^2\beta^2 H_0^2}
\end{aligned}
\right\} \quad (H_0 \| z') \quad (8\text{-}89)
$$

These Zeeman effect energy level shifts are shown in Figs. 8-10 to 8-13.

An example of the transitions observed with H_0 parallel to the z' axis is given in Fig. 8-14 which shows the symmetric splitting of an ESR singlet at high field by the presence of a D interaction, and the asymmetry that results when an E interaction is added.

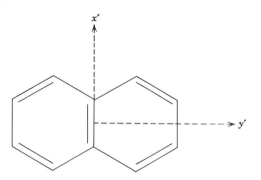

Fig. 8-15. Naphthalene molecule, showing the principal axes x' and y'.

8-13 TRIPLET STATES

A few molecules such as NO, ClO_2, and O_2 have triplet ground states, and a large number have triplet excited states. These triplet states with $S = 1$ are paramagnetic and may be observed by ESR. Many of them have rather large zero field splittings that are extremely anisotropic and that prevent observation in powder samples. Excited-state triplets are frequently relatively long lived. They may be excited optically by pulsed methods and studied by ESR.

A typical aromatic molecule with a triplet state is naphthalene.[15-17] It has the principal axes indicated in Fig. 8-15 where z' is perpendicular to the molecular plane. The principal values of D and E and the g-factor are as follows:

$$D = 0.1012 \text{ cm}^{-1} = 3034 \text{ MHz}$$
$$E = 0.0141 \text{ cm}^{-1} = 423 \text{ MHz} \tag{8-90}$$
$$g = 2.0030$$

The sign of E is the same as the sign of D in accordance with Eq. 8-71. The sign of D may be obtained at liquid helium temperature from the change in Boltzmann ratios and their effect on the line intensities.[17]

8-14 TRANSITION-METAL IONS

Many transition-metal ions have zero field splittings.[9-11] Typical cases from the first transition series are

$$
\begin{array}{ll}
3d^2 \text{ V}^{3+}, \text{ Cr}^{4+} & (S = 1) \\
3d^3 \text{ V}^{2+}, \text{ Cr}^{3+}, \text{ Mn}^{4+} & (S = \tfrac{3}{2}) \\
3d^4 \text{ Cr}^{2+} & (S = 2) \\
3d^5 \text{ Cr}^+, \text{ Mn}^{2+}, \text{ Fe}^{3+} & (S = \tfrac{5}{2})
\end{array}
\tag{8-91}
$$

$$3d^6 \; Fe^{2+} \qquad\qquad\qquad (S = 2)$$
$$3d^7 \; Fe^+, Co^{2+}, Ni^{3+} \qquad (S = \tfrac{3}{2}) \qquad\qquad \text{(8-91 cont.)}$$
$$3d^8 \; Co^+, Ni^{2+}, Cu^{3+} \qquad (S = 1)$$

Since space does not permit an elaboration of each case, a typical one is discussed in detail as an example.

The ions Mn^{2+} and Fe^{3+} have a zero field splitting even in a cubic field. It arises from the a term in the equivalent spin Hamiltonian:

$$\mathscr{H} = \beta \vec{H} \cdot \overset{\leftrightarrow}{g} \cdot \vec{S} + \tfrac{1}{6} a [S_\xi^4 + S_\eta^4 + S_\zeta^4 - \tfrac{1}{5} S(S+1)(3S^2 + 3S - 1)]$$
$$+ D[S_z^2 - \tfrac{1}{3} S(S+1)] + \vec{S} \cdot \overset{\leftrightarrow}{T} \cdot \vec{I}$$
$$+ \frac{F}{180} \{35 S_z^4 - 30 S(S+1) S_z^2 + 25 S_z^2 + 3 S(S+1)[S(S+1) - 2]\}$$

$$(8\text{-}92)$$

where the $\xi\eta\zeta$ coordinate system refers to three mutually perpendicular axes which are fourfold axes of the crystal field. For axial symmetries one must use the D and F terms, where for trigonal distortion the z axis is the [111] axis of the $\xi\eta\zeta$ system, and for tetragonal distortion the xyz axes coincide with $\xi\eta\zeta$. The g-factor is either axially symmetric or isotropic.

The Hamiltonian matrix may be formed by multiplying various spin matrices, such as

$$S_\xi^4 = S_\xi S_\xi S_\xi S_\xi \qquad\qquad (8\text{-}93)$$

and then transforming them all to the same coordinate system. The resulting 6×6 Hamiltonian matrix is quite complex.

The three strongly allowed transitions were found to be (see Bleaney and Trenam[18]) for the $\pm \tfrac{5}{2} \leftrightarrow \pm \tfrac{3}{2}$ transitions

$$g\beta H = g\beta H_0 \mp [2D(3 \cos^2 \theta - 1) + 2(1 - 5\phi)a + Fq] - 8\delta_1 + \delta_2 + \epsilon_1$$

$$(8\text{-}94)$$

for the $\pm \tfrac{3}{2} \leftrightarrow \pm \tfrac{1}{2}$ transitions

$$g\beta H = g\beta H_0 \mp [D(3 \cos^2 \theta - 1) - \tfrac{3}{2}(1 - 5\phi)a - \tfrac{5}{4} Fq] + \delta_1 - \tfrac{5}{4}\delta_2 + \epsilon_2$$

$$(8\text{-}95)$$

and for the $+ \tfrac{1}{2} \leftrightarrow - \tfrac{1}{2}$ transition

$$g\beta H = g\beta H_0 + 4\delta_1 - 2\delta_2 + \epsilon_3 \qquad\qquad (8\text{-}96)$$

where the external field H makes the angle θ with the trigonal axis, ϕ refers to the direction cosines of H in the $\xi\eta\zeta$ system,

$$\phi = \cos^2 \theta_{H\xi} \cos^2 \theta_{H\eta} + \cos^2 \theta_{H\eta} \cos^2 \theta_{H\zeta} + \cos^2 \theta_{H\zeta} \cos^2 \theta_{H\xi}$$

and the various parameters are defined by

$$\epsilon_1 = \frac{5}{3}\frac{a^2\phi}{g\beta H_0}(1-7\phi)$$

$$\epsilon_2 = -\frac{5}{48}\frac{a^2}{g\beta H_0}(3+178\phi-625\phi^2)$$

$$\epsilon_3 = \frac{10}{3}\frac{a^2\phi}{g\beta H_0}(7-25\phi)$$

$$q = \tfrac{1}{6}(35\cos^4\theta-30\cos^2\theta+3) \qquad\qquad (8\text{-}97)$$

$$\delta_1 = \frac{4D^2}{g\beta H_0}\sin^2\theta\cos^2\theta$$

$$\delta^2 = \frac{4D^2}{g\beta H_0}\sin^4\theta$$

$$\hbar\omega = g\beta H_0$$

Abragam and Bleaney[9] may be consulted for further details. A typical Mn^{2+} energy level diagram with a small cubic splitting (a), a large Zeeman splitting ($g\beta H$), and a small hyperfine splitting (T) is shown in Fig. 8-16.

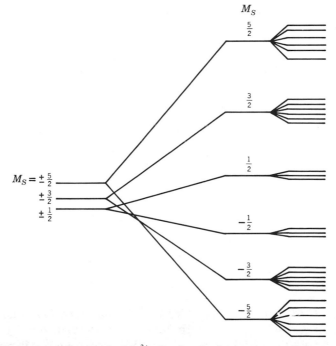

Fig. 8-16. Energy level diagram for Mn^{2+}, showing the fine structure splitting on the left and the hyperfine structure levels on the right. The figure is drawn for a zero field splitting much less than the Zeeman interaction. (Adapted from Ref. 19.)

The preceding discussion dealt with the first transition series. Rare earths have much larger and more pronounced zero field effects since their main crystal field splittings are much closer to the Zeeman energies than is the case for the first transition series. The spin–orbit coupling dominates the spectrum since it tends to exceed the crystal field splittings. The orbital angular momentum, which is quenched in the first transition series $(L \sim 0)$, is a good quantum number for rare earths, so that one can define $\vec{J} = \vec{L} + \vec{S}$ to describe the spectrum.

Now that a specific spin Hamiltonian has been presented, a few words about transition-metal ions in general is helpful. The requirement of time-reversal symmetry excludes odd powers of spin operators from the Hamiltonian. Operators of degree higher than $2S$ can be omitted. In other words, for $S = 1$ and $\frac{3}{2}$ only second-degree terms are permitted (S_x^2, etc.). For $S = 2$ and $\frac{5}{2}$ quadratic terms can occur (S_x^4, etc.), and for $S = 3$ and $\frac{7}{2}$ sixth powers are permitted.

8-15 HYPERFINE OR SPIN–SPIN ZERO FIELD LIMIT

It was mentioned in the introduction to this chapter that the Hamiltonian term $\vec{I}_1 \cdot \vec{T} \cdot \vec{I}_2$ produces a splitting in zero field. It is of interest to formulate this problem in terms of the low-field wavefunctions.

The Hamiltonian written in the conventional manner

$$\mathcal{H} = g_1 \beta H I_{1z} + g_2 \beta H I_{2z} + T \vec{I}_1 \cdot \vec{I}_2 \tag{8-98}$$

is in a form convenient for use with the high-field wavefunctions $|m_1 m_2\rangle$. It is desired to express this Hamiltonian in terms of the total spin operator

$$F = I_1 + I_2 \tag{8-99}$$

and its components F_x, F_y, F_z. To do this, one should observe that

$$\begin{aligned} F^2 &= (\vec{I}_1 + \vec{I}_2) \cdot (\vec{I}_1 + \vec{I}_2) \\ &= I_1^2 + I_2^2 + 2\vec{I}_1 \cdot \vec{I}_2 \end{aligned} \tag{8-100}$$

since I_1 and I_2 commute. As a result

$$\vec{I}_1 \cdot \vec{I}_2 = \tfrac{1}{2}[F(F+1) - I_1(I_1+1) - I_2(I_2+1)] \tag{8-101}$$

which permits the Hamiltonian of Eq. 8-98 to be written in the form

$$\begin{aligned} \mathcal{H} = \tfrac{1}{2}\beta H(g_1 + g_2)F_z &+ \tfrac{1}{2}\beta H(g_1 - g_2)(I_{1z} - I_{2z}) \\ &+ \tfrac{1}{2}T[F(F+1) - I_1(I_1+1) - I_2(I_2+1)] \end{aligned} \tag{8-102}$$

In the basis $|FM_F\rangle$ the first and third Hamiltonian terms are diagonal, and only the $\beta H(g_1 - g_2)(I_{1z} - I_{2z})$ term remains to be evaluated. The energies for the zero field limit of the $S = \frac{1}{2}$, $I = 1$ ESR case are given in Section 8-2. Section 12-10 of the first edition gives a more detailed discussion of this topic.

REFERENCES

1. M. H. Cohen and F. Reif, *Solid State Phys.*, **5**, 322 (1957).

2. T. P. Das and E. L. Hahn, *Nuclear Quadrupole Resonance Spectroscopy*, Academic Press, New York, 1958.

3. R. Livingston, "Nuclear Quadrupole Resonance," in *Methods of Experimental Physics*, Vol. 3, D. Williams, ed., Academic Press, New York, 1962, p. 501.

4. E. A. Lucken, *Nuclear Quadrupole Coupling Constants*, Academic Press, New York, 1969.

5. E. Scrocco, *Phys. Rev.*, **95**, 736 (1954); **96**, 951 (1954).

6. J. A. Smith, ed., *Advances in Nuclear Quadrupole Resonance*, Vol. 1 (1974) to Vol. 5 (1983), Academic Press, New York. A continuing series.

7. R. M. Sternheimer, *Phys. Rev.*, **95**, 736 (1954); **96**, 951 (1954).

8. T. P. Das and R. Bersohn, *Phys. Rev.*, **102**, 733 (1956).

9. A. Abragam and B. Bleaney, *Electron Paramagnetic Resonance of Transition Ions*, Clarendon Press, Oxford, 1979.

10. S. A. Al'tshuler and B. M. Kozyrev, *Electron Paramagnetic Resonance*, C. P. Poole, Jr., translation ed., Academic Press, New York, 1964.

11. A. Carrington and A. D. McLachlin, *Introduction to Magnetic Resonance*, Harper & Row, New York, 1967.

12. H.S. Jarrett, *Solid State Phys.*, **14**, 215 (1963).

13. M. Bersohn and J. C. Baird, *An Introduction to Electron Paramagnetic Resonance*, Benjamin, New York, 1966.

14. J. E. Wertz and J. R. Bolton, *Electron Spin Resonance*, McGraw-Hill, New York, 1972.

15. C. A. Hutchison, Jr. and B. W. Mangum, *J. Chem. Phys.*, **34**, 908 (1961).

16. R. W. Brandon, G. L. Closs, and C. A. Hutchison, Jr., *J. Chem. Phys.*, **37**, 1878 (1962).

17. J. H. van der Waals and M. S. DeGroot, *Mol. Phys.*, **2**, 333 (1959); **6**, 545 (1963).

18. B. Bleaney and R. S. Trenam, *Proc. R. Soc. A*, **223**, 1 (1954).

9

MÖSSBAUER RESONANCE

9-1 NATURE OF MÖSSBAUER EFFECT

The Mössbauer effect is the recoilless transition from an excited nuclear level denoted by N^* to a nuclear ground state N. It is exhibited by certain atoms in crystal lattices. This transition is of particular interest to magnetic resonance workers when the excited and ground states have different nuclear spins I^* and I, respectively. Instead of discussing this topic in general terms, it is illustrated for ^{57}Fe, which is by far the most widely studied nucleus. The generalization to other nuclei is straightforward.

The nuclear interactions involved in the ^{57}Fe Mössbauer effect are shown in Fig. 9-1. A cobalt nucleus ^{57}Co captures an electron (e capture) and thereby is transformed to a nuclear excited state of ^{57}Fe. The excited state is one of odd parity with a nuclear spin of $\frac{5}{2}$. It is 136 keV above the stable ground state of ^{57}Fe and may decay to the ground state by one of two processes. It passes directly to the ground state 9% of the time by emitting a 136-keV γ-ray, while 91% of the time it transforms to the $I^* = \frac{3}{2}$ first excited state, which has a half-life of 1.4×10^{-7} sec, before spontaneously decaying to the ground state by emitting a 14.4-keV γ-ray. The Mössbauer effect is the recoilless emission of the 14.4-keV γ-ray by a source and its absorption by an absorber.

The $I^* = \frac{3}{2}$ state at 14.4 keV and the ground state have the following characteristics which correspond to an ideal Mössbauer isotope (2.19% abundant)

Energy of transition	$E_0 = 14.4 \times 10^3$ eV
Natural linewidth	$\Delta E = 4.7 \times 10^{-9}$ eV
Shift due to recoil	$2E_R = 2.9 \times 10^{-3}$ eV

$$(9\text{-}1)$$

Doppler velocity due to natural linewidth	$v = \left(\dfrac{\Delta E}{E_0}\right)c = 0.095$ mm/sec

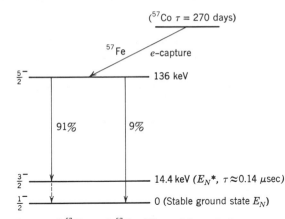

Fig. 9-1. Decay scheme of ^{57}Co and ^{57}Fe. The solid vertical arrows correspond to γ-ray emissions, and the dotted vertical arrow is the recoilless γ-ray or Mössbauer transition (from Ref. 3, p. 198).

Typical Debye temperature $\quad\quad \Theta_D = 420°C$

Recoilless fraction $\quad\quad\quad\quad F = \begin{cases} 0.92 \text{ at } 0°C \\ 0.79 \text{ at } 300°C \end{cases}$
(with $\Theta_D = 420°C$)

(9-1 cont.)

Nuclear spins $\quad\quad\quad\quad\quad I^* = \frac{3}{2},\ I = \frac{1}{2}$

Nuclear g-factors $\quad\quad\quad\quad g_N^* = -0.103,\ g_N = 0.18$

Gyromagnetic ratios $\quad\quad \begin{cases} \gamma^*/2\pi = 82 \text{ Hz}/G \\ \gamma/2\pi = 138 \text{ Hz}/G \end{cases}$

The magnetic moments are given by $\mu = g_N\beta_N I$, $\mu^* = g_N^*\beta_N I$, where $\gamma\hbar = g_N\beta_N$ and the asterisk denotes the excited state. The significance of these characteristics will become evident later in the chapter.

The Mössbauer experiment is carried out with a source which emits γ-rays (14.4 eV) and an absorber which absorbs them. The absorber is moved in relation to the source at a variable velocity v, and the resulting Doppler shift in energy Ev/c shows minima in transmitted γ-rays at velocities where the source and absorber energy levels differ by the relative velocity (i.e., by Ev/c). If the source lacks structure and the absorber has several energy levels of the type shown in Fig. 9-2 or 9-3, then a graph of γ-ray count versus relative velocity plots out the hyperfine or quadrupole pattern of the ^{57}Fe in the absorber, as shown in Fig. 9-4.

Space does not permit an elaboration of the experimental method employed in Mössbauer work, and other texts may be consulted for such details.[1,2] One should note that energies are measured in the units $E_\gamma v/c$, where c is the velocity of light. A number of reviews[3-7] of Mössbauer spectroscopy are available.

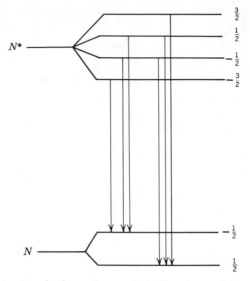

Fig. 9-2. Hyperfine structure in the nuclear excited N^* and ground N states of ^{57}Fe, showing the six hyperfine transitions. The quadrupole splitting is neglected in this figure.

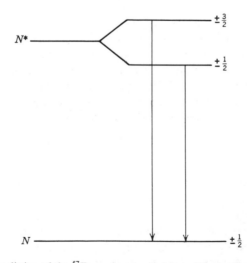

Fig. 9-3. Quadrupole splitting of the ^{57}Fe nuclear excited state N^*, showing the two Mössbauer transitions. The hyperfine splitting is neglected in this figure.

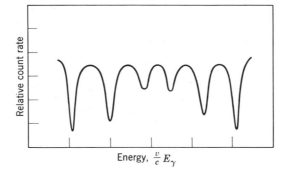

$$\text{Energy, } \frac{v}{c} E_\gamma$$

Fig. 9-4. Typical hyperfine pattern for ^{57}Fe.

9-2 HAMILTONIAN TERMS

In this chapter we discuss both the energy and the transition probability matrix for the full hyperfine pattern shown in Fig. 9-4. We calculate using the ket vectors $|N^*I^*m^*\rangle$, where $-\frac{3}{2} \leq m^* \leq \frac{3}{2}$ for the nuclear excited state, and $|NIm\rangle$, where $m = \pm\frac{1}{2}$ for the nuclear ground state. The Hamiltonian is

$$\mathcal{H} = \mathcal{H}_N - \vec{\mu} \cdot \vec{H}$$
$$= \mathcal{H}_N - g'_N\beta_N(H_x I'_x + H_y I'_y + H_z I'_z) \tag{9-2}$$

where $\mathcal{H}_N$ is the nuclear Hamiltonian, μ is the nuclear magnetic moment, I'_i is the spin operator for the ground (I_i) or excited (I_i^*) state, and $\vec{H}$ is the hyperfine field at the nucleus. For iron, $H \sim 300{,}000$ G.

We begin by calculating matrix elements for the Hamiltonian $\mathcal{H}_N$. This nuclear Hamiltonian is diagonal in the main nuclear quantum number N (and I) and is independent of m:

$$\langle N^*m^*|\mathcal{H}_N|N^*m^{*\prime}\rangle = E_{N^*}\delta_{m^*m^{*\prime}}$$
$$\langle Nm|\mathcal{H}_N|Nm'\rangle = E_N\delta_{mm'} \tag{9-3}$$
$$\langle N^*m^*|\mathcal{H}_N|Nm\rangle = \langle Nm|\mathcal{H}_N|N^*m^*\rangle = 0$$

This gives the zero-order energies of Fig. 9-1.

For the second term in the Hamiltonian we require the matrix elements of the magnetic moment. They are of the general form given by the Wigner–Eckart theorem:

$$\langle N^*I^*m^*|\vec{\mu}|NIm'\rangle = (N^*I^*|\vec{\mu}|NI)f(I^*Im^*m) \tag{9-4}$$

where $(N^*I^*|\vec{\mu}|NI)$ is the reduced matrix element which characterizes the intrinsic probability of $\vec{\mu}$ inducing a transition, and $f(I^*Im^*m)$ is the parameter which gives the relative intensity of each line. For ^{57}Fe the nuclear spins have the values given by Eq. 9-1, and $m^* = \pm\frac{1}{2}, \pm\frac{3}{2}, m = \pm\frac{1}{2}$ for the excited and ground states, respectively.

The matrices for the second term in the Hamiltonian of 9-2 may be written down with the aid of matrices 2-57 to 2-59 and the following expressions for the nonvanishing cross-terms (cf. Ref. 8, p. 63):

$$\langle N^*I + 1m \pm 1|\mu_x|NIm\rangle = \mp \tfrac{1}{2}\beta_N g_{12}\sqrt{(I \pm m + 1)(I \pm m + 2)}$$

$$\langle N^*I + 1m \pm 1|\mu_y|NIm\rangle = \tfrac{1}{2}i\beta_N g_{12}\sqrt{(I \pm m + 1)(I \pm m + 2)} \qquad (9\text{-}5)$$

$$\langle N^*I + 1m|\mu_z|NIm\rangle = \beta_N g_{12}\sqrt{(I + 1)^2 - m^2}$$

where for the present case $I = \tfrac{1}{2}$. The matrices for the three components of $\vec{\mu}$ have the following explicit forms:

$$\frac{\mu_x}{\beta_N} =
\begin{array}{c}
\quad \quad |N^* \tfrac{3}{2}\rangle \quad |N^* \tfrac{1}{2}\rangle \quad |N^* - \tfrac{1}{2}\rangle \; |N^* - \tfrac{3}{2}\rangle \quad |N \tfrac{1}{2}\rangle \quad |N - \tfrac{1}{2}\rangle \\[4pt]
\begin{array}{l}
\langle N^* \tfrac{3}{2}| \\[14pt]
\langle N^* \tfrac{1}{2}| \\[14pt]
\langle N^* - \tfrac{1}{2}| \\[14pt]
\langle N^* - \tfrac{3}{2}| \\[14pt]
\langle N \tfrac{1}{2}| \\[14pt]
\langle N - \tfrac{1}{2}|
\end{array}
\left(
\begin{array}{cccc|cc}
0 & \frac{\sqrt{3}}{2}g_N^* & 0 & 0 & -\frac{\sqrt{6}}{2}g_{12} & 0 \\[6pt]
\frac{\sqrt{3}}{2}g_N^* & 0 & g_N^* & 0 & 0 & -\frac{\sqrt{2}}{2}g_{12} \\[6pt]
0 & g_N^* & 0 & \frac{\sqrt{3}}{2}g_N^* & \frac{\sqrt{2}}{2}g_{12} & 0 \\[6pt]
0 & 0 & \frac{\sqrt{3}}{2}g_N^* & 0 & 0 & \frac{\sqrt{6}}{2}g_{12} \\[6pt]
\hline
-\frac{\sqrt{6}}{2}g_{12} & 0 & \frac{\sqrt{2}}{2}g_{12} & 0 & 0 & \tfrac{1}{2}g_N \\[6pt]
0 & -\frac{\sqrt{2}}{2}g_{12} & 0 & \frac{\sqrt{6}}{2}g_{12} & \tfrac{1}{2}g_N & 0
\end{array}
\right)
\end{array}
$$

$$(9\text{-}6)$$

$$\frac{\mu_y}{\beta_N} =
\left(
\begin{array}{cccc|cc}
0 & -\frac{i\sqrt{3}}{2}g_N^* & 0 & 0 & \frac{i\sqrt{6}}{2}g_{12} & 0 \\[6pt]
\frac{i\sqrt{3}}{2}g_N^* & 0 & -ig_N^* & 0 & 0 & \frac{i\sqrt{2}}{2}g_{12} \\[6pt]
0 & ig_N^* & 0 & -\frac{i\sqrt{3}}{2}g_N^* & \frac{i\sqrt{2}}{2}g_{12} & 0 \\[6pt]
0 & 0 & \frac{i\sqrt{3}}{2}g_N^* & 0 & 0 & \frac{i\sqrt{6}}{2}g_{12} \\[6pt]
\hline
-i\frac{\sqrt{6}}{2}g_{12} & 0 & -\frac{i\sqrt{2}}{2}g_{12} & 0 & 0 & -\tfrac{i}{2}g_N \\[6pt]
0 & -\frac{i\sqrt{2}}{2}g_{12} & 0 & -\frac{i\sqrt{6}}{2}g_{12} & \tfrac{i}{2}g_N & 0
\end{array}
\right)
$$

$$(9\text{-}7)$$

$$\frac{\mu_z}{\beta_N} = \begin{pmatrix} \frac{3}{2}g_N^* & 0 & 0 & 0 & 0 & 0 \\ 0 & \frac{1}{2}g_N^* & 0 & 0 & \sqrt{2}g_{12} & 0 \\ 0 & 0 & -\frac{1}{2}g_N^* & 0 & 0 & \sqrt{2}g_{12} \\ 0 & 0 & 0 & -\frac{3}{2}g_N^* & 0 & 0 \\ \hline 0 & \sqrt{2}g_{12} & 0 & 0 & \frac{1}{2}g_N & 0 \\ 0 & 0 & \sqrt{2}g_{12} & 0 & 0 & -\frac{1}{2}g_N \end{pmatrix}$$ (9-8)

Similar matrices may be constructed easily for the quadrupole interaction operators I_x^2, $I_x I_z$, and so on.

9-3 ENERGY LEVELS

To deduce the energies it is useful to select the quantization axis along the hyperfine field direction, which gives the simplified Hamiltonian

$$\mathcal{H} = \mathcal{H}_N - \mu_z H_z \tag{9-9}$$

In this system the Hamiltonian is effectively diagonal, and the corresponding matrix has the form of 9-11. Off-diagonal matrix elements connecting the states $|N^*m^*\rangle$ to $|N\,\frac{1}{2}\rangle$ and $|N^*m^*\rangle$ to $|N -\frac{1}{2}\rangle$ may be neglected since in second-order perturbation theory they have the energy denominator $E_N^* - E_N$ and

$$\frac{(\mu_z H_z)^2}{E_N^* - E_N} \sim 10^{-11}\mu_z H_z \tag{9-10}$$

which is rather small. Under these conditions the energy matrix has the form

	$\lvert N^*\,\tfrac{3}{2}\rvert$	$\lvert N^*\,\tfrac{1}{2}\rangle$	$\lvert N^* -\tfrac{1}{2}\rangle$	$\lvert N^* -\tfrac{3}{2}\rangle$	$\lvert N\,\tfrac{1}{2}\rangle$	$\lvert N -\tfrac{1}{2}\rangle$
$\langle N^*\,\tfrac{3}{2}\rvert$	$E_N^* + \tfrac{3}{2}g_N^*\beta_N H$	0	0	0	0	0
$\langle N^*\,\tfrac{1}{2}\rvert$	0	$E_N^* + \tfrac{1}{2}g_N^*\beta_N H$	0	0	0	0
$\langle N^* -\tfrac{1}{2}\rvert$	0	0	$E_N^* - \tfrac{1}{2}g_N^*\beta_N H$	0	0	0
$\langle N^* -\tfrac{3}{2}\rvert$	0	0	0	$E_N^* - \tfrac{3}{2}g_N^*\beta_N H$	0	0
$\langle N\,\tfrac{1}{2}\rvert$	0	0	0	0	$E_N + \tfrac{1}{2}g_N\beta_N H$	0
$\langle N -\tfrac{1}{2}\rvert$	0	0	0	0	0	$E_N - \tfrac{1}{2}g_N\beta_N H$

(9-11)

corresponding to the energy level diagram of Fig. 9-2.

Energy matrix 9-11 corresponds to the case of an ^{57}Fe ion in a cubic field site where the gradient of the electric field tensor is zero. When this gradient does not vanish, the quadrupole interaction must be added to the Hamiltonian (cf. Section 8-4)

$$\mathscr{H} = \mathscr{H}_N - g_N \beta_N H I' + \vec{I}' \cdot \vec{\vec{Q}} \cdot \vec{I}' \tag{9-12}$$

where the nuclear g-factor is isotropic for both the nuclear ground and excited states. The quantization axis is taken along the z' principal direction of the quadrupole tensor, to give a Hamiltonian matrix where $\vec{\vec{Q}}$ is diagonal. With the aid of Eqs. 8-35 and 9-6 to 9-8 one may write the Hamiltonian matrix by using the notation

$$A = \frac{e^2 qQ}{6}, \qquad \vec{G}_N = g_N \beta_N \vec{H}, \qquad \vec{G}_N^* = g_N^* \beta_N \vec{H} \tag{9-13}$$

and the symbol η for the asymmetry parameter. The result with $\vec{H}$ in the yz plane is as follows:

$$
\left(
\begin{array}{cccc|cc}
E_N^* + \frac{3}{2}G_{Nz}^* + 3A & -\frac{i\sqrt{3}}{2}G_{Ny}^* - 3\eta A & 0 & 0 & 0 & 0 \\[2mm]
\frac{i\sqrt{3}}{2}G_{Ny}^* - 3\eta A & E_N^* + \frac{1}{2}G_{Nz}^* - 3A & -iG_{Ny}^* & 0 & 0 & 0 \\[2mm]
0 & iG_{Ny}^* & E_N^* - \frac{1}{2}G_{Nz}^* - 3A & -\frac{i\sqrt{3}}{2}G_{Ny}^* - 3\eta A & 0 & 0 \\[2mm]
0 & 0 & \frac{i\sqrt{3}}{2}G_{Ny}^* - 3\eta A & E_N^* - \frac{3}{2}G_{Nz}^* + 3A & 0 & 0 \\[2mm]
\hline
0 & 0 & 0 & 0 & E_N + \frac{1}{2}G_{nz} & -\frac{1}{2}iG_{Ny} \\[2mm]
0 & 0 & 0 & 0 & \frac{1}{2}iG_{Ny} & E_N - \frac{1}{2}G_{Nz}
\end{array}
\right)
$$

$$\tag{9-14}$$

The matrix elements connecting the ground and excited nuclear states are neglected as a result of Eq. 9-10, and of course the quadrupole interaction vanishes in the nuclear ground state where $I = \frac{1}{2}$. The energy levels are shown in Fig. 9-5 for the hyperfine splitting alone, and with the addition of a weak and a medium quadrupole interaction. Figure 9-6 gives the observed spectra for several ratios of A/G^* and Fig. 9-7 shows details of a spectrum with combined hyperfine and quadrupole contributions. The ground nuclear state has no quadrupole effect; hence its energies are, as usual (omitting E_N),

$$E = \pm \frac{1}{2} g_N \beta_N H \tag{9-15}$$

Hamiltonian matrix 9-14 has no solution in closed form. The $I = \frac{3}{2}$ part is

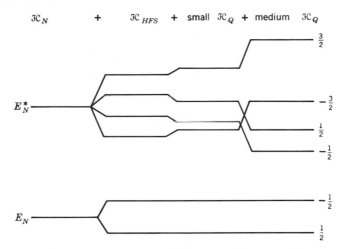

Fig. 9-5. Energy level diagram of ground and nuclear excited states of ^{57}Fe, showing from left to right the effect of adding a hyperfine and a quadrupole interaction.

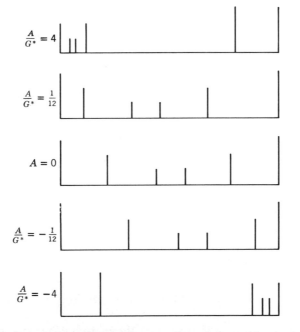

Fig. 9-6. Mössbauer structure patterns for the range from a large positive (top) to zero (center) to a large negative (bottom) quadrupole contribution to the hyperfine pattern. The spectra are normalized to a constant overall spread in energy.

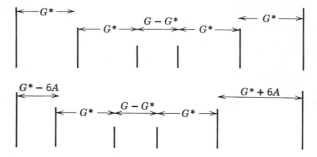

Fig. 9-7. Mössbauer spectrum for ^{57}Fe, showing hyperfine structure (top) and hyperfine structure plus a weak quadrupole effect (bottom).

equivalent to the case discussed in Section 11-7 of the first edition (vide Eq. 11-73 there). If the hyperfine magnetic field is aligned along one of the principal directions of the quadrupole tensor, then all off-diagonal terms G^*_{Ny} vanish, and the $I^* = \frac{3}{2}$ part of the Hamiltonian reduces to two 2×2 submatrices with the solutions (omitting E_N)

$$E = g^*_N \beta_N H \pm \tfrac{1}{2} e^2 q Q \sqrt{\left(1 + \frac{g^*_N \beta_N H}{e^2 q Q}\right)^2 + \eta^2} \qquad (m = \tfrac{3}{2}, -\tfrac{1}{2})$$

$$\tag{9-16}$$

$$= -g^*_N \beta_N H \pm \tfrac{1}{2} e^2 q Q \sqrt{\left(1 - \frac{g^*_N \beta_N H}{e^2 q Q}\right)^2 + \eta^2} \qquad (m = -\tfrac{3}{2}, \tfrac{1}{2})$$

These expressions permit the sign of $e^2 q Q$ to be determined. If the field gradient is axially symmetric with the symmetry axis oriented at an angle θ with respect to the hyperfine field direction, then the asymmetry parameter η vanishes. For the case where $\eta = 0$ and the Zeeman term is much greater than the quadrupole term one may use the procedures of Section 11-2 of the first edition to give

$$E = \pm \tfrac{3}{2} g^*_N \beta_N H + \frac{e^2 q Q}{8} (3 \cos^2 \theta - 1) \qquad (m = \pm \tfrac{3}{2})$$

$$\tag{9-17}$$

$$= \pm \tfrac{1}{2} g^*_N \beta_N H + \frac{e^2 q Q}{8} (3 \cos^2 \theta - 1) \qquad (m = \pm \tfrac{1}{2})$$

The product $e^2 q Q (3 \cos^2 \theta - 1)/8$ constitutes the observed quadrupole splitting.

Thus far we have taken into account the main nuclear, the hyperfine, and the quadrupolar interactions. If additional Hamiltonian terms are appreciable, they can complicate the spectrum.

9-4 HYPERFINE FIELD

Throughout the book the hyperfine interaction Hamiltonian has been written

$$\mathcal{H}_{HF} = \vec{S} \cdot \vec{\vec{T}} \cdot \vec{I} \qquad (9\text{-}18)$$

This may also be written as a Zeeman interaction:

$$\mathcal{H}_{HF} = g_N \omega_N \vec{H}_{HF} \cdot \vec{I} \qquad (9\text{-}19)$$

When T is isotropic, the magnitude of the hyperfine field or effective magnetic field at the nucleus is

$$H_{HF} = \frac{ST}{g_N \beta_N} = \frac{ST}{\gamma_N \hbar} \qquad (9\text{-}20)$$

For ^{57}Fe in the nuclear ground state the hyperfine coupling constant T is 49 MHz or 17.5 G. One may convert T in gauss to T in megahertz by the expression

$$T_{(G)} = \frac{T_{(MHz)}}{g\beta} \qquad (9\text{-}21)$$

where for an electronic g-factor of 2 the quantity $g\beta = 2.8$ MHz/G. Hence, if the hyperfine coupling constant in megahertz is divided by $\gamma_N / 2\pi$ from Eq. 9-1, one obtains the hyperfine field

$$H_{HF} = \frac{ST_{(MHz)}}{\gamma_N / 2\pi} = \left(\frac{g\beta}{g_N \beta_N} \right) ST_{(G)} \qquad (9\text{-}22)$$

$$\sim 300{,}000 \text{ G} \qquad (9\text{-}23)$$

for ^{57}Fe nuclei, where $S = \frac{5}{2}$.

The hyperfine field at the nucleus arises from the Fermi contact interaction and is proportional to the square of the S-state wavefunction at the nucleus:

$$H_{HF} = \frac{8\pi}{3} g\beta S |\psi_{S(0)}|^2 \qquad (9\text{-}24)$$

For closed shells the wavefunction $\psi_{S(0)}$ at the nuclear site is slightly different for spin up and spin down. For the first transition series each $1S$ electronic wavefunction produces hyperfine fields of about 2.5×10^9 G, $2S$ electrons produce fields of 2.2×10^8 G, and $3S$ electrons produce 3×10^7 G. The difference fields of the spin up and spin down electrons in each of the ns levels are the order of hundreds of kilogauss, since the wavefunctions in each closed shell are almost (but not quite) identical, and their fields do not cancel.

9-5 ISOMER SHIFT

One additional effect is often measured in Mössbauer experiments and that is the isomer shift which is sometimes referred to as the chemical shift. This arises from a shift in the positions of the unsplit nuclear energy levels of the type shown in Fig. 9-8. It results from the finite nuclear radius R.

If the nuclear ground and excited states had zero radius, their energy difference would be $E_{N0}^* - E_{N0}$. The effect of a finite nuclear radius R is to raise each energy level in relation to its zero radius value, E_{N0}^* or E_{N0}, to give the energies

$$E_N^* = E_{N0}^* + \frac{2\pi}{5} Ze^2 R^{*2} |\psi_{(0)}|^2$$

$$E_N = E_{N0} + \frac{2\pi}{5} Ze^2 R^2 |\psi_{(0)}|^2 \tag{9-25}$$

where Z is the nuclear charge, and $|\psi_{(0)}|^2$ is the total S-state electron density at the nucleus. This might be referred to as an electronic monopole interaction to contrast it with the magnetic dipole (hyperfine) and electric quadrupole interactions. In the absence of the other two interactions the γ-ray will have the energy

$$E_\gamma = E_N^* - E_N \tag{9-26}$$

$$= (E_{N0}^* - E_{N0}) + \frac{2\pi}{5} Ze^2 |\psi_{(0)}|^2 (R^{*2} - R^2) \tag{9-27}$$

The magnitude of $|\psi_{(0)}|^2$ depends somewhat on the electronic configuration, chemical environment, and other factors and varies from one compound to another. An experiment usually determines the shift in γ-ray energy between the absorber and the emitter and gives the isomer shift δ:

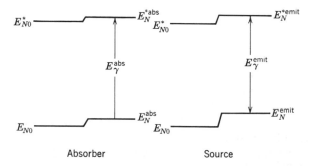

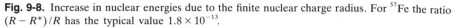

Fig. 9-8. Increase in nuclear energies due to the finite nuclear charge radius. For ^{57}Fe the ratio $(R - R^*)/R$ has the typical value 1.8×10^{-13}.

$$\delta = E_\gamma^{abs} - E_\gamma^{emit} \tag{9-28}$$

$$= \tfrac{2}{5}\pi Ze^2(R^{*2} - R^2)(|\psi_{(0)}|_{abs}^2 - |\psi_{(0)}|_{emit}^2) \tag{9-29}$$

The isomer shift is observed as a shift in energy (velocity) of the entire hyperfine or quadrupole pattern relative to a standard sample spectrum.

REFERENCES

1. R. L. Cohen, *Applications of Mössbauer Spectroscopy*, Vol. 2, Academic Press, New York, 1980.

2. M. Leopold, ed., *An Introduction of Mössbauer Spectroscopy*, Plenum, New York, 1971.

3. G. K. Shenoy and F. E. Wagner, eds., *Mössbauer Isomer Shifts*, Elsevier, New York, 1978.

4. J. G. Stevens, ed., *Cumulative Index to the Mössbauer Spectroscopy Data Indices*, Plenum, New York, 1979. (See also the continuing series *Mössbauer Effect Data Indices*.)

5. A. Vertes, L. Korecz, and K. Burger, *Mössbauer Spectroscopy*, Elsevier, New York, 1980.

6. U. Gonsev, ed., *Mössbauer Spectroscopy II: The Exotic Side of the Methods*, Springer-Verlag, Berlin, 1981.

7. J. G. Stevens and G. P. Shenoy, *Mössbauer Spectroscopy and Its Chemical Applications*, American Chemical Society, Washington, DC, 1981.

8. E. U. Condon and G. H. Shortley, *The Theory of Atomic Spectra*, Cambridge University Press, Cambridge, 1953. Chapter 3 discusses angular momentum.

ATOMIC SPECTRA AND CRYSTAL FIELD THEORY

10-1 INTRODUCTION

Almost all research in atomic spectroscopy has dealt with atoms and radicals in their ground electronic states. In recent years flash photolysis and high-power pulsed laser techniques have been employed to study excited electronic states, and sometimes magnetic resonance is detected by changes that occur in optical transitions.[1-6] To gain some insight into the relationship between optical levels and magnetic field splittings it is appropriate to discuss several topics from atomic spectroscopy.[7-9] The usual direct product method is employed in developing the theory. The first edition treats this topic more thoroughly. The second half of this chapter deals with crystal field theory.[10-13]

10-2 SPIN–ORBIT COUPLING

The spin–orbit interaction may be written[1]

$$\mathcal{H}_{SO} = \lambda \vec{L} \cdot \vec{S} \tag{10-1}$$

where λ is the spin–orbit coupling constant. The strength of this interaction is sometimes expressed in terms of the positive definite parameter ζ:

$$\lambda = \pm \frac{\zeta}{2S} \tag{10-2}$$

where the negative sign refers to a shell that is more than half filled, and the positive sign corresponds to one that is less than half filled. Some authors refer to ζ as the spin–orbit coupling constant. Figure 10-1 shows the variation ζ with atomic number for several transition series.

Instead of discussing the spin–orbit interaction in general, it is convenient to calculate the spin–orbit splittings for the specific case $L = 1$, $S = \frac{1}{2}$, using the $|LSm_L m_S\rangle = |m_L m_S\rangle$ representation.

The Hamiltonian

$$\mathcal{H}_{SO} = \lambda[L_x S_x + L_y S_y + L_z S_z] \tag{10-3}$$

is expanded in direct products, using the matrices of Eqs. 2-57 to 2-59. This gives the Hamiltonian matrix:

	$\lvert 1\,\tfrac{1}{2}\rangle$	$\lvert 1-\tfrac{1}{2}\rangle$	$\lvert 0\,\tfrac{1}{2}\rangle$	$\lvert 0-\tfrac{1}{2}\rangle$	$\lvert -1\,\tfrac{1}{2}\rangle$	$\lvert -1-\tfrac{1}{2}\rangle$
$\langle 1\,\tfrac{1}{2}\rvert$	$\tfrac{1}{2}\lambda$	0	0	0	0	0
$\langle 1-\tfrac{1}{2}\rvert$	0	$-\tfrac{1}{2}\lambda$	$\dfrac{1}{\sqrt{2}}\lambda$	0	0	0
$\langle 0\,\tfrac{1}{2}\rvert$	0	$\dfrac{1}{\sqrt{2}}\lambda$	0	0	0	0
$\langle 0-\tfrac{1}{2}\rvert$	0	0	0	0	$\dfrac{1}{\sqrt{2}}\lambda$	0
$\langle -1\,\tfrac{1}{2}\rvert$	0	0	0	$\dfrac{1}{\sqrt{2}}\lambda$	$-\tfrac{1}{2}\lambda$	0
$\lvert\langle -1-\tfrac{1}{2}\rvert$	0	0	0	0	0	$\tfrac{1}{2}\lambda$

$$\tag{10-4}$$

which has the eigenvalues

$$\tfrac{1}{2}\lambda,\ \tfrac{1}{2}\lambda,\ \tfrac{1}{2}\lambda,\ \tfrac{1}{2}\lambda,\ -\lambda,\ -\lambda \tag{10-5}$$

as illustrated in Fig. 10-2 for the ^2P state.

The above energies could have been obtained by writing the Hamiltonian directly in the $|Jm\rangle$ representation with the aid of the expression

$$J^2 = (\vec{L} + \vec{S})^2 = L^2 + S^2 + 2\vec{L}\cdot\vec{S} \tag{10-6}$$

where $\vec{L}$ and $\vec{S}$ commute $(\vec{L}\cdot\vec{S} = \vec{S}\cdot\vec{L})$.

Substituting $\vec{L}\cdot\vec{S}$ from Eq. 10-6 into the Hamiltonian (10-1) gives

$$\mathcal{H}_{SO} = \tfrac{1}{2}\lambda(J^2 - L^2 - S^2) \tag{10-7}$$

which has the following eigenvalues in the $|Jm\rangle$ representation:

$$E_{SO} = \tfrac{1}{2}\lambda[J(J+1) - L(L+1) - S(S+1)] \tag{10-8}$$

$$= \tfrac{1}{2}\lambda[J(J+1) - \tfrac{11}{4}] \tag{10-9}$$

$$= \begin{cases} \tfrac{1}{2}\lambda & (J = \tfrac{3}{2}) \\ -\lambda & (J = \tfrac{1}{2}) \end{cases} \tag{10-10}$$

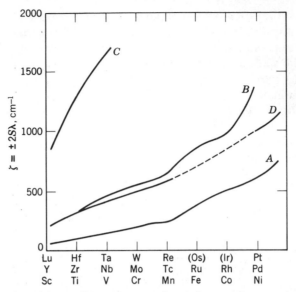

Fig. 10-1. The spin–orbit coupling constant ζ for d-electrons in neutral atoms: (A) first transition series, (B) second series, (C) third series, (D) third series scaled down by a factor of 4 (from Ref. 8). Reprinted from J.S. Griffith, The Theory of Transition—Metal Ions. Copyright 1961 by Cambridge University Press.

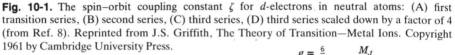

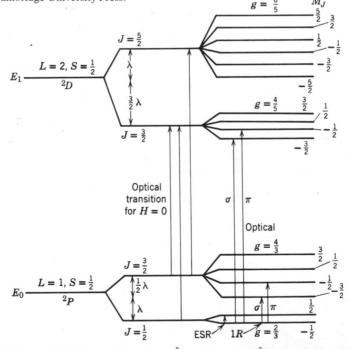

Fig. 10-2. Energy level diagram for a lower ^2P and an upper ^2D state (left), with spin–orbit coupling added (center) and with magnetic field added (right). The levels are labeled with their quantum numbers. Several optical, two infrared, and one ESR transitions are indicated. The Zeeman spacings are exaggerated.

in agreement with Eq. 10-5. Equation 10-8 gives the spin–orbit splitting for various configurations such as S states $(L = 0)$, P states $(L = 1)$, D states $(L = 2)$, F states $(L = 3)$, G states $(L = 4)$, H states $(L = 5)$, and I, J, K, and so on states. Figure 10-2 shows two doublet $(S = \frac{1}{2}, 2S + 1 = 2)$ states, one a P and the other a D state. The P state is assumed to have the electronic energy E_0, and the next excited state has the energy E_1. The total angular momentum J will, of course, obey the relation

$$|L - S| \le J \le L + S \tag{10-11}$$

to give the values $J = \frac{3}{2}, \frac{5}{2}$ for the D state of Fig. 10-2.

10-3 ZEEMAN EFFECT

The Zeeman effect is the magnetic field splitting that occurs for weak magnetic fields where the magnetic moment or Zeeman interaction $g\beta\vec{H}$ is much less than the spin–orbit interaction $\lambda\vec{L}\cdot\vec{S}$. The Hamiltonian has the form

$$\mathcal{H} = \mathcal{H}_{SO} + \mathcal{H}_{Ze} \tag{10-12}$$

$$= \lambda\vec{L}\cdot\vec{S} + \beta(\vec{L} + 2\vec{S})\cdot\vec{H} \tag{10-13}$$

where the g-factors for orbital and spin motion are 1 and 2, respectively. The Hamiltonian may be rearranged to the form

$$\mathcal{H} = \tfrac{1}{2}\lambda[J(J + 1) - L(L + 1) - S(S + 1)] + \tfrac{3}{2}\beta J_z H + \tfrac{1}{2}\beta(S_z - L_z)H \tag{10-14}$$

where the magnetic field $\vec{H}$ is in the z direction. This is the same Hamiltonian that was treated in Section 12-10 of the first edition and therefore the energy can be written immediately:

$$E = \tfrac{1}{2}\lambda[J(J + 1) - L(L + 1) - S(S + 1)] + g\beta H m_J \tag{10-15}$$

where the Landé g-factor is given by

$$g = \frac{3}{2} + \frac{S(S + 1) - L(L + 1)}{2J(J + 1)} \tag{10-16}$$

This expression may be obtained by letting $I_1 = L$, $I_2 = S$, $F = J$, and $g_2 = 2g_1 = 2$ in Eq. 12-85 of the first edition. Each main spin–orbit level is characterized by a total angular momentum J, and the magnetic field splits the J level into $2J + 1$ sublevels with the shifts in energy $g\beta H m_J$ from the unsplit position. One should note that the center of gravity of the levels remains invariant under the influence of both the spin–orbit and the Zeeman interactions. This occurs because the corresponding Hamiltonian matrices have zero trace.

10-4 ZEEMAN TRANSITION PROBABILITIES

In the optical region transitions are observed between different main optical states, as illustrated in Fig. 10-2. The transition probability arises from the square of the matrix elements of the electric dipole operator $e\vec{r}$ connecting the various Zeeman sublevels. The matrix elements of this operator are proportional to the orbital angular momentum matrix elements

$$\langle E_iLSJm|e\vec{r}|E_gL'S'J'm'\rangle = K_{(E_iE_gLL'SS')}\langle E_iJm|\vec{L}|E_gJ'm'\rangle$$

$$(10\text{-}17)$$

which may be written in the simpler notation

$$\langle E_iJm|e\vec{r}|E_gJ'm'\rangle = K_{(E_iE_gJJ')}\langle E_iJm|\vec{L}|E_gJ'm'\rangle \qquad (10\text{-}18)$$

where the symbol E_i refers to a state with L, S, and E_g refers to one with L', S' orbital and spin angular momenta. The vector

$$\vec{L} = L_x\hat{i} + L_y\hat{j} + L_z\hat{k} \qquad (10\text{-}19)$$
$$= \tfrac{1}{2}L^+(\hat{i} - i\hat{j}) + \tfrac{1}{2}L^-(\hat{i} + i\hat{j}) + L_z\hat{k} \qquad (10\text{-}20)$$

may be treated in the manner discussed in Section 13-4 of the first edition. The matrix elements $\langle E_iJm|\vec{L}|E_gJ'm'\rangle$ between the m magnetic sublevels of the main optical level E_g, and the excited-state counterparts (m, E_i), may be arranged in a probability amplitude matrix. These matrix elements are of the following forms:[7]

$$\langle E_iJm|e\vec{r}|E_gJ+1m\pm 1\rangle = \mp K_{(E_iE_gJJ+1)}\tfrac{1}{2}\sqrt{(J\pm m+1)(J\pm m+2)}(\hat{i}\pm i\hat{j})$$

$$\langle E_iJm]e\vec{r}|E_gJ+1m\rangle \;\;= K_{(E_iE_gJJ+1)}\sqrt{(J+1)^2 - m^2}\hat{k}$$

$$\langle E_iJm|e\vec{r}|E_gJm\pm 1\rangle \;\;= K_{(E_iE_gJJ)}\tfrac{1}{2}\sqrt{(J\mp m)(J\pm m+1)}(\hat{i}\pm i\hat{j})$$

$$\langle E_iJm|e\vec{r}|E_gJm\rangle \;\;= K_{(E_iE_gJJ)}m\hat{k}$$

$$\langle E_iJm|e\vec{r}|E_gJ-1m\pm 1\rangle \;\;= \pm K_{(E_iE_gJJ-1)}\tfrac{1}{2}\sqrt{(J\mp m)(J\mp m-1)}(\hat{i}\pm i\hat{j})$$

$$\langle E_iJm|e\vec{r}|E_gJ-1m\rangle \;\;= K_{(E_iE_gJJ-1)}\sqrt{J^2 - m^2}\hat{k}$$

$$(10\text{-}21)$$

where the Hermitian property

$$\langle E_iJm|L_n|E_gJ'm'\rangle^* = \langle E_gJ'm'|L_n|E_iJm\rangle \qquad (10\text{-}22)$$

renders the matrix elements L_x symmetric about the diagonal whereas those of L_y change from i to $-i$ across the diagonal of the transition probability matrix. The reduced matrix element $K_{(E_iE_gJJ')}$ varies from one set of transitions to another, while the relative intensity within a group of lines is given by the function of J and M to the right of the reduced matrix elements in Eq. 10-21. (See also Sections 9-2 and 10-9.)

Equations 10-21 were written in Section 9-2 for the Mössbauer case, and the matrices of μ_x, μ_y, μ_z were given explicitly for $J = \frac{1}{2}$, $J' = \frac{3}{2}$. The same matrices arise here for this J case, but with different reduced matrix elements.

The energy levels for the $^2P \rightarrow {}^2D$ transitions are shown in Fig. 10-2. Typical ESR, infrared, and optical transitions are indicated, where the spin–orbit splitting is usually in the infrared region, and the electronic splitting $(E_1 - E_0)$ is in the optical region. A similar diagram may correspond to E_0 and E_1 with the same principal quantum number, in which case $E_1 - E_0$ would probably be an infrared transition. However, optical Zeeman studies are usually carried out between groups of lines with different principal quantum numbers, as assumed above.

All allowed optical Zeeman transitions are shown in Fig. 10-3. The three sets of lines in this figure correspond to the three "optical" boxes in the following matrix:

2×2 $E_0 - \lambda$ $J = \frac{1}{2}$	2×4 Infrared $K_{(E_0 E_0\, 1/2\, 3/2)}$	2×4 Optical $K_{(E_0 E_1\, 1/2\, 3/2)}$	Forbidden
	4×4 $E_0 + \frac{1}{2}\lambda$ $J = \frac{3}{2}$	4×4 Optical $K_{(E_0 E_1\, 3/2\, 3/2)}$	4×6 Optical $K_{(E_0 E_1\, 3/2\, 5/2)}$
		4×4 $E_1 - \frac{3}{2}\lambda$ $J = \frac{3}{2}$	4×6 infrared $K_{(E_1 E_1\, 3/2\, 5/2)}$
			6×6 $E_1 + \lambda$ $J = \frac{5}{2}$

$$(10\text{-}23)$$

Since the matrix is Hermitian, only the diagonal and upper right boxes are filled in. The $J = \frac{1}{2} \leftrightarrow J = \frac{5}{2}$ transitions are forbidden by the selection rule

$$\Delta J = 0, \pm 1 \qquad (10\text{-}24)$$

The infrared transitions within a multiplet arise from boxes labeled "infrared" in the matrix. The ordinary ESR transitions are entirely within the diagonal boxes.

It is appropriate to give each of the optical transition submatrices calculated from Eq. 10-21. For $\frac{1}{2} \rightarrow \frac{3}{2}$ each element of the following matrix should be multiplied by $K_{(E_0 E_1\, 1/2\, 3/2)}$ to give the probability amplitude $\langle E_0 \frac{1}{2} m | e\vec{r} | E_1 \frac{3}{2} m \rangle$:

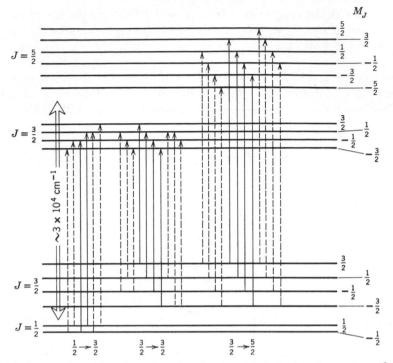

Fig. 10-3. Zeeman transitions for σ (–––) and π (——) polarizations from a lower ^2P to an upper ^2D state. The large arrow on the left gives the single transition in the absence of an applied field.

$$
\begin{array}{c}
& \left|\frac{3}{2}\,\frac{3}{2}\right\rangle & \left|\frac{3}{2}\,\frac{1}{2}\right\rangle & \left|\frac{3}{2}-\frac{1}{2}\right\rangle & \left|\frac{3}{2}-\frac{3}{2}\right\rangle \\[6pt]
\left\langle\frac{1}{2}\,\frac{1}{2}\right| & \begin{pmatrix} -\dfrac{\sqrt{3}}{2}(\hat{i}+i\hat{j}) & \sqrt{2}\hat{k} & \dfrac{1}{\sqrt{2}}(\hat{i}-i\hat{j}) & 0 \\[12pt]
\left\langle\frac{1}{2}-\frac{1}{2}\right| & 0 & -\dfrac{1}{\sqrt{2}}(\hat{i}+i\hat{j}) & \sqrt{2}\hat{k} & \dfrac{\sqrt{3}}{2}(\hat{i}-i\hat{j}) \end{pmatrix}
\end{array}
\tag{10-25}
$$

in agreement with Eq. 13-19 of the first edition.

For the $J=\frac{3}{2}\to J=\frac{3}{2}$ case the following matrix elements are multiplied by the common factor $K_{(E_0E_1\ 3/2\ 3/2)}$:

$$
\begin{array}{c}
& \left|\frac{3}{2}\,\frac{3}{2}\right\rangle & \left|\frac{3}{2}\,\frac{1}{2}\right\rangle & \left|\frac{3}{2}-\frac{1}{2}\right\rangle & \left|\frac{3}{2}-\frac{3}{2}\right\rangle \\[6pt]
\left\langle\frac{3}{2}\,\frac{3}{2}\right| & \begin{pmatrix} \dfrac{3}{2}\hat{k} & \dfrac{\sqrt{3}}{2}(\hat{i}-i\hat{j}) & 0 & 0 \\[12pt]
\left\langle\frac{3}{2}\,\frac{1}{2}\right| & \dfrac{\sqrt{3}}{2}(\hat{i}+i\hat{j}) & \dfrac{1}{2}\hat{k} & (\hat{i}-i\hat{j}) & 0 \\[12pt]
\left\langle\frac{3}{2}-\frac{1}{2}\right| & 0 & (\hat{i}+i\hat{j}) & -\dfrac{1}{2}\hat{k} & \dfrac{\sqrt{3}}{2}(\hat{i}-i\hat{j}) \\[12pt]
\left\langle\frac{3}{2}-\frac{3}{2}\right| & 0 & 0 & \dfrac{\sqrt{3}}{2}(\hat{i}+i\hat{j}) & -\dfrac{3}{2}\hat{k} \end{pmatrix}
\end{array}
\tag{10-26}
$$

The $\frac{3}{2} \leftrightarrow \frac{5}{2}$ transitions are in accordance with the following probability amplitudes, where the common factor is $K_{(E_0 E_1\, 3/2\, 5/2)}$. Squaring these amplitudes gives the relative intensities.

$$
\begin{array}{ccccccc}
 & |\tfrac{5}{2}\,\tfrac{5}{2}\rangle & |\tfrac{5}{2}\,\tfrac{3}{2}\rangle & |\tfrac{5}{2}\,\tfrac{1}{2}\rangle & |\tfrac{5}{2}-\tfrac{1}{2}\rangle & |\tfrac{5}{2}-\tfrac{3}{2}\rangle & |\tfrac{5}{2}-\tfrac{5}{2}\rangle \\[4pt]
\langle\tfrac{3}{2}\,\tfrac{3}{2}| & -\sqrt{5}(\hat{i}+i\hat{j}) & 2\hat{k} & \dfrac{1}{\sqrt{2}}(\hat{i}-i\hat{j}) & 0 & 0 & 0 \\[10pt]
\langle\tfrac{3}{2}\,\tfrac{1}{2}| & 0 & -\sqrt{3}(\hat{i}+i\hat{j}) & \sqrt{6}\hat{k} & \dfrac{\sqrt{3}}{2}(\hat{i}-i\hat{j}) & 0 & 0 \\[10pt]
\langle\tfrac{3}{2}-\tfrac{1}{2}| & 0 & 0 & -\dfrac{\sqrt{3}}{2}(\hat{i}+i\hat{j}) & \sqrt{6}\hat{k} & \sqrt{3}(\hat{i}-i\hat{j}) & 0 \\[10pt]
\langle\tfrac{3}{2}-\tfrac{3}{2}| & 0 & 0 & 0 & -\dfrac{1}{\sqrt{2}}(\hat{i}+i\hat{j}) & 2\hat{k} & +\sqrt{5}(\hat{i}-i\hat{j})
\end{array}
$$

$$(10\text{-}27)$$

10-5 PASCHEN–BACH EFFECT

In the low-field Zeeman effect the dominance of the spin–orbit interaction made the Hamiltonian of Eq. 10-13 follow the $|Jm\rangle$ coupling scheme. At the other extreme, where the Zeeman term far exceeds the spin–orbit interaction, it is appropriate to employ the representation $|m_L m_S\rangle$, where $\mathcal{H}_{Ze}$ is diagonal and $\mathcal{H}_{SO}$ is not. This limit is referred to as the Paschen–Bach effect, and the energies become (assuming $\mathcal{H}_{SO} \sim 0$)

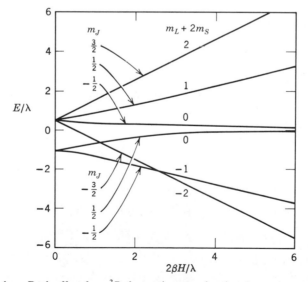

Fig. 10-4. Paschen–Bach effect for a ^2P electronic state, showing the quantum numbers for the low- and high-field cases.

$$E = \beta H(m_L + 2m_S) \tag{10-28}$$

corresponding to levels being approached on the extreme right side of Fig. 10-4.

At intermediate fields one has the incomplete Paschen–Bach effect. The energy levels are determined most easily from the spin Hamiltonian of Eq. 10-13, written in the $|m_L m_S\rangle$ representation. Selecting H in the z direction gives

$$\mathcal{H} = \lambda(L_x S_x + L_y S_y + L_z S_z) + \beta H(L_z + 2S_z) \tag{10-29}$$

The Hamiltonian matrix for the $L = 1$, $S = \frac{1}{2}$ configuration may be written easily with the aid of the spin–orbit part (10-3) as follows:

	$\lvert 1\,\tfrac{1}{2}\rangle$	$\lvert 1-\tfrac{1}{2}\rangle$	$\lvert 0\,\tfrac{1}{2}\rangle$	$\lvert 0-\tfrac{1}{2}\rangle$	$\lvert -1\,\tfrac{1}{2}\rangle$	$\lvert -1-\tfrac{1}{2}\rangle$
$\langle 1\,\tfrac{1}{2}\rvert$	$\tfrac{1}{2}\lambda + 2\beta H$	0	0	0	0	0
$\langle 1-\tfrac{1}{2}\rvert$	0	$-\tfrac{1}{2}\lambda$	$\dfrac{1}{\sqrt{2}}\lambda$	0	0	0
$\langle 0\,\tfrac{1}{2}\rvert$	0	$\dfrac{1}{\sqrt{2}}\lambda$	βH	0	0	0
$\langle 0-\tfrac{1}{2}\rvert$	0	0	0	$-\beta H$	$\dfrac{1}{\sqrt{2}}\lambda$	0
$\langle -1\,\tfrac{1}{2}\rvert$	0	0	0	$\dfrac{1}{\sqrt{2}}\lambda$	$-\tfrac{1}{2}\lambda$	0
$\langle -1-\tfrac{1}{2}\rvert$	0	0	0	0	0	$\tfrac{1}{2}\lambda - 2\beta H$

$$\tag{10-30}$$

There are two 2×2 submatrices to diagonalize. The resulting secular equation provides the six energies:

$$
\begin{aligned}
E_1 &= \tfrac{1}{2}\lambda + 2\beta H \\
E_2 &= -\tfrac{1}{4}\lambda + \tfrac{1}{2}\beta H + \tfrac{1}{2}(\beta^2 H^2 - \lambda\beta H + \tfrac{9}{4}\lambda^2)^{1/2} \\
E_3 &= -\tfrac{1}{4}\lambda - \tfrac{1}{2}\beta H + \tfrac{1}{2}(\beta^2 H^2 + \lambda\beta H + \tfrac{9}{4}\lambda^2)^{1/2} \\
E_4 &= -\tfrac{1}{4}\lambda + \tfrac{1}{2}\beta H - \tfrac{1}{2}(\beta^2 H^2 - \lambda\beta H + \tfrac{9}{4}\lambda^2)^{1/2} \\
E_5 &= -\tfrac{1}{4}\lambda - \tfrac{1}{2}\beta H - \tfrac{1}{2}(\beta^2 H^2 + \lambda\beta H + \tfrac{9}{4}\lambda^2)^{1/2} \\
E_6 &= \tfrac{1}{2}\lambda - 2\beta H
\end{aligned}
\tag{10-31}
$$

The corresponding energy level diagram is shown in Fig. 10-4 with the low-field lines labeled by m_J and the high-field lines labeled by $m_L + 2m_S$.

10-6 CRYSTAL FIELDS

In preceding chapters the direct product method of treating anisotropic Hamiltonian parameters was discussed. The angular dependence of ESR spectra was found to arise mainly from anisotropic g-factors and hyperfine coupling constants. These in turn result from the crystalline electric fields and chemical bonds in solids. Therefore it is appropriate to apply the direct product method to the calculation of crystal field effects.

A first-transition-series ion in a solid experiences a very strong Stark effect on its $3d^n$ unfilled electronic shell of many thousands of wavenumbers (several electron volts) because of the crystalline electric fields of the surrounding ions. This is comparable to the basic electronic energies of the ion, and it far exceeds the spin–orbit interaction. Rare-earth ions, on the other hand, have the outermost $5s^2 5p^6$ filled shells shielding the inner paramagnetic $4f^n$ electrons, and therefore the crystalline Stark effect splitting amounts to only hundreds of reciprocal centimeters, less than the spin–orbit energy. These energy splittings far exceed the electronic Zeeman interaction of about $-0.3\,\text{cm}^{-1}$, and hence they are detected directly in optical spectroscopy. However, they strongly influence the magnitude and anisotropy of the spin Hamiltonian parameters, and therefore it is appropriate to consider them here.

10-7 CRYSTAL FIELD POTENTIAL

In most solids the crystalline electric field at a positive ion site has a dominant component arising from a basically octahedral, tetrahedral, or cubic array of surrounding negative charges or ligands, as illustrated in Fig. 10-5. The nearest-neighbor ions dominate in determining the crystal potential. Ordinarily, the surrounding charges are axially distorted with a tetragonal axis of symmetry, as shown in Fig. 10-6. Lower symmetry distortions are also common.

The electric potential near the center of an array of N negative charges q_i may be written

$$V = \sum_{i=1}^{N} \frac{q_i}{r_i} \tag{10-32}$$

where r_i is the distance from the ith ion. It is more convenient to select the center of the array as the reference point, and to express the potential at an arbitrary point p in terms of the radius vector $\vec{r}$ from the origin

$$V = \sum_{i=1}^{N} \left(\frac{q_i}{R_i}\right)\left[1 + \left(\frac{r}{R_i}\right)^2 - 2\frac{r}{R_i}\cos\alpha_i\right]^{-1/2} \tag{10-33}$$

where α_i is the angle between $\vec{R}_i$ and $\vec{r}$ for each ion, as defined in Fig. 10-7.

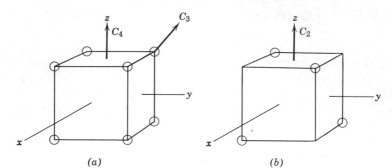

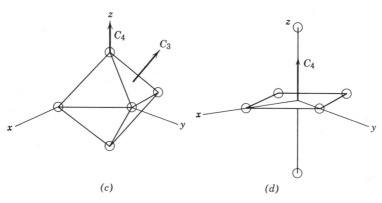

Fig. 10-5. Charge distributions: (*a*) cubic, (*b*) tetrahedral, (*c*) octahedral, and (*d*) square planar plus axial (tetragonally distorted octahedral).

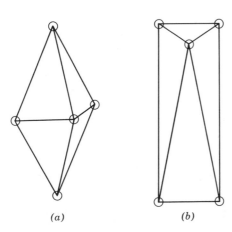

Fig. 10-6. Tetragonally (*a*) and trigonally (*b*) distorted octahedral array of charges.

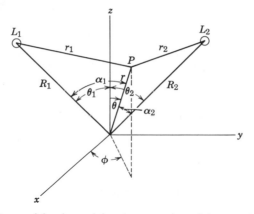

Fig. 10-7. Coordinates used for determining the expansion of the crystal field potential at an arbitrary point $P = P_{(r,\theta,\phi)}$ arising from more distant $(R_1, R_2 > r)$ ligands L_1 and L_2.

In crystal field studies one is generally interested in points P closer to the origin than the ligand distance R_1, and so a power series expansion is made in terms of the ratio r/R_i:

$$V = \sum_{i=1}^{N} \left(\frac{q_i}{R_i}\right) \sum_{l=0}^{\infty} \left(\frac{r}{R_i}\right)^l P_l(\cos \alpha_i) \qquad (10\text{-}34)$$

where $P_l(\cos \alpha_i)$ is the lth-order Legendre polynomial. This may be expressed in polar coordinates with the aid of the spherical harmonic addition theorem:

$$P_l(\cos \alpha_i) = \frac{4\pi}{2l+1} \sum_{m=-l}^{l} (-1)^m Y_{l(\theta_i,\phi_i)}^{-m} Y_{l(\theta,\phi)}^{m} \qquad (10\text{-}35)$$

and the real spherical harmonics $Z_{lm(\theta,\phi)}$, called tesseral harmonics, which are related to the usual spherical harmonics in the following manner:

$$Z_{l0} = Y_{l0}$$

$$Z_{lm}^c = \frac{1}{\sqrt{2}} [Y_l^{-m} + (-1)^m Y_l^m] \qquad (10\text{-}36)$$

$$Z_{lm}^s = \frac{1}{\sqrt{2}} [Y_l^{-m} - (-1)^m Y_l^m]$$

where $m > 0$, and the superscripts c and s denote cosine and sine, respectively. This follows since

$$Z_{l0} = \sqrt{\frac{2l+1}{4\pi}} \; P_l(\cos\theta)$$

$$Z_{lm}^c = \sqrt{\frac{(2l+1)(l-m)!}{2\pi(l+m)!}} \; P_l^m(\cos\theta)\cos m\phi \qquad (10\text{-}37)$$

$$Z_{lm}^s = \sqrt{\frac{(2l+1)(l-m)!}{2\pi(l+m)!}} \; P_l^m(\cos\theta)\sin m\phi$$

where the associated Legendre polynomials P_l^m are functions of θ. The tesseral harmonics most often used in crystal field calculations are listed in Table 10-1.

Using these relations, one obtains for the crystal field potential 10-33

$$V_{(r,\theta,\phi)} = \sum_{l=0}^{\infty} \sum_{m=0}^{l} r^l A_{lm} Z_{lm(\theta,\phi)} \qquad (10\text{-}38)$$

where there are terms Z_{l0}, Z_{lm}^c, and Z_{lm}^s for all $l>0$. The corresponding coefficients A_{lm} have the following forms for a discrete charge distribution of N ligands:

$$A_{l0} = \frac{4\pi}{2l+1} \sum_{i=1}^{N} q_i \frac{Z_{l0}(\theta_i\phi_i)}{R_i^{l+1}}$$

$$A_{lm}^c = \frac{4\pi}{2l+1} \sum_{i=1}^{N} q_i \frac{Z_{lm}^c(\theta_i\phi_i)}{R_i^{l+1}} \qquad (10\text{-}39)$$

$$A_{lm}^s = \frac{4\pi}{2l+1} \sum_{i=1}^{N} q_i \frac{Z_{lm}^s(\theta_i\phi_i)}{R_i^{l+1}}$$

For many of the cases considered in this book there are N identical nearest-neighbor ligand charges located the same distance R from the transition metal ion site, to give

$$A_{lm}^c = \frac{4\pi}{2l+1} \frac{q}{R^{l+1}} \sum_{i=1}^{N} Z_{lm}^c(\theta_i\phi_i) \qquad (10\text{-}40)$$

where the summation constitutes a geometric factor, and similar expressions may be written for A_{lm}^s and A_{l0}. The summation in Eq. 10-38 in an infinite sum over all l, but for reasonably high symmetry most terms will vanish. In addition, one needs to retain only terms which satisfy the inequality $l \le 2l'$, where l' equals 2 for d-electrons and 3 for f-electrons. For a continuous charge distribution the summation over i in Eq. 10-33 may be replaced by an integration over a charge density $\rho_{(\bar{r}_i)}$. When the coefficients A_{lm} are listed below, the factor $[4\pi/(2l+1)]^{1/2}$ will be explicitly written down. Various authors adopt somewhat different definitions for these quantities A_{lm}, and one should exercise caution in comparing formulas.[8-13]

The crystal field potential for a particular array of charges is determined by inserting the ligand coordinates into Eq. 10-39. This is referred to as the point charge model, and it is a good first approximation to the potential. The $l = 0$ term is a constant times q/R which shifts all the levels equally and hence may be omitted.

For the cubic array of point charges Ze shown in Fig. 10-5a, one has

$$A_{00} = 8 \frac{Ze}{R} \sqrt{4\pi}$$

$$A_{40} = -\frac{28}{9} \frac{Ze}{R^5} \sqrt{\frac{4\pi}{9}} \qquad (10\text{-}41)$$

$$A_{60} = \frac{16}{9} \frac{Ze}{R^7} \sqrt{\frac{4\pi}{13}}$$

and A_{44}^c and A_{64}^c are obtained from the relations

$$A_{44}^c = \sqrt{\tfrac{5}{7}} A_{40}, \qquad A_{64}^c = -\sqrt{7} A_{60} \qquad (10\text{-}42)$$

For a tetrahedral array of the type shown in Fig. 10-5b, the corresponding potential coefficients A_{lm} are one-half of the cubic values in Eqs. 10-41. For the octahedron of Fig. 10-5c the coefficients are

$$A_{00} = \frac{6Ze}{R} \sqrt{4\pi}$$

$$A_{40} = \frac{7}{2} \frac{Ze}{R^5} \sqrt{\frac{4\pi}{9}} \qquad (10\text{-}43)$$

$$A_{60} = \frac{3}{4} \frac{Ze}{R^7} \sqrt{\frac{4\pi}{13}}$$

with A_{44}^c and A_{64}^c still given by Eq. 10-42. For d-electrons only the A_{40} and A_{44} terms contribute to the energy splitting, and the cubic coefficients A_{4m} are $\frac{8}{9}$ times as large as the octahedral coefficients. Thus, for the same distances and charges, the octahedral group is slightly more effective (by a factor of $\frac{9}{8}$) than the eight cubic charges in splitting the energy levels.

The formulas in this section are valid for a point whose distance from the origin r is less than the nearest ligand distance. For the opposite case of a point that is beyond the furthest charge in the array, the roles of r and R_i in Eq. 10-33 are reversed. As a result the expansion should be carried out in powers of $(R_i/r)^l$, thereby interchanging the quantities r and R. Subsequent formulas are valid here also. However, in this chapter we confine our interest to the practical case wherein $r < R_i$ for all R_i.

TABLE 10-1 Commonly Occurring Tesseral Harmonics in Cartesian Coordinates (Column 2) and Spherical Coordinates (Column 3). Each Polynomial Is in Square Brackets Preceded by a Normalization Constant N_{kq} Used in Eq. 10-37

Tesseral Harmonic	Cartesian Representation	Spherical Representation
Z_{20}	$\dfrac{1}{2}\sqrt{\dfrac{5}{4\pi}}\left[\dfrac{3z^2-r^2}{r^2}\right]$	$\dfrac{1}{2}\sqrt{\dfrac{5}{2\pi}}\,[3\cos^2\theta-1]$
Z_{22}^{c}	$\dfrac{1}{2}\sqrt{\dfrac{15}{4\pi}}\left[\dfrac{x^2-y^2}{r^2}\right]$	$\dfrac{1}{2}\sqrt{\dfrac{15}{4\pi}}\,[\sin^2\theta\cos 2\phi]$
Z_{40}	$\dfrac{1}{8}\sqrt{\dfrac{9}{4\pi}}\left[\dfrac{35z^4-30z^2r^2+3r^4}{r^4}\right]$	$\dfrac{1}{8}\sqrt{\dfrac{9}{4\pi}}\,[35\cos^4\theta-30\cos^2\theta+3]$
Z_{42}^{c}	$\dfrac{3}{4}\sqrt{\dfrac{5}{4\pi}}\left[\dfrac{(7z^2-r^2)(x^2-y^2)}{r^4}\right]$	$\dfrac{3}{4}\sqrt{\dfrac{5}{4\pi}}\,[\sin^2\theta(7\cos^2\theta-1)\cos 2\phi]$
Z_{43}^{c}	$\dfrac{3}{4}\sqrt{\dfrac{70}{4\pi}}\left[\dfrac{xz(x^2-3y^2)}{r^4}\right]$	$\dfrac{3}{4}\sqrt{\dfrac{70}{4\pi}}\,[\sin^3\theta\cos\theta\cos 3\phi]$
Z_{43}^{s}	$\dfrac{3}{4}\sqrt{\dfrac{70}{4\pi}}\left[\dfrac{yz(3x^2-y^2)}{r^4}\right]$	$\dfrac{3}{4}\sqrt{\dfrac{70}{4\pi}}\,[\sin^3\theta\cos\theta\sin 3\phi]$

Z^c_{44} $\quad \dfrac{3}{8}\sqrt{\dfrac{35}{4\pi}}\left[\dfrac{x^4-6x^2y^2+y^4}{r^4}\right] \qquad\qquad \dfrac{3}{8}\sqrt{\dfrac{35}{4\pi}}\left[\sin^4\theta\cos4\phi\right]$

Z^s_{44} $\quad \dfrac{3}{8}\sqrt{\dfrac{35}{4\pi}}\left[\dfrac{4xy(x^2-y^2)}{r^4}\right] \qquad\qquad \dfrac{3}{8}\sqrt{\dfrac{35}{4\pi}}\left[\sin^4\theta\sin4\phi\right]$

Z_{60} $\quad \dfrac{1}{16}\sqrt{\dfrac{13}{4\pi}}\left[\dfrac{231z^6-315z^4r^2+105z^2r^4-5r^6}{r^6}\right] \qquad \dfrac{1}{16}\sqrt{\dfrac{13}{4\pi}}\left[231\cos^6\theta-315\cos^4\theta+105\cos^2\theta-5\right]$

Z^c_{62} $\quad \dfrac{1}{32}\sqrt{\dfrac{2730}{4\pi}}\left[\dfrac{[16z^4-16(x^2+y^2)z^2+(x^2+y^2)^2](x^2-y^2)}{r^6}\right] \qquad \dfrac{1}{32}\sqrt{\dfrac{2730}{4\pi}}\left[\sin^2\theta(33\cos^4\theta-18\cos^2\theta+1)\cos2\phi\right]$

Z^c_{63} $\quad \dfrac{1}{16}\sqrt{\dfrac{2730}{4\pi}}\left[\dfrac{xz(11z^2-3r^2)(x^2-3y^2)}{r^6}\right] \qquad \dfrac{1}{16}\sqrt{\dfrac{2730}{4\pi}}\left[\sin^3\theta(11\cos^3\theta-3\cos\theta)\cos3\phi\right]$

Z^c_{64} $\quad \dfrac{21}{16}\sqrt{\dfrac{13}{28\pi}}\left[\dfrac{(11z^2-r^2)(x^4-6x^2y^2+y^2)}{r^6}\right] \qquad \dfrac{21}{16}\sqrt{\dfrac{13}{28\pi}}\left[\sin^4\theta(11\cos^2\theta-1)\cos4\phi\right]$

Z^c_{66} $\quad \dfrac{1}{32}\sqrt{\dfrac{6006}{4\pi}}\left[\dfrac{x^6-15x^4y^2+15x^2y^4-y^6}{r^6}\right] \qquad \dfrac{1}{32}\sqrt{\dfrac{6006}{4\pi}}\left[\sin^6\theta\cos6\phi\right]$

10-8 CUBIC AND LOWER SYMMETRY POTENTIALS

The preceding section presented some general formulas for the electrostatic potential due to an array of charges Ze. Now it is appropriate to write down the explicit expressions for particular charge arrangements of the types shown in Fig. 10-5. The specific formulas are obtained by inserting the charge coordinates. For the present we limit our attention to first-transition-series ions $(3d^n)$, which have $l = 2$.

Eight charges at the vertices of a cube in the positions ± 1, ± 1, ± 1 as shown in Fig. 10-5a have the potential

$$V_c = \frac{8Ze}{R} - \frac{56\sqrt{\pi}}{27}\left(\frac{Ze}{R^5}\right)r^4(Z_{40} + \sqrt{\tfrac{5}{7}}Z_{44}^c) \tag{10-44}$$

If half of the charges are removed from the cubes to form the regular tetrahedron of Fig. 10-5b, one has

$$V_t = \frac{4Ze}{R} - \frac{28\sqrt{\pi}}{27}\left(\frac{Ze}{R^5}\right)r^4(Z_{40} + \sqrt{\tfrac{5}{7}}Z_{44}^c) \tag{10-45}$$

which is exactly one-half of the cubic potential. For an octahedral charge distribution corresponding to Fig. 10-5c with charges along the coordinate axes, the potential is

$$V_0 = \frac{6Ze}{R} + \frac{7\sqrt{\pi}}{3}\left(\frac{Ze}{R^5}\right)r^4(Z_{40} + \sqrt{\tfrac{5}{7}}Z_{44}^c) \tag{10-46}$$

Two charges located along the z axis as shown in Fig. 10-5d give

$$V_z = \frac{2Ze}{R} + 4\sqrt{\frac{\pi}{5}}\left(\frac{Ze}{R^3}\right)r^2 Z_{20} + \frac{4\sqrt{\pi}}{3}\left(\frac{Ze}{R^5}\right)r^4 Z_{40} \tag{10-47}$$

and a square planar array of four charges along the coordinate axes in the xy plane corresponds to

$$V_{xy} = \frac{4Ze}{R} - 4\sqrt{\frac{\pi}{5}}\left(\frac{Ze}{R^3}\right)r^2 Z_{20} + \sqrt{\pi}\left(\frac{Ze}{R^5}\right)r^4\left(Z_{40} + \frac{\sqrt{35}}{3}Z_{44}^c\right) \tag{10-48}$$

One should note that the octahedral potential is the sum of the axial and square planar ones when all six ligands are at the same distance from the origin:

$$V_0 = V_z + V_{xy} \tag{10-49}$$

The angularly dependent part of the cubic potential (excluding the A_{00} term) is minus $\tfrac{8}{9}$ times the octahedral potential. The negative sign has the effect of inverting the crystal field energy levels.

A tetragonally distorted octahedron is formed by adding the potentials for V_{xy} and V_z when the two axial ions are at a different distance from the square planar ions. One has

$$V = PV_z + QV_{xy} \tag{10-50}$$

where the dimensionless constants P and Q are normalized:

$$P + Q = 2 \tag{10-51}$$

so that for a perfect octahedron $P = Q = 1$. The slightly distorted potential in Cartesian coordinates is a polynomial of the form

$$+ K(P - Q)(3z^2 - r^2) + x^4 + y^4 + z^4 - \tfrac{3}{5}r^4 \tag{10-52}$$

where K is a constant. The tetragonal distortion term $(P - Q)$ is positive for stretching and negative for compression along the z axis. For cubic, tetrahedral, or octahedral symmetry one has a polynomial of the type

$$\pm(x^4 + y^4 + z^4 - \tfrac{3}{5}r^4) \tag{10-53}$$

where the positive sign denotes octahedral symmetry, and the negative sign is for the cubic and tetrahedral cases.

The tetragonal distortion under discussion occurs along the z axis and hence is easy to handle. A distortion may also occur in directions other than along a symmetry axis, in which case the mathematics becomes more complex. To treat trigonal distortions it is convenient to know the potential for an octahedron with the z axis coinciding with a threefold axis; it has the explicit form

$$V = \frac{6Ze}{R} - \frac{14\sqrt{\pi}}{9}\left(\frac{Ze}{R^5}\right)r^4(Z_{40} + 2\sqrt{\tfrac{5}{7}}Z_{43}^s) \tag{10-54}$$

Other special cases can also be written.

10-9 IRREDUCIBLE TENSOR OPERATORS

The crystal field potential expressed in spherical coordinates as in Eqs. 10-47 and 10-48 or in Cartesian coordinates as in Eq. 10-52 is not in a form that is convenient for use with the direct product method. Therefore it is helpful to transform the potentials so as to make use of angular momentum operators. This may be accomplished by the use of the equivalent operator method, which is based on the Wigner–Eckart theorem. The present section provides particular equivalent operators, and the following sections use them in typical calculations. General cases are not discussed. Abragam and Bleaney[11] give a thorough treatment of this topic from another point of view.

The Wigner–Eckart theorem relates the matrix element $\langle \alpha jm | T_k^q | \alpha'j'm' \rangle$ of the qth component of a kth-order irreducible tensor operator to a Clebsch–Gordon coefficient $\langle j'km'q | jm \rangle$ as follows:

$$\langle \alpha jm | T_k^q | \alpha'j'm' \rangle = \frac{(\alpha j \| T_k \| \alpha'j')}{\sqrt{2J+1}} \langle j'km'q | jm \rangle \qquad (10\text{-}55)$$

where $(\alpha j \| T_k \| \alpha'j')$ is the reduced matrix element which is independent of m and the particular component of T_k. The reduced matrix element is an intrinsic property of the individual tensor operator T_k. An irreducible tensor operator T_k may be defined as an operator whose $2k+1$ components T_k^q obey the following commutation laws with respect of the total angular momentum J:

$$[J_\pm, T_k^q] = \sqrt{k(k+1) - q(q \pm 1)} \, T_k^{q \pm 1} \qquad (10\text{-}56)$$

$$[J_z, T_k^q] = q T_k^q \qquad (10\text{-}57)$$

This is equivalent to the property that the qth component of an irreducible tensor operator transforms under rotation into a sum of one or more of the $2k+1$ components of T_k^q:

$$P_R T_k^q P_R^{-1} = \sum_{q'=-k}^{k} T_k^{q'} D_{k(R)_{qq'}} \qquad (10\text{-}58)$$

where P_R is a $(2k+1) \times (2k+1)$ general rotation matrix, and the $D_{k(R)_{qq'}}$ are coefficients characteristic of the particular rotation and k.

A discussion of irreducible tensor operators is beyond the scope of this book. Insead of dealing with the matrix elements in Eq. 10-55, it is more convenient to use operator equivalents. For this purpose it is best to express all vector operators V_x, V_y, V_z in irreducible tensor form:

$$V_1^1 = -\frac{V_x + iV_y}{\sqrt{2}}$$

$$V_1^0 = V_z \qquad (10\text{-}59)$$

$$V_1^{-1} = \frac{V_x - iV_y}{\sqrt{2}}$$

As a special case of this one may define the angular momentum vector operator J_1 with the components

$$J_1^1 = -\frac{J_x + iJ_y}{\sqrt{2}} = -\frac{J_+}{\sqrt{2}}$$

$$J_1^0 = J_z \qquad (10\text{-}60)$$

$$J_1^{-1} = \frac{J_x - iJ_y}{\sqrt{2}} = \frac{J_-}{\sqrt{2}}$$

Within a given manifold of J a vector operator V may be expressed in the form

$$V = CJ \tag{10-61}$$

with the components

$$V_1^1 = -\frac{C}{\sqrt{2}} J_+ = CJ_1^1$$

$$V_1^0 = CJ_z \tag{10-62}$$

$$V_1^{-1} = \frac{C}{\sqrt{2}} J_- = CJ_1^{-1}$$

As a result the components of the operator V are proportional to those of the total angular momentum J within the manifold J. The proportionality constant C depends on J and α and is related to the reduced matrix elements of Eq. 10-55.

In analogy to the vector operator case one can build up a second-rank irreducible angular momentum tensor J_2 with the components J_2^q given by

$$J_2^0 = 3J_z^2 - J(J+1)$$

$$J_2^{\pm 1} = \frac{\sqrt{6}}{2}(J_z J_\pm + J_\pm J_z) \tag{10-63}$$

$$J_2^{\pm 2} = \frac{\sqrt{6}}{2} J_\pm^2$$

Inside the manifold J all second-order irreducible tensors have matrix elements proportional to those of J_2^q

$$T_2^q = C'J_2^q \tag{10-64}$$

where C' is also a function of J and α but differs in value from C of Eq. 10-61.

When Cartesian components are used for a typical crystal field potential term arising from the summation over N ligands, one may define Cartesian components T_{ij} of T_2^q where $i, j = x, y, z$. For the position operator T_{ij} has the form

$$T_{ij} = 3x_i x_j - r^2 \delta_{ij} \tag{10-65}$$

and it is proportional to the corresponding term of J_2^q

$$T_{ij} = C'J_{ij} \tag{10-66}$$

where

$$J_{ij} = \tfrac{3}{2}(J_i J_j + J_j J_i) - J(J+1)\delta_{ij} \tag{10-67}$$

which reduces to the well-known expression $[3J_z^2 - J(J+1)]$ for $i = j$.

10-10 EQUIVALENT OPERATORS

The formalism developed in Section 10-9 is very helpful in evaluating crystal field energy levels. The crystal field potential (10-38) corresponds to polynomials in powers of r, $\sin \theta$, $\cos \theta$, $\sin \phi$, $\cos \phi$, which is equivalent to a polynomial in powers of x, y, z, and r. The vector r may be selected as the vector V of Eq. 10-59 to give, from Eq. 10-61,

$$x = CJ_x , \qquad y = CJ_y , \qquad z = CJ_z \qquad (10\text{-}68)$$

Higher-order polynomials can be formed by using the identification $z^2 = C'J_z^2$ and $r^2 = C'J(J+1)$, but terms like xy require more careful treatment because J_x and J_y do not commute. To take this lack of commutation into account one should replace products like $x^m y^n z^p$ by a symmetrized product of $J_x^m J_y^n J_z^p$. In other words,

$$xy = \tfrac{1}{2} C'(J_x J_y + J_y J_x) \qquad (10\text{-}69)$$

where the $\frac{1}{2}$ is a normalization constant. The symmetrized product is an average of all the products where J_x, J_y, and J_z can appear, respectively, m, n, and p times. This average can be simplified by using the commutation relations such as

$$[J_x, J_y] = J_x J_y - J_y J_x = iJ_z \qquad (10\text{-}70)$$

The notation

$$\{A, B\}_s = \tfrac{1}{2}(AB + BA) \qquad (10\text{-}71)$$

will be used for a symmetrized product.

Particular combinations of angular momentum operators are referred to as equivalent operators O_k^q. The ones that are most often needed in crystal field theory studies are listed in Table 10-2.

A charge distribution produces the potential $V_{(r\theta\phi)}$ of Eq. 10-38 and for several electrons j this gives the crystal field Hamiltonian

$$\mathcal{H}_{CF} = e \sum_j V_{(r_j \theta_j \phi_j)} = \sum_{k=0}^{\infty} \sum_{q=-k}^{k} B_k^q O_k^q \qquad (10\text{-}72)$$

where each operator O_k^q has a coefficient B_k^q associated with it, and the angular momentum operator is the orbital momentum L. Most of the B_k^q will be zero for crystal fields with appreciable symmetry. The more symmetric the atomic site, the smaller the number of potential terms in expansion 10-72. If there is a center of inversion at the ionic site, there will be no odd-k terms in the expansion. A tetrahedron lacks a center of inversion, and so it has terms in O_k^q with odd k. However, these terms do not contribute to the energy within the L manifold. They will, nevertheless, admix excited states with the ground state. If the z axis is an m-fold axis, the potential will contain terms of the type O_k^m.

TABLE 10-2 Equivalent Operators O_p^q (from Abragam and Bleaney[11]). The Notation $\{A, B\}_s$ Is Used as a Shorthand for $\frac{1}{2}(AB + BA)$. Thus $O_4^3 = \frac{1}{2}\{J_z(J_+^3 + J_-^3)\}_s$ Denotes $O_4^3 = \frac{1}{4}\{J_z(J_+^3 + J_-^3) + (J_+^3 + J_-^3)J_z\}$

$O_2^0 = 3J_z^2 - J(J + 1)$
$O_2^2 = \frac{1}{2}(J_+^2 + J_-^2) - (J_x^2 - J_y^2)$

$O_4^0 = 35J_z^4 - 30J(J + 1)J_z^2 + 25J_z^2 - 6J(J + 1) + 3J^2(J + 1)^2$
$O_4^2 = \frac{1}{2}\{[7J_z^2 - J(J + 1) - 5](J_+^2 + J_-^2)\}_s$
$O_4^3 = \frac{1}{2}\{J_z(J_+^3 + J_-^3)\}_s$
$O_4^4 = \frac{1}{2}(J_+^4 + J_-^4)$

$O_6^0 = 231J_z^6 - 315J(J + 1)J_z^4 + 735J_z^4 + 105J^2(J + 1)^2J_z^2 - 525J(J + 1)J_z^2 + 294J_z^2$
$\qquad - 5J^3(J + 1)^3 + 40J^2(J + 1)^2 - 60J(J + 1)$
$O_6^3 = \frac{1}{2}\{[11J_z^3 - 3J(J + 1)J_z - 59J_z](J_+^3 + J_-^3)\}_s$
$O_6^4 = \frac{1}{2}\{[11J_z^2 - J(J + 1) - 38](J_+^4 + J_-^4)\}_s$
$O_6^6 = \frac{1}{2}(J_+^6 + J_-^6)$

The number of terms required in the expansion is limited by the electronic configuration of the transition metal that is interacting with the crystal field. The maximum required k is equal to $2l$, where l is the individual electron's orbital quantum number. For d-electrons $l = 2$, and for f-electrons $l = 3$. It is not the total $L = \Sigma\, l_i$ that limits the summation; hence a $3d^3$ configuration with a ^4F ground state and $L = 3$ is characterized by the cutoff of $2l = 4$, and not by the limit of 6.

For fairly symmetric charge distributions one has

$$\mathcal{H}_{CF} = \sum_{k=2,4} \sum_{q=0}^{k} B_k^q O_k^q \quad (d\text{-electrons})$$

$$\mathcal{H}_{CF} = \sum_{k=2,4,6} \sum_{q=0}^{k} B_k^q O_k^q \quad (f\text{-electrons})$$

$$(10\text{-}73)$$

where the constant term $B_0^0 O_0^0$ is omitted since it only affects the zero point of energy. For octahedral or cubic coordination the potential becomes

$$\mathcal{H}_{CF} = B_4(O_4^0 + 5O_4^4) \qquad\qquad (d\text{-electrons})$$
$$\mathcal{H}_{CF} = B_4(O_4^0 + 5O_4^4) + B_6(O_6^0 - 21O_6^4) \quad (f\text{-electrons})$$

$$(10\text{-}74)$$

where the notation is simplified by using B_4 and B_6. If the symmetry is predominantly cubic or octahedral with a tetragonal or trigonal distortion, the equivalent operator Hamiltonian has the general form

$$\mathcal{H}_{CF} = \sum_{k=2,4} \sum_{q=0}^{k} [B_{k\,cub}^q + B_{kD}^q] O_k^q$$

which separates the cubic or octahedral terms $B^q_{k\,cub}$ from the distortion ones B^q_{kD}. The former appear in Eqs. 10-74 as B_4 and B_6 instead of B^0_4 and B^0_6, respectively. For tetragonal distortion one has[11]

$$\mathcal{H}_{CF} = B_4(O^0_4 + 5O^4_4) + B^0_{2D}O^0_2 + B^0_{4D}O^0_4 \quad \text{(d-electrons)} \quad (10\text{-}75)$$

A predominantly cubic environment with trigonal distortion gives[11]

$$\mathcal{H}_{CF} = -\tfrac{2}{3}B_4(O^0_4 + 20\sqrt{2}O^3_4) + B^2_{2D}O^0_2 + B^0_{4D}O^0_4 \quad \text{(d-electrons)}$$
$$(10\text{-}76)$$

Expressions 10-75 and 10-76 are for d-electrons where the maximum value of k is 4. Additional terms with $k = 6$ must be added for f-electrons.

10-11 ENERGIES AND WAVEFUNCTIONS FOR d-ELECTRONS

For the octahedral case with d-electrons Eq. 10-74 gives (see Eq. 8-92)

$$\mathcal{H}_{CF} = B_4[35L^4_z - 30L(L+1)L^2_z + 25L^2_z - 6L(L+1)$$
$$+ 3L^2(L+1)^2 + \tfrac{5}{2}(L^4_+ + L^4_-)] \quad (10\text{-}77)$$

where J was replaced by L in the equivalent operators O^0_4 and O^4_4 of Table 10-2. The usual ground states for d-electronic configurations d^n are D and F states. The D-state case is discussed in detail. This section ignores the sin–orbit coupling and spin since the spin–orbit coupling constant $|\lambda|$ is much less than the crystal field splittings for the first transition series.

For a D state $L = 2$, and Eq. 10-77 becomes

$$\mathcal{H}_{CF} = B_4[35L^4_z - 155L^2_z + 72I + \tfrac{5}{2}(L^4_+ + L^4_-)] \quad (10\text{-}78)$$

where I is the 5×5 unit matrix. The basic matrices may be formed from the prescriptions given in Section 2-6:

$$L_+ = \begin{pmatrix} 0 & 2 & 0 & 0 & 0 \\ 0 & 0 & \sqrt{6} & 0 & 0 \\ 0 & 0 & 0 & \sqrt{6} & 0 \\ 0 & 0 & 0 & 0 & 2 \\ 0 & 0 & 0 & 0 & 0 \end{pmatrix}, \quad L_- = \begin{pmatrix} 0 & 0 & 0 & 0 & 0 \\ 2 & 0 & 0 & 0 & 0 \\ 0 & \sqrt{6} & 0 & 0 & 0 \\ 0 & 0 & \sqrt{6} & 0 & 0 \\ 0 & 0 & 0 & 2 & 0 \end{pmatrix}$$

$$L_z = \begin{pmatrix} 2 & 0 & 0 & 0 & 0 \\ 0 & 1 & 0 & 0 & 0 \\ 0 & 0 & 0 & 0 & 0 \\ 0 & 0 & 0 & -1 & 0 \\ 0 & 0 & 0 & 0 & -2 \end{pmatrix} \quad (10\text{-}79)$$

and matrix multiplication gives the four matrices L^4_+, L^4_-, L^2_z, and L^4_z, which way be inserted into Eq. 10-78 to give

$$\mathcal{H}_{CF} = 12B_4 \begin{pmatrix} 1 & 0 & 0 & 0 & 5 \\ 0 & -4 & 0 & 0 & 0 \\ 0 & 0 & 6 & 0 & 0 \\ 0 & 0 & 0 & -4 & 0 \\ 5 & 0 & 0 & 0 & 1 \end{pmatrix} \qquad (10\text{-}80)$$

This matrix is diagonal except for the 2×2 $m_L = \pm 2$ part with the corresponding secular equation

$$\begin{vmatrix} 12B_4 - E & 60B_4 \\ 60B_4 & 12B_4 - E \end{vmatrix} = 0 \qquad (10\text{-}81)$$

which has the roots

$$E = 72B_4, \qquad -48B_4 \qquad (10\text{-}82)$$

each of which is degenerate with another root. Thus we see that the crystal field interaction splits the D state into an orbital triplet and an orbital doublet with the energies

$$E = \begin{cases} 72B_4 & \text{(Doublet)} \\ -48B_4 & \text{(Triplet)} \end{cases} \qquad (10\text{-}83)$$

The center of gravity of the original D level remains invariant, as expected. The usual notation for the separation between these energy levels is Δ or $10Dq$ to give

$$\Delta = 10Dq = 120B_4 \qquad (10\text{-}84)$$

The fact that the D term splits into a triplet and a doublet could have been found independently by group theory, but the determination of the magnitude of the energy level separation requires a crystal field calculation.

For a tetragonally distorted octahedron or cube the lower symmetry terms of Eq. 10-75 may be employed. The $B_{2D}^0 O_2^0$ term requires the use of the function $(3L_z^2 - 6)$ from Table 10-2 and the $B_{4D}^0 O_4^0$ term is identical with the first half of the octahedral field potential that was just calculated, with $|B_{4D}^0| \ll |B_4|$. Both are diagonal and together they give the following tetragonal distortion matrix:

$$\mathcal{H}_{\text{tetrag}} = \begin{pmatrix} 6(B_{2D}^0 + 12B_{4D}^0) & 0 & 0 & 0 & 0 \\ 0 & -3(B_{2D}^0 + 16B_{4D}^0) & 0 & 0 & 0 \\ 0 & 0 & -6(B_{2D}^0 - 12B_{4D}^0) & 0 & 0 \\ 0 & 0 & 0 & -3(B_{2D}^0 + 16B_{4D}^0) & 0 \\ 0 & 0 & 0 & 0 & 6(B_{2D}^0 - 8B_{4D}^0) \end{pmatrix}$$

$$(10\text{-}85)$$

This matrix is added to 10-80 to provide the crystal field energy levels in a tetragonally distorted octahedron. The manner in which the tetragonal terms shift the energy levels is illustrated in Fig. 10-8. The energies comprise one doublet and three singlets.

The wavefunctions $|m_L\rangle = |1\rangle$, $|0\rangle$, and $|-1\rangle$ are unchanged by the crystal field potential, but the $|\pm 2\rangle$ ket vectors are mixed by the off-diagonal matrix elements of O_4^4. Therefore the eigenfunctions must be obtained from the 2×2 submatrix corresponding to determinant 10-81. They are of the forms

$$|2^s\rangle = \frac{1}{\sqrt{2}}[|2\rangle + |-2\rangle] \qquad (E = 72B_4)$$

$$|2^a\rangle = \frac{1}{\sqrt{2}}[|2\rangle - |-2\rangle] \qquad (E = -48B_4)$$

$$(10\text{-}86)$$

The energy levels in Fig. 10-8 are labeled with their corresponding wavefunctions.

Similar calculations may be carried out for trigonal distortions and F states, and typical results are presented in Figs. 10-9 to 10-11.

In conclusion, it helpful to say a few words about cubic type levels in general. For the configurations $3d^1$ and $3d^9$ the energy levels are of the type shown in Fig. 10-12. The levels become inverted in going from an octahedral to a cubic or tetrahedral field. In addition, the $3d^1$ and $3d^9$ configurations are inverted in relation to each other because $3d^9$ may be treated as a $3d^1$ positive hole in a filled $(3d^{10})$ shell. The F-state energy levels also invert under the same conditions. These inversions result from the sign changes that occur in various coefficients that were derived in Section 15-7 of the first edition.

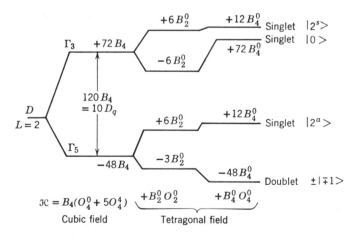

Fig. 10-8. Splitting of a $3d^9$, D state in cubic (left) and tetragonal (center) crystalline electric fields. The wavefunctions are shown on the right in terms of $|m_L\rangle$ and the symmetric and antisymmetric combinations of $|\pm 2\rangle$ (adapted from Ref. 11).

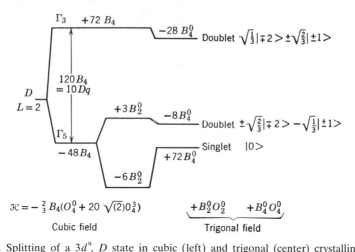

$$\mathcal{K} = -\tfrac{2}{3} B_4 (O_4^0 + 20 \sqrt{(2)} O_4^3)$$

Cubic field $+B_2^0 O_2^0 \quad +B_4^0 O_4^0$

         Trigonal field

Fig. 10-9. Splitting of a $3d^9$, D state in cubic (left) and trigonal (center) crystalline electric fields, using the notation of Fig. 10-8 (adapted from Ref. 11).

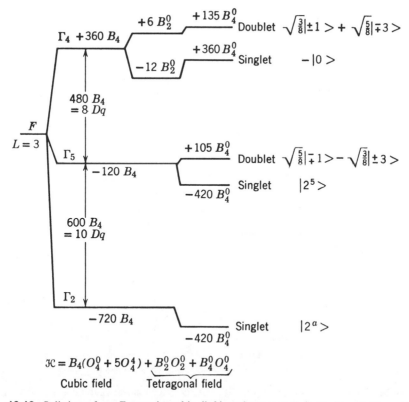

$$\mathcal{K} = B_4 (O_4^0 + 5 O_4^4) + B_2^0 O_2^0 + B_4^0 O_4^0$$

Cubic field   Tetragonal field

Fig. 10-10. Splitting of an F state in cubic (left) and tetragonal (center) crystalline electric fields, using the notation of Fig. 10-8 (from Ref. 11).

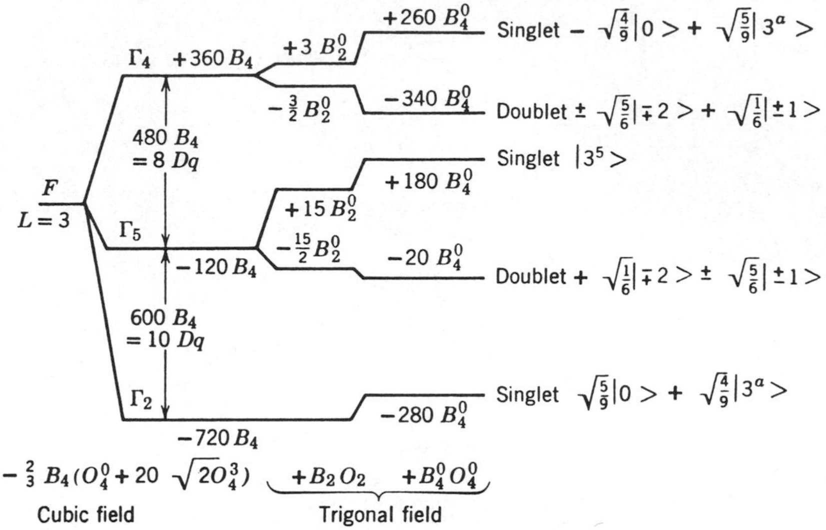

$+260\,B_4^0$ Singlet $-\sqrt{\frac{4}{9}}|0> + \sqrt{\frac{5}{9}}|3^a >$

$\Gamma_4 \quad +360\,B_4$ $+3\,B_2^0$

$-340\,B_4^0$ Doublet $\pm \sqrt{\frac{5}{6}}|\mp 2> + \sqrt{\frac{1}{6}}|\pm 1>$

$-\frac{3}{2}B_2^0$

$480\,B_4$
$= 8\,Dq$

Singlet $|3^5>$

F $+180\,B_4^0$

$L = 3$ Γ_5 $+15\,B_2^0$

$-\frac{15}{2}B_2^0$ $-20\,B_4^0$

$-120\,B_4$ Doublet $+ \sqrt{\frac{1}{6}}|\mp 2> \pm \sqrt{\frac{5}{6}}|\pm 1>$

$600\,B_4$
$= 10\,Dq$

Γ_2 Singlet $\sqrt{\frac{5}{9}}|0> + \sqrt{\frac{4}{9}}|3^a >$

$-280\,B_4^0$

$-720\,B_4$

$\mathcal{H} = -\frac{2}{3}B_4\left(O_4^0 + 20\,\sqrt{2}O_4^3\right) \underbrace{\quad +B_2O_2 \quad +B_4^0O_4^0 \quad}$

Cubic field Trigonal field

Fig. 10-11. Splitting of an F state in cubic (left) and trigonal (center) crystalline electric fields, using the notation of Fig. 10-8 (from Ref. 11).

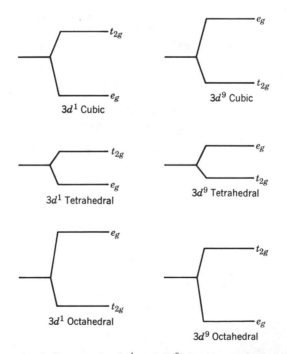

Fig. 10-12. Energy level diagrams for $3d^1$ and $3d^9$ electron configurations in cubic (top), tetrahedral (center), and octahedral (bottom) sites. The ligand distance R is constant for this set of levels.

10-12 IRREDUCIBLE REPRESENTATIONS AND KRAMERS' THEOREM

Crystal field energy levels for orbital electrons in octahedral, cubic, or tetrahedral environments have five symmetry types called irreducible representations. These may be designated by one of the following two systems of notation:

$$
\left.
\begin{array}{llll}
\text{Singlet} & A_1 & \Gamma_1 \\
\text{Singlet} & A_2 & \Gamma_2 \\
\text{Doublet} & E & \Gamma_3 \\
\text{Triplet} & T_1 & \Gamma_4 \\
\text{Triplet} & T_2 & \Gamma_5
\end{array}
\right\}
\begin{array}{c}
\text{Integral} \\
J
\end{array}
\tag{10-87}
$$

$$
\left.
\begin{array}{llll}
\text{Doublet} & E' & \Gamma_6 \\
\text{Doublet} & E'' & \Gamma_7 \\
\text{Quartet} & U & \Gamma_8
\end{array}
\right\}
\begin{array}{c}
\text{Half-} \\
\text{integral} \\
J
\end{array}
\tag{10-88}
$$

For d-electrons one often uses the notation A_{1g}, T_{2g}, Γ_{2g}, and so on, where g stands for *gerade* and means that the wavefunction is asymmetric under inversion. Odd or *ungerade* (u) levels have wavefunctions which change sign under inversion. Octahedral, cubic, and tetrahedral crystal fields split levels with half-integral angular momenta into combinations or irreducible representations which are doublets (Γ_6 and Γ_7) and quartets (Γ_8). Table 10-3 lists the number and types of energy levels for various magnitudes of angular momentum. The results presented in this table are most easily obtained from group theory.

TABLE 10-3 Irreducible Representations for Spectroscopic States J in Cubic (O_h), Tetrahedral (T_d), Octahedral (O_h), Tetragonal (D_4), and Trigonal (D_3) Crystalline Electric Fields

J	O_h, T_d	D_4	D_3
0	A_1	A_1	A_1
1	T_1	$A_2 + E$	$A_2 + E$
2	$E + T_2$	$A_1 + B_1 + B_2 + E$	$A_1 + 2E$
3	$A_2 + T_1 + T_2$	$A_2 + B_1 + B_2 + 2E$	$A_1 + 2A_2 + 2E$
4	$A_1 + E + T_1 + T_2$	$2A_1 + A_2 + B_1 + B_2 + 2E$	$2A_1 + A_2 + 3E$
5	$E + 2T_1 + T_2$	$A_1 + 2A_2 + B_1 + B_2 + 3E$	$A_1 + 2A_2 + 4E$
6	$A_1 + A_2 + E + T_1 + 2T_2$	$2A_1 + A_2 + 2B_1 + 2B_2 + 3E$	$3A_1 + 2A_2 + 4E$
$\frac{1}{2}$	E'	E'	E'
$\frac{3}{2}$	U	$E' + E''$	$E' + E''$
$\frac{5}{2}$	$E'' + U$	$E' + 2E''$	$2E' + E''$
$\frac{7}{2}$	$E' + E'' + U$	$2E' + 2E''$	$3E' + E''$
$\frac{9}{2}$	$E' + 2U$	$3E' + 2E''$	$3E' + 2E''$

In tetragonal environments integral spins have five irreducible represent-ations—four singlet (A_1, A_2, B_1, B_2) and one doublet (E), and half-integral spins have two doublet (E', E'') representations. Trigonal symmetry is characterized by two singlet (A_1, A_2) and one doublet (E) for integral spin, and two doublets (E', E'') for half-integral J. The combinations of representations that characterize each J are listed in Table 10-3.

A number of years ago Kramers proved that in an arbitrary electrostatic field all states of a half-integral electron system are at least twofold degenerate. As a result all irreducible representations of such an odd electron system have even degeneracies. Their irreducible representations are denoted E', E'', and U in Table 10-3. Rare earths with half-integral J are called Kramers ions, and those with integral J are non-Kramers ions. Electrons spin resonance is, in principle, always observable in the former. The latter sometimes have Stark effect splittings which separate the non-degenerate levels beyond the range of ESR so that no resonances can be observed.

10-13 CALCULATION OF g-FACTORS

The electron spin resonance spectrum of a transition-series ion in free space exhibits an energy level separation which depends on the Landé g-factor with the total angular momentum J constituting a good quantum number, as explained in Section 10-3. In a solid, the crystal field splittings produce large energy separations which strongly influence the g-factor. For rare earths, the spin–orbit coupling far exceeds the crystal field strength and J remains a good quantum number. However, a crystal field splitting can be quite large, to produce an effective g-factor which deviates considerably from the Landé value. For the first transition series the crystal field interaction dominates, so that J is no longer a good quantum number. As a result L is uncoupled from S and is quantized along the crystalline electric fields. The orbital angular momentum is said to be quenched. Therefore the spin S is quantized along the external magnetic field, and one obtains g-factors close to 2.

If the crystal field ground state is an orbital doublet or triplet, it is possible to define an "effective orbital angular momentum" L of the corresponding multiplicity. This couples to the spin S and produces the Zeeman splitting. Quite often the ground state is an orbital singlet, and only the M_s splittings occur, with deviations from the free-electron g-factor arising from the ratio of the spin–orbit coupling constant to an orbital energy separation. This latter case is considerably simpler and is discussed here in detail.

For illustrative purposes the spectroscopic state ^2D with $L = 2$, $S = \frac{1}{2}$ is considered. This occurs for the $3d^1$ and $3d^9$ transition-metal ions, and the crystal field results are described in Section 10-11. Our attention is confined to the case of dominant octahedral symmetry with a small tetragonal

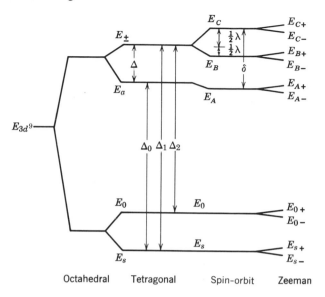

Fig. 10-13. Energy level diagram for a $3d^9$ electronic configuration in an octahedral crystal field with tetragonal symmetry, showing the effect of the first-order spin–orbit interaction and the Zeeman effect. The E_0 and E_s levels exhibit a second-order splitting in λ which is not shown.

distortion, where the tetragonal crystal field is assumed to far exceed the magnitude of the Zeeman splitting. The pertinent energy level diagram is shown in Fig. 10-8. This figure corresponds to the $3d^1$ configuration in an octahedral site, or to $3d^9$ in a cubic or tetrahedral site. It is inverted for $3d^1$ in a cubic or tetrahedral site, and for $3d^9$ in an octahedral site, as indicated in Fig. 10-13. In addition, the signs of the tetragonal distortion terms B_{2D}^0 and B_{4D}^0 depend on whether or not the octahedron (or cube) is stretched or compressed along the C_4 axis. As a result the energy levels can order themselves in several ways. Since $120B_4$ far exceeds $10B_{2D}^0$ and $60B_{4D}^0$ in magnitude, the energy levels occur in pairs (E_s, E_0) and $(E_a, E_\pm)$ as shown in Fig. 10-13. If one level of a pair is the ground state, the other is the lowest excited state. The particular configuration will have an effect on the g-factor.

The spin–orbit coupling matrix

$$\langle \psi_i | \lambda \vec{L} \cdot \vec{S} | \psi_j \rangle = \lambda \langle \psi_i | L_x S_x + L_y S_y + L_z S_z | \psi_j \rangle \qquad (10\text{-}89)$$

discussed in Section 10-2 mixes the wavefunctions of the various levels, and the amount of mixing may be calculated by perturbation theory. This matrix is as follows:

	$\lvert 2^s +\rangle$	$\lvert 2^s -\rangle$	$\lvert 0 +\rangle$	$\lvert 0 -\rangle$	$\lvert 2^a +\rangle$	$\lvert 2^a -\rangle$	$\lvert 1 +\rangle$	$\lvert -1 -\rangle$	$\lvert 1 -\rangle$	$\lvert -1 +\rangle$
$\langle 2^s +\rvert$	E_s	0	0	0	λ	0	0	$\dfrac{\lambda}{\sqrt{2}}$	0	0
$\langle 2^s -\rvert$	0	E_s	0	0	0	$-\lambda$	$\dfrac{\lambda}{\sqrt{2}}$	0	0	0
$\langle 0 +\rvert$	0	0	E_0	0	0	0	0	0	$\dfrac{\sqrt{3}}{2}\lambda$	0
$\langle 0 -\rvert$	0	0	0	E_0	0	0	0	0	0	$\dfrac{\sqrt{3}}{2}\lambda$
$\langle 2^a +\rvert$	λ	0	0	0	E_a	0	0	$-\dfrac{\lambda}{\sqrt{2}}$	0	0
$\langle 2^a -\rvert$	0	$-\lambda$	0	0	0	E_a	$\dfrac{\lambda}{\sqrt{2}}$	0	0	0
$\langle 1 +\rvert$	0	$\dfrac{\lambda}{\sqrt{2}}$	0	0	0	$\dfrac{\lambda}{\sqrt{2}}$	$E_\pm + \tfrac{1}{2}\lambda$	0	0	0
$\langle -1 -\rvert$	$\dfrac{\lambda}{\sqrt{2}}$	0	0	0	$-\dfrac{\lambda}{\sqrt{2}}$	0	0	$E_\pm + \tfrac{1}{2}\lambda$	0	0
$\langle 1 -\rvert$	0	0	$\dfrac{\sqrt{3}}{2}\lambda$	0	0	0	0	0	$E_\pm - \tfrac{1}{2}\lambda$	0
$\langle -1 +\rvert$	0	0	0	$\dfrac{\sqrt{3}}{2}\lambda$	0	0	0	0	0	$E_\pm - \tfrac{1}{2}\lambda$

$$(10\text{-}90)$$

The levels E_s and E_0 do not mix, and hence nondegenerate perturbation theory may be directly employed to obtain their eigenfunctions correct to first order in accordance with Eq. 2-106. Since the levels $\lvert -1, +\rangle$ and $\lvert +1, -\rangle$ which form E_B only couple to the much lower E_0 sublevel, perturbation theory applies to them also. The remaining four levels form two pairs $[\lvert 2^a +\rangle, \lvert -1 -\rangle]$ and $[\lvert 2^a, -\rangle, \lvert 1, +\rangle]$. Each of these pairs has the same secular equation

$$\begin{vmatrix} E_a - E & \dfrac{1}{\sqrt{2}}\lambda \\[2mm] \dfrac{1}{\sqrt{2}}\lambda & E_\pm + \tfrac{1}{2}\lambda - E \end{vmatrix} = 0 \qquad (10\text{-}91)$$

with the eigenvalues E denoted by E_A and E_C, as follows:

$$E_A = \tfrac{1}{2}(E_a + E_\pm) + \tfrac{1}{4}\lambda - \tfrac{1}{2}\sqrt{\Delta^2 + \Delta\lambda + \tfrac{9}{4}\lambda^2}$$
$$E_B = E_\pm - \tfrac{1}{2}\lambda \tag{10-92}$$
$$E_C = \tfrac{1}{2}(E_a + E_\pm) + \tfrac{1}{4}\lambda + \tfrac{1}{2}\sqrt{\Delta^2 + \Delta\lambda + \tfrac{9}{4}\lambda^2}$$

This corresponds to a splitting of the $E_\pm$ energy level and a shift of E_a, as shown in Fig. 10-13. The sign before the radical in Eq. 10-92 is selected so that $E_A = E_a$ and $E_C = E_B = E_\pm$ in the limit when $\lambda \to 0$. For convenience the following energy level separations can be defined:

$$\delta = E_C - E_A = \sqrt{\Delta^2 + \Delta\lambda + \tfrac{9}{4}\lambda^2}$$
$$\Delta = E_\pm - E_a$$
$$\Delta_0 = E_a - E_s \tag{10-93}$$
$$\Delta_1 = E_\pm - E_s$$
$$\Delta_2 = E_\pm - E_0$$

as indicated in Fig. 10-13. These parameters enter directly into the g-factor shifts from the free-electron value.

The wavefunctions for the four levels mixed by the secular equation may be referred to as $|\psi_{A\pm}\rangle$ and $|\psi_{C\pm}\rangle$ with the phases chosen so that $|\psi_{A\pm}\rangle \to |2^a\ \pm\rangle$ in the limit $\lambda \to 0$. If $\lambda \ll \Delta$ as is usually the case, then it is not necessary to solve Eq. 10-91. Instead the wavefunctions may be calculated in first order as follows, using the labeling scheme of Fig. 10-13:

$$\left.\begin{aligned}
\psi_{S+} &= |2^s\ +\rangle - \frac{\lambda}{\Delta_0}|2^a\ +\rangle - \frac{\lambda}{\sqrt{2}\Delta_1}|-1,\ -\rangle \\[2mm]
\psi_{S-} &= |2^s\ -\rangle + \frac{\lambda}{\Delta_0}|2^a\ -\rangle + \frac{\lambda}{\sqrt{2}\Delta_1}|+1,\ +\rangle
\end{aligned}\right\}E_s$$

$$\left.\begin{aligned}
\psi_{0+} &= |0,\ +\rangle - \frac{\lambda}{\Delta_2}\frac{\sqrt{3}}{2}|+1,\ -\rangle \\[2mm]
\psi_{0-} &= |0,\ -\rangle - \frac{\lambda}{\Delta_2}\frac{\sqrt{3}}{2}|-1,\ +\rangle
\end{aligned}\right\}E_0 \tag{10-94}$$

$$\left.\begin{aligned}
\psi_{A+} &= |2^a,\ +\rangle + C_1|2^S\ +\rangle + C_2|-1,\ -\rangle \\
\psi_{A-} &= |2^a,\ -\rangle + C_3|2^S\ -\rangle + C_4|+1,\ +\rangle
\end{aligned}\right\}E_A$$

$$\left.\begin{aligned}
\psi_{B+} &= |-1,\ +\rangle + C_5|0\ -\rangle \\
\psi_{B-} &= |+,\ -\rangle + C_6|0\ +\rangle
\end{aligned}\right\}E_B$$

$$\left.\begin{aligned}
\psi_{C+} &= |+1,\ +\rangle + C_7|2^S\ -\rangle + C_8|2^a,\ -\rangle \\
\psi_{C-} &= |-1,\ -\rangle + C_9|2^S\ +\rangle + C_{10}|2^a,\ +\rangle
\end{aligned}\right\}E_C$$

Each of the doublets can be characterized by an effective spin $\tfrac{1}{2}$, and therefore the wavefunctions may be denoted by $|\mp\rangle$ and $|\approx\rangle$, where the sign

$\sim$ indicates an effective spin. In accordance with this notation one may write, for example,

$$|\mp\rangle_A = \psi_{A+}$$
$$|\approx\rangle_A = \psi_{A-} \tag{10-95}$$

and similarly for $|\mp\rangle_B$ and $|\mp\rangle_C$. The concept of effective spin is also useful for characterizing triplets $[|\hat{1}\rangle, |\hat{0}\rangle, |-\hat{1}\rangle]$ and higher multiplets.

Now that the wavefunctions have been obtained, it is appropriate to calculate the g-factor. This is accomplished by assuming the identity

$$\beta\vec{H}\cdot(\vec{L}+2\vec{S}) = \beta\vec{H}\cdot\vec{\tilde{g}}\cdot\vec{\tilde{S}} \tag{10-96}$$

where $\vec{\tilde{S}}$ is the effective spin, and $\vec{S}$ is the actual spin. This is equivalent to

$$\vec{L}+2\vec{S} = \vec{\tilde{g}}\cdot\vec{\tilde{S}} \tag{10-97}$$

with the components

$$(L_x+2S_x)\hat{i} + (L_y+2S_y)\hat{j} + (L_z+2S_z)\hat{k} = g_\perp(\tilde{S}_x\hat{i} + \tilde{S}_y\hat{j}) + g_\parallel\tilde{S}_z\hat{k} \tag{10-98}$$

The matrices of the left- and right-hand components are compared to determine the quantities $g_\perp$ and $g_\parallel$. The effective spin eigenfunctions $|\mp\rangle$ are employed to determine the matrix for $\vec{\tilde{g}}\cdot\vec{\tilde{S}}$:

$$\vec{\tilde{g}}\cdot\vec{\tilde{S}} = \begin{matrix} & |\mp\rangle & |\approx] \\ \langle\mp| & \begin{pmatrix} \frac{1}{2}g_\parallel\hat{k} & \frac{1}{2}g_\perp(\hat{i}-i\hat{j}) \\ \frac{1}{2}g_\perp(\hat{i}+i\hat{j}) & -\frac{1}{2}g_\parallel\hat{k} \end{pmatrix} \end{matrix} \tag{10-99}$$

which is valid for all the doublet pairs $\psi_{j\pm}$ of Eq. 10-94. The matrix of $(\vec{L}+2\vec{S})$ is calculated with the actual wavefunctions $\psi_{j\pm}$ of Eq. 10-94. This will be done for several cases. The effective spin $\vec{\tilde{S}}$ is a valid concept if the matrix of $(\vec{L}+2\vec{S})$ assumes the form of Eq. 10-99.

If the lowest orbital level is $|0\rangle$, then in the absence of the spin–orbit interaction one has

$$\langle 0|L_i|0\rangle = 0 \tag{10-100}$$

independent of spin. As a result

$$(\vec{L}+2\vec{S}) = \begin{matrix} & |+\rangle_0 & |-\rangle_0 \\ \langle+|_0 & \begin{pmatrix} \hat{k} & (\hat{i}-i\hat{j}) \\ (\hat{i}+i\hat{j}) & -\hat{k} \end{pmatrix} \end{matrix} \tag{10-101}$$

and a comparison with Eq. 10-99 gives

$$g_\parallel = g_\perp = 2 \tag{10-102}$$

for the completely quenched case $(\lambda = 0)$.

If the spin–orbit interaction is taken into account for the ψ_0 level, the wavefunctions $|\mp\rangle_0$ and $|\approx\rangle_0$ of Eqs. 10-94 and 10-95 give the matrix

$$
(\vec{L} + 2\vec{S}) = \begin{array}{c} \\ \langle\mp|_0 \\ \\ \langle\approx|_0 \end{array}
\begin{array}{cc} |\mp\rangle_0 & \quad |\approx\rangle_0 \\
\left(\begin{array}{cc}
\hat{k} & \frac{1}{2}(\hat{i} - i\hat{j})\left(2 - \frac{6\lambda}{\Delta_2}\right) \\
\frac{1}{2}(\hat{i} + i\hat{j})\left(2 - \frac{6\lambda}{\Delta_2}\right) & -\hat{k}
\end{array} \right) \end{array}
\tag{10-103}
$$

As a result the g-factors are

$$
g_\| = 2
$$
$$
\tag{10-104}
$$
$$
g_\perp = 2 - \frac{6\lambda}{\Delta_2}
$$

If the $|2^s\rangle$ level lies lowest the observed g-factors in the absence of spin–orbit coupling are $g_\| = g_\perp = 2$. When spin–orbit coupling is taken into account, the matrix of $(\vec{L} + 2\vec{S})$

$$
(\vec{L} + 2\vec{S}) = \begin{array}{c} \\ \langle\mp|_s \\ \\ \langle\approx|_s \end{array}
\begin{array}{cc} |\mp\rangle_s & \quad\quad |\approx\rangle_s \\
\left(\begin{array}{cc}
\frac{1}{2}\hat{k}\left(2 - \frac{8\lambda}{\Delta_0}\right) & \frac{1}{2}(\hat{i} - i\hat{j})\left(2 - \frac{2\lambda}{\Delta_1}\right) \\
\frac{1}{2}(\hat{i} + i\hat{j})\left(2 - \frac{2\lambda}{\Delta_1}\right) & -\frac{1}{2}\hat{k}\left(2 - \frac{8\lambda}{\Delta_0}\right)
\end{array} \right) \end{array}
\tag{10-105}
$$

may be compared with 10-99 to give the g-factors:

$$
g_\| = 2 - \frac{8\lambda}{\Delta_0}
$$
$$
\tag{10-106}
$$
$$
g_\perp = 2 - \frac{2\lambda}{\Delta_1}
$$

For Cu^{2+} $(3d^9)$ the spin–orbit coupling constant is negative. A typical value, $\lambda/\Delta \sim -0.05$, gives $g_\| = 2.4$ and $g_\perp = 2.1$.

The general methods described in this section may be employed to calculate anisotropic hyperfine coupling constants and quadrupole effects. They are also applicable to the case of rare-earth ions with $4f^n$ electronic configurations and to other transition series.

REFERENCES

1. R. Blinc, ed., "Magnetic Resonance and Relaxation," in *Proceedings of the 14th Colloque Ampere*, Ljubljana, Sessions 13 and 25, North-Holland, Amsterdam, 1967.

2. C. P. Poole, Jr. and H. A. Farach, *Relaxation in Magnetic Resonance*, Wiley, New York, 1971; cf. reference list in the appendix.

3. M. Blume, R. Orbach, A. Kiel, and S. Geschwind, *Phys. Rev.*, **139**, A314 (1965).

4. M. S. de Groot, I. A. M. Hesselmann, J. Schmidt, and J. H. Van der Waals, *Mol. Phys.*, **15**, 7 (1968).

5. P. H. Fisher and A. B. Denison, *Mol. Phys.*, **17**, 297 (1969).

6. M. Sharnoff, *J. Chem. Phys.*, **51**, 451 (1969).

7. E. U. Condon and G. H. Shortley, *The Theory of Atomic Spectra*, Cambridge University Press, Cambridge, 1953.

8. J. S. Griffith, *The Theory of Transition Metal Ions*, Cambridge University Press, Cambridge, 1961.

9. J. C. Slater, *Quantum Theory of Atomic Structure*, McGraw-Hill, New York, 1960, Vols. 1 and 2.

10. M. T. Hutchings, *Solid State Phys.*, **16**, 227 (1964).

11. A. Abragam and B. Bleaney, *Electron Paramagnetic Resonance of Transition Ions*, Oxford University Press, London, 1970.

12. C. J. Ballhausen, *Introduction of Ligand Field Theory*, McGraw-Hill, New York, 1962.

13. B. N. Figgis, *Introduction to Ligand Fields*, Interscience, New York, 1966.

11

LINESHAPES

11-1 INTRODUCTION

Since the formalism adopted in this book has been the use of Hamiltonian matrices to obtain energy levels, we will proceed to introduce the linewidth into the matrices themselves. The present chapter begins with a discussion of the Anderson theory of exchange narrowing, which employed imaginary quantities for the parameters affecting the linewidth. The remainder of the chapter is devoted to a discussion of lineshapes from a more conventional point of view.

The shapes of the majority of the resonant lines encountered in magnetic resonance are either Gaussian or Lorentzian, or a combination of both, and so it is appropriate to present these basic shape functions. The manner in which these shapes combine depends on whether the spin system is a gas, liquid, solid, or glass, whether the broadening arises from dipole–dipole, exchange, or other causes, whether the line is homogeneous or inhomogeneous, and so on. The theory and applications of many of these cases were developed elsewhere, and therefore they are not extensively treated here.

11-2 ANDERSON THEORY OF EXCHANGE NARROWING

Rapid motion in general and exchange interactions in particular are known to produce a narrowing of spectral lines.[1-4] Anderson[5] treated the case of random exchange by considering it as a stationary Markov process; this method has been summarized by Abragam.[1] Anderson made use of a relaxation function to derive a group of matrices called exchange matrices, each of which may be combined with an appropriate energy matrix to explain exchange-narrowing phenomena. The present treatment begins with these matrices since they fit quite well into the formalism of the book.

Anderson[5] or Abragam[2] may be consulted for a justification of the exchange matrices.

Anderson made use of real exchange matrices and purely imaginary Hamiltonian-type matrices. We find it more convenient to employ purely imaginary exchange matrices, since this leaves the remainder of our formalism unchanged. It is merely necessary to multiply Anderson's matrices by $-i$ to convert them to the form adopted here.

The exchange process under discussion can correspond to one of several physical models. For example, there may be a rapid interchange of a spin between two magnetic environments, or mutual spin flips may occur between nearby magnetic moments. The exchange process is assumed to occur randomly at the average rate ω_e; hence ω_e is referred to as the exchange frequency, and $\hbar\omega_e$ is the exchange energy.

The exchange matrix is constructed from the elements $\pm i\omega_e$ and rational fractions thereof.[6] The diagonal elements are multiplied by $+i$ and the off-diagonal ones by $-i$. The matrices are normalized so that the sum of the elements in every row and every column equals zero. Two limiting cases are considered. In case a all hyperfine lines are equally coupled by exchange, so all off-diagonal elements are equal to $-i\omega_e/(N-1)$ for an $N \times N$ matrix. In case b only adjacent spin states are coupled via exchange, so all elements adjacent to the diagonal are equal to $-i\omega_e/2$ and all other off-diagonal elements vanish. In case a all diagonal elements are $i\omega_e$, while in case b all are $i\omega_e$ except the two end ones, which assume half this magnitude, $i\omega_e/2$.

Several exchange matrices have the following explicit forms:

$$\begin{pmatrix} i\omega_e & -i\omega_e \\ -i\omega_e & i\omega_e \end{pmatrix} \quad \text{(Cases } a \text{ and } b) \tag{11-1}$$

$$\begin{pmatrix} i\omega_e & -\tfrac{1}{2}i\omega_e & -\tfrac{1}{2}i\omega_e \\ -\tfrac{1}{2}i\omega_e & i\omega_e & -\tfrac{1}{2}i\omega_e \\ -\tfrac{1}{2}i\omega_e & -\tfrac{1}{2}i\omega_e & i\omega_e \end{pmatrix} \quad \text{(Case } a) \tag{11-2}$$

$$\begin{pmatrix} \tfrac{1}{2}i\omega_e & -\tfrac{1}{2}i\omega_e & 0 \\ -\tfrac{1}{2}i\omega_e & i\omega_e & -\tfrac{1}{2}i\omega_e \\ 0 & -\tfrac{1}{2}i\omega_e & \tfrac{1}{2}i\omega_e \end{pmatrix} \quad \text{(Case } b) \tag{11-3}$$

$$\begin{pmatrix} i\omega_e & -\tfrac{1}{3}i\omega_e & -\tfrac{1}{3}i\omega_e & -\tfrac{1}{3}i\omega_e \\ -\tfrac{1}{3}i\omega_e & i\omega_e & -\tfrac{1}{3}i\omega_e & -\tfrac{1}{3}i\omega_e \\ -\tfrac{1}{3}i\omega_e & -\tfrac{1}{3}i\omega_e & i\omega_e & -\tfrac{1}{3}i\omega_e \\ -\tfrac{1}{3}i\omega_e & -\tfrac{1}{3}i\omega_e & -\tfrac{1}{3}i\omega_e & i\omega_e \end{pmatrix} \quad \text{(Case } a) \tag{11-4}$$

$$\begin{pmatrix} \tfrac{1}{2}i\omega_e & -\tfrac{1}{2}i\omega_e & 0 & 0 \\ -\tfrac{1}{2}i\omega_e & i\omega_e & -\tfrac{1}{2}i\omega_e & 0 \\ 0 & -\tfrac{1}{2}i\omega_e & i\omega_e & -\tfrac{1}{2}i\omega_e \\ 0 & 0 & -\tfrac{1}{2}i\omega_e & \tfrac{1}{2}i\omega_e \end{pmatrix} \quad \text{(Case } b) \tag{11-5}$$

Higher-order matrices for each case are easily written by analogy with these. Matrices intermediate between cases a and b may be constructed by replacing the zeros of case b by elements between the values of 0 and $-i\omega_e/(N-1)$. For these intermediate cases the remaining off-diagonal elements and the two end diagonal elements must be adjusted so that the sums over rows and columns vanish, as mentioned above.

11-3 EXCHANGE FOR THE SPIN-$\frac{1}{2}$ CASE

Anderson considered the explicit case of a spin-$\frac{1}{2}$ spectrum with two absorption lines at the positions ω_1 and ω_2 separated by the distance $2\,A$, as shown in Fig. 11-1a. In the absence of exchange the separation of the lines is given by

$$2A = (\omega_2 - \omega_1) \tag{11-6}$$

with the average position located at

$$\omega_0 = \tfrac{1}{2}(\omega_2 + \omega_1) \tag{11-7}$$

as shown in Fig. 11-1b.

The corresponding position matrix is

$$\begin{pmatrix} \omega_1 & 0 \\ 0 & \omega_2 \end{pmatrix} = \begin{pmatrix} \omega_0 + A & 0 \\ 0 & \omega_0 - A \end{pmatrix} \tag{11-8}$$

This is added to the exchange matrix (11-1) to give

$$\begin{pmatrix} \omega_0 + A + i\omega_e & -i\omega_e \\ -i\omega_e & \omega_0 - A + i\omega_e \end{pmatrix} \tag{11-9}$$

which has the eigenvalues

$$\lambda = i\omega_e + \omega_0 \pm (A^2 - \omega_e^2)^{1/2} \tag{11-10}$$

This may be written in the general form

$$\lambda = i\Delta\omega + P \tag{11-11}$$

where the imaginary part $\Delta\omega$ is the width, and P is the position of the spectral line. For $A > \omega_e$ one has

$$\Delta\omega = \omega_e$$
$$P = \omega_0 \pm A_{\mathrm{eff}} \tag{11-12}$$

where the effective line spacing parameter A_{eff} has the explicit form

$$A_{\mathrm{eff}} = (A^2 - \omega_e^2)^{1/2} \tag{11-13}$$

which decreases continuously with increasing exchange, as shown in Fig.

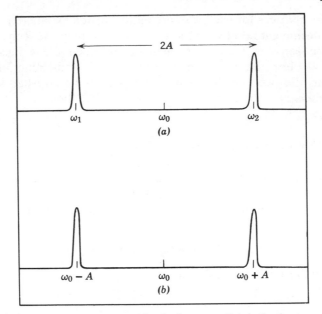

Fig. 11-1. A symmetric doublet separated by the frequency $2A$, indicating two conventions for labeling the line positions.

11-2 (where $2A\hbar = T$), and the linewidth increases linearly with exchange in accordance with Eq. 11-12, as shown in Fig. 11-3. For exchange rates ω_e exceeding A one has

$$\Delta\omega = \omega_e \pm (\omega_e^2 - A^2)^{1/2}$$
$$P = \omega_0 \tag{11-14}$$

corresponding to an imaginary A_{eff} which now contributes only to the linewidth. The doublet collapses to a singlet centered at the average position ω_0, as shown in Fig. 11-4. In the high exchange limit $\omega_e > A$ there are two superimposed lines; the linewidth $\omega_e + (\omega_e^2 - A^2)^{1/2}$ of one rapidly broadens beyond detection, and the other,

$$\Delta\omega = \omega_e\left[1 - \left(1 - \frac{A^2}{\omega_e^2}\right)^{1/2}\right] \tag{11-15}$$

narrows to the limiting value

$$\Delta\omega = \frac{A^2}{2\omega_e} \tag{11-16}$$

for very high exchange rates $\omega_e \gg A$. This is the well-known exchange-narrowing formula.

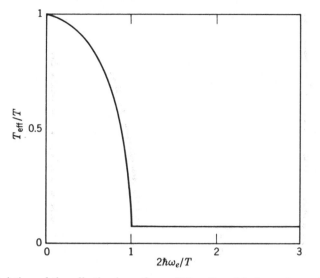

Fig. 11-2. Variation of the effective hyperfine splitting T_{eff} with the exchange frequency ω_e. Both the abscissa and ordinate are normalized relative to the true hyperfine coupling constant T.

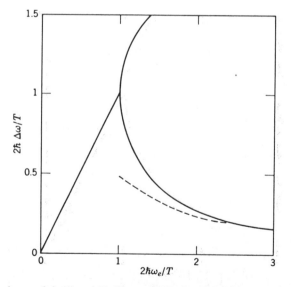

Fig. 11-3. Dependence of the linewidth $\Delta\omega$ on the exchange frequency ω_e for the case of zero intrinsic linewidth ($\Delta\omega_0 = 0$). Both the ordinate and abscissa are normalized relative to the true hyperfine coupling constant $T = 2\hbar A$. The dotted line corresponds to the asymptotic exchange-narrowing formula (Eq. 11-16).

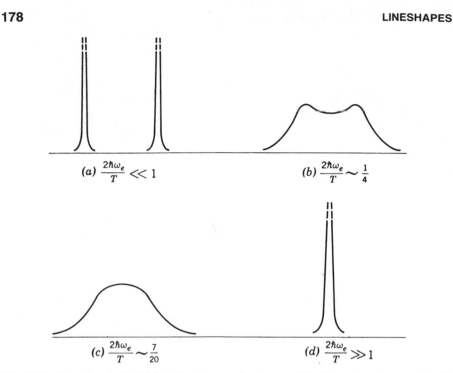

Fig. 11-4. Hyperfine doublet (*a*) in the absence of exchange, (*b*) with weak exchange, (*c*) with moderate exchange, and (*d*) with strong exchange.

The preceding discussion was for the case of a zero intrinsic width or delta function linewidth. If the intrinsic lineshape is Lorentzian in the absence of exchange, then the corresponding intrinsic linewidth $\Delta\omega_0$ may be added to each of the diagonal matrix elements. In other words, one starts with the position matrix

$$\begin{pmatrix} \omega_0 + A + i\Delta\omega_0 & 0 \\ 0 & \omega_0 - A + i\Delta\omega_0 \end{pmatrix} \tag{11-17}$$

corresponding to 11-8, which has the eigenvalues

$$\lambda' = i\Delta\omega_0 + \omega_0 \pm A \tag{11-18}$$

This is combined with the exchange matrix to give

$$\begin{pmatrix} (\omega_0 + A + i\omega_e + i\Delta\omega_0) & -i\omega_e \\ -i\omega_e & (\omega_0 - A + i\omega_e + i\Delta\omega_0) \end{pmatrix} \tag{11-19}$$

The eigenvalues λ' of Eq. 11-18 are the same as the λ of Eq. 11-9 with the added width $\Delta\omega_0$:

$$\lambda' = \lambda + i\Delta\omega_0 \tag{11-20}$$
$$= i(\Delta\omega_0 + \omega_e) + \omega_0 \pm (A^2 - \omega_e^2)^{1/2} \tag{11-21}$$

Equations 11-11 to 11-16 are easily modified to take into account the intrinsic width. For example, in the exchange-narrowing region $(\omega_e \gg A)$ one has

$$\Delta\omega = \Delta\omega_0 + \frac{A^2}{2\omega_e} \tag{11-22}$$

as expected.

The lineshape of an exchange-narrowed line with zero intrinsic width $\Delta\omega_0 = 0$ is

$$Y_{(\omega-\omega_0)} = \frac{8\omega_e A^2}{(\omega - \omega_0)^4 + 2(\omega - \omega_0)^2(2\omega_e^2 - A^2) + A^4} \tag{11-23}$$

For small exchange $\omega_e \ll A$ this reduces to

$$Y_{(\omega-\omega_0)} \approx \frac{2\omega_e}{(\omega - \omega_0 \pm A)^2 + \omega_e^2} \tag{11-24}$$

which is a Lorentzian shape centered at the two positions

$$\omega = \omega_0 \pm A \tag{11-25}$$

For strong exchange, $\omega_e \gg A$, the lineshape in the center, where $|\omega - \omega_0| < A$, becomes

$$Y_{(\omega-\omega_0)} \approx \frac{2A^2/\omega_e}{(\omega - \omega_0)^2 + (A^2/2\omega_e)^2} \tag{11-26}$$

which is a singlet of width $A^2/2\omega_e$ centered at the position $\omega = \omega_0$. This shape is Lorentzian also.

From a somewhat more general viewpoint one may say that the Anderson theory of exchange narrowing combines a Hermitian Hamiltonian matrix (H_{ij}) with an imaginary non-Hermitian linewidth matrix $(i\Delta\omega)$ into a secular equation. The method of combining these matrices provides a simple model for exchange which constitutes a first approximation to more sophisticated formulations. One may also generalize the Bloch equations to take into account exchanging spins, as shown elsewhere.[7-13]

This section treated the spin-$\frac{1}{2}$ case in general. In the next two sections the approach employed here is applied to two specific physical situations, and in the section immediately thereafter the intrinsic linewidth is discussed.

11-4 EXCHANGE NARROWING WITH HYPERFINE STRUCTURE

Section 11-3 discussed the effect of exchange on two resonant lines separated by the interval $2A$ about a center frequency ω_0. Anderson's explicit expressions were derived for the gradual collapse and subsequent exchange narrowing of the spectrum as the exchange rate ω_e increases beyond $\omega_e/A = 1$. Since the formalism in this book emphasizes the energy levels, it is appropriate to suggest how these levels can change in order to produce the observed spectral changes. In other words, the observed variations in the widths and positions of the spectral lines arise from alterations in the effective widths and magnitudes of the energy levels themselves. A very rapid exchange associated with two energy levels produces an observed level at an intermediate energy, and moderate exchange causes these energy levels to appear broader and closer together.

Since exchange affects the energy levels themselves and is dependent on probabilities of transitions between particular states, it is important to diagonalize the Hamiltonian in the absence of exchange and then to add exchange to the resulting energy matrix. For the hyperfine case only first-order or diagonal hyperfine terms are taken into account. The terms $\pm \frac{1}{2}i\omega_e$ are used in the energy matrix instead of $\pm i\omega_e$ to obtain the matrix 11-19 for the spectral line positions, where the hyperfine interaction constant $\frac{1}{2}T$ equals the line position parameter $A\hbar$.

For this ESR case, exchange is considered with respect to the simplified Hamiltonian

$$\mathscr{H} = g\beta HS_z + TS_zI_z \tag{11-27}$$

where the nuclear Zeeman term and the off-diagonal hyperfine terms are neglected. The Hamiltonian matrix with exchange is

$$
\begin{pmatrix}
\frac{1}{2}g\beta H + \frac{1}{4}T \\ + \frac{1}{2}i\hbar(\omega_e + \Delta\omega_0) & -\frac{1}{2}\hbar i\omega_e & 0 & 0 \\[2ex]
-\frac{1}{2}i\hbar\omega_e & \begin{array}{c}\frac{1}{2}g\beta H - \frac{1}{4}T \\ + \frac{1}{2}i\hbar(\omega_e + \Delta\omega_0)\end{array} & 0 & 0 \\[2ex]
0 & 0 & \begin{array}{c}-\frac{1}{2}g\beta H - \frac{1}{4}T \\ + \frac{1}{2}i\hbar(\omega_e + \Delta\omega_0)\end{array} & -\frac{1}{2}i\hbar\omega_e \\[2ex]
0 & 0 & -\frac{1}{2}i\hbar\omega_e & \begin{array}{c}-\frac{1}{2}g\beta H + \frac{1}{4}T \\ + \frac{1}{2}i\hbar(\omega_e + \Delta\omega_0)\end{array}
\end{pmatrix}
$$

$$\tag{11-28}$$

where the use of the intrinsic width term $\frac{1}{2}i\Delta\omega_0$ will be justified below.

The four eigenvalues of this matrix are

$$
\begin{aligned}
\lambda_1 &= \tfrac{1}{2}g\beta H + \tfrac{1}{2}i\hbar(\omega_e + \Delta\omega_0) + \tfrac{1}{4}[T^2 - (2\hbar\omega_e)^2]^{1/2} \\
\lambda_2 &= \tfrac{1}{2}g\beta H + \tfrac{1}{2}i\hbar(\omega_e + \Delta\omega_0) - \tfrac{1}{4}[T^2 - (2\hbar\omega_e)^2]^{1/2} \\
\lambda_3 &= -\tfrac{1}{2}g\beta H + \tfrac{1}{2}i\hbar(\omega_e + \Delta\omega_0) + \tfrac{1}{4}[T^2 - (2\hbar\omega_e)^2]^{1/2} \\
\lambda_4 &= -\tfrac{1}{2}g\beta H + \tfrac{1}{2}i\hbar(\omega_e + \Delta\omega_0) - \tfrac{1}{4}[T^2 - (2\hbar\omega_e)^2]^{1/2}
\end{aligned}
\tag{11-29}
$$

All these eigenvalues have the form

$$\lambda_i = \pm \tfrac{1}{2} g\beta H \pm \tfrac{1}{4} T_{\text{eff}} + \tfrac{1}{2} i\hbar \Delta\omega \tag{11-30}$$

where the first and second terms on the right-hand side are real, and all linewidth contributions are incorporated into the third term $\tfrac{1}{2} i\hbar \Delta\omega$. For exchange rates $\hbar\omega_e < \tfrac{1}{2} T$ the effective hyperfine coupling constant decreases continuously with increasing ω_e, where

$$T_{\text{eff}} = T\left[1 - \left(\frac{2\hbar\omega_e}{T}\right)^2\right]^{1/2} \tag{11-31}$$

as illustrated in Fig. 11-2. The linewidth increases linearly with ω_e

$$\tfrac{1}{2}\Delta\omega = \tfrac{1}{2}(\omega_e + \Delta\omega_0) \tag{11-32}$$

as shown in Fig. 11-3. At the point $\hbar\omega_e = \tfrac{1}{2} T$ the hyperfine doublets collapse into singlets:

$$T_{\text{eff}} = 0 \tag{11-33}$$

and the linewidth for $\hbar\omega_e > \tfrac{1}{2} T$ becomes

$$\tfrac{1}{2}\Delta\omega = \tfrac{1}{2}\Delta\omega_0 + \tfrac{1}{2}\omega_e\left\{1 \pm \left[1 - \left(\frac{T}{2\hbar\omega_e}\right)^2\right]^{1/2}\right\} \tag{11-34}$$

For strong exchange, $\hbar\omega_e \gg \tfrac{1}{2} T$, the lower sign gives an exchange-narrowed line characterized by the width

$$\tfrac{1}{2}\Delta\omega = \tfrac{1}{2}\Delta\omega_0 + \frac{T^2}{16\hbar^2\omega_e} \tag{11-35}$$

and the second root corresponding to the upper sign of Eq. 11-34 gives an unobservable contribution to the width, as was explained in the discussion after Eq. 11-14. The effect of exchange on the energy levels may be pictured as occurring in the manner illustrated in Fig. 11-5.

To obtain the transition frequencies $\omega_{ij} = (\lambda_j - \lambda_i)/\hbar$ from the levels

$$\begin{aligned} \lambda_i &= \tfrac{1}{2} i\hbar\Delta\omega_i + E_i \\ \lambda_j &= \tfrac{1}{2} i\hbar\Delta\omega_j + E_j \end{aligned} \tag{11-36}$$

the energies are subtracted algebraically $(E_j - E_i)$ and the widths of the levels are added to give $i\hbar\Delta\omega$. In other words, a transition between two levels with the respective linewidths $\tfrac{1}{2}\Delta\omega_i$ and $\tfrac{1}{2}\Delta\omega_j$ is assumed to produce an observed spectral line with the width

$$\Delta\omega = \tfrac{1}{2}(\Delta\omega_i + \Delta\omega_j) \tag{11-37}$$

This addition of the widths is a reasonable approximation on physical grounds. When these prescriptions are followed in the range $\hbar\omega_e < \tfrac{1}{2} T$, the two allowed complex transition frequencies are

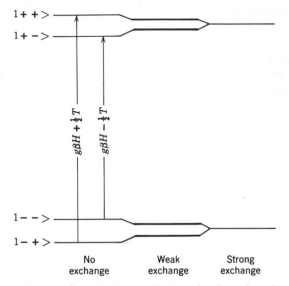

Fig. 11-5. Behavior of $S = I = \frac{1}{2}$ coupled energy levels under the action of exchange. The two allowed transitions with $\Delta m_S = \pm 1$ and $\Delta m_I = 0$ are indicated.

$$\hbar \omega_{13} = i\hbar(\Delta \omega_0 + \omega_e) + g\beta H + [(\tfrac{1}{2}T)^2 - \hbar^2 \omega_e^2]^{1/2}$$
$$\hbar \omega_{24} = i\hbar(\Delta \omega_0 + \omega_e) + g\beta H - [(\tfrac{1}{2}T)^2 - \hbar^2 \omega_e^2]^{1/2} \qquad (11\text{-}38)$$

in agreement with Eq. 11-21 when one makes the identifications

$$\lambda \rightarrow \hbar \omega_{ij}, \qquad T \rightarrow 2A\hbar \qquad (11\text{-}39)$$

In other words,

$$\omega_{ij} = i\Delta \omega + P \qquad (11\text{-}40)$$

where the real part P gives the spectral line position and the imaginary part $\Delta \omega$ is the width, corresponding to Eq. 11-11. By using identifications 11-39 the various formulas 11-10 to 11-26 may be applied directly to this case.

11-5 EXCHANGE EFFECTS ON CHEMICAL SHIFTS

When a nucleus exchanges rapidly between two sites with slightly different gyromagnetic ratios γ_1 and γ_2, the spectral lines broaden and approach each other, merge, and then exchange-narrow with increasing exchange rates in the manner illustrated in Fig. 11-4. Using ω_0 as the mean line position and δ as the relative chemical shift,

$$\omega_0 = \tfrac{1}{2}(\gamma_1 + \gamma_2)H_0$$

$$\delta = \frac{\gamma_2 - \gamma_1}{\tfrac{1}{2}(\gamma_2 - \gamma_1)} \qquad (11\text{-}41)$$

$$2A = \delta \omega_0 = (\gamma_2 - \gamma_1)H_0$$

the effect of exchange is taken into account by the use of Eq. 11-19 with $2A$ replaced by $\delta\omega_0$:

$$\begin{pmatrix} (\omega_0 + \frac{1}{2}\delta\omega_0 + i\omega_e + i\Delta\omega_0) & -i\omega_e \\ -i\omega_e & (\omega_0 - \frac{1}{2}\delta\omega_0 + i\omega_e + i\Delta\omega_0) \end{pmatrix} \quad (11\text{-}42)$$

Therefore all the equations of Section 11-3 are applicable with A replaced by $\frac{1}{2}\delta\omega_0$. For example, one can define an effective chemical shift for $\omega_e < \frac{1}{2}\delta\omega_0$,

$$\delta_{\text{eff}} = \pm\delta\left[1 - \left(\frac{2\omega_e}{\delta\omega_0}\right)^2\right]^{1/2} \quad (11\text{-}43)$$

and write down the exchange-narrowing formula for $\omega_e \gg \frac{1}{2}\delta\omega_0$,

$$\Delta\omega = \Delta\omega_0 + \frac{\delta^2\omega_0^2}{8\omega_e} \quad (11\text{-}44)$$

This Anderson treatment is equivalent to assuming that the nucleus spends half of its time at each site, corresponding to the observation of two equal intensity lines in the absence of exchange. More sophisticated theories have been devised to take into account unequal site occupation times.[14-16] Many of these theories are based on the modified Bloch equations.

11-6 INTRINSIC LINEWIDTH

Dissipative processes are ordinarily introduced into real differential equations as imaginary terms. For example, the dissipative impedance or resistance in an electrical circuit is imaginary, whereas the nondissipative impedance or reactance arising from an inductance or capacitance is real. Friction terms added to transport equations, dissipative contributions to dielectric constants, and so on are also imaginary. Therefore it is not surprising that line-broadening quantities may be added to Hamiltonian matrices as purely imaginary terms. These quantities constitute non-Hermitian matrix elements since for them

$$\mathcal{H}_{ij} = -\mathcal{H}^*_{ji} \quad (11\text{-}45)$$

in contrast to the Hermitian property of ordinary Hamiltonian matrix elements:

$$\mathcal{H}_{ij} = \mathcal{H}^*_{ji} \quad (11\text{-}46)$$

The non-Hermitian line-broadening matrix thus has purely imaginary diagonal elements and identical imaginary off-diagonal elements at corresponding locations, ij and ji.

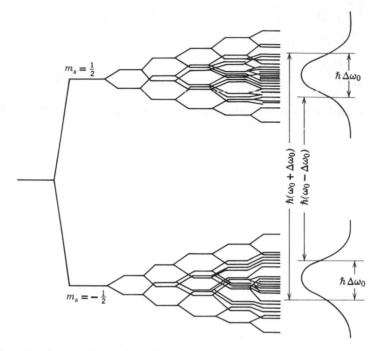

Fig. 11-6. Broadening of spin $S = \frac{1}{2}$ Zeeman energy levels by the successive applications of small $I = \frac{1}{2}$ dipolar interactions. The lineshape of each level is shown at the right.

The intrinsic linewidth may be viewed as arising from a large number of mechanisms such as dipolar interactions. Each individual interaction splits all levels by a small amount, as shown in Fig. 11-6. As a result, each energy level acquires a half-width $\frac{1}{2}\hbar\Delta\omega_0$. The figure shows the extreme transitions corresponding to parts separated by the closest and the most remote half-amplitude points on the lines. These occur at

$$E_{max} = \hbar(\omega_0 + \Delta\omega_0)$$
$$E_{min} = \hbar(\omega_0 - \Delta\omega_0)$$

(11-47)

corresponding to a spectral line half-width of $\Delta\omega_0$. Thus we see that the spectral line tends to be wider than the individual levels, and Eq. 11-37 may be taken as an approximation to this increased width.

11-7 GAUSSIAN AND LORENTZIAN SHAPES

The preceding sections considered the linewidths of energy levels and ended with a discussion of exchange-narrowed lineshapes. These next three sections are devoted to various lineshapes.

The analytical expression for a Gaussian absorption lineshape as a function of the magnetic field strength H is

$$Y^G_{(H)} = y_m \exp\left[-0.693\left(\frac{H-H_0}{\frac{1}{2}\Delta H_{1/2}}\right)^2\right]$$ (11-48)

and the corresponding expression for a Lorentzian line is

$$Y^L_{(H)} = \frac{y_m}{1 + [(H-H_0)/\frac{1}{2}\Delta H_{1/2}]^2}$$ (11-49)

where y_m denotes the maximum amplitude, obtained in both cases at the center of the line, where $H = H_0$. The quantity $\Delta H_{1/2}$ is the full linewidth between half-amplitude points, and $\ln 2 = 0.693$.

In field modulation spectrometers one ordinarily detects the first derivative counterparts of Eqs. 11-48 and 11-49; these have the forms

$$Y'^G_{(H)} = 1.649 y'_m \left(\frac{H-H_0}{\frac{1}{2}\Delta H_{pp}}\right) \exp\left[-\frac{1}{2}\left(\frac{H-H_0}{\frac{1}{2}\Delta H_{pp}}\right)^2\right]$$ (11-50)

and

$$Y'^L_{(H)} = \frac{16 y'_m[(H-H_0)/\frac{1}{2}\Delta H_{pp}]}{[3 + \{(H-H_0)/\frac{1}{2}\Delta H_{pp}\}^2]^2}$$ (11-51)

For a Gaussian line the amplitude y'_m has the explicit form

$$y'_m = \frac{y_m}{\frac{1}{2}\Delta H_{pp} e^{1/2}}$$ (11-52)

and the full peak-to-peak linewidth is related to the full half-amplitude linewidth in the following manner:

$$\Delta H_{1/2} = (2 \ln 2)^{1/2} \Delta H_{pp} = 1.1774 \Delta H_{pp}$$ (11-53)

The corresponding expressions for the Lorentzian case are

$$y'_m = \frac{3}{4} y_m (\Delta H_{pp})^{-1}$$

$$\Delta H_{1/2} = \sqrt{3}\Delta H_{pp}$$ (11-54)

These lineshapes are normalized in terms of the amplitude, in contrast to the practice of some other authors who normalize to a unit area.

The number of spins in the sample is proportional to the area

$$A = \int_0^\infty Y_{(H-H_0)} dH_0 = \int_0^\infty (H-H_0) Y'_{(H-H_0)} dH$$ (11-55)

The various line-broadening mechanisms are characterized by the moments of the line $\langle (H - H_0)^n \rangle$, which are defined by

$$\langle (H - H_0)^n \rangle = \langle H^n \rangle = \int_0^\infty (H - H_0)^n Y_{(H-H_0)} dH$$

$$= \frac{1}{n+1} \int_0^\infty (H - H_0)^{n+1} Y'_{(H-H_0)} dH \qquad (11\text{-}56)$$

where a negative sign is omitted from the two integrals containing $Y'_{(H-H_0)}$. One should note that the area is a zeroth moment. For the symmetric shapes under discussion all odd-order moments ($n = 1, 3$, etc.) vanish.

The parameters which characterize the lines are defined in Fig. 11-7, and the absorption and derivative lineshapes are sketched in Figs. 11-8 and 11-9. The area and moments are related to other parameters in the manner given in Table 11-1. The subject has been reviewed elsewhere.[4,17]

The Lorentzian and Gaussian lineshapes described in this section are associated with the absorption of radiofrequency energy that occurs at resonance, and hence they are referred to as absorption lineshapes. The passage through resonance is also accompanied by a change in the frequency of the cavity or coil, and this produces what is referred to as a dispersion signal. It has a shape that is similar in appearance to an absorption derivative. Most ESR spectrometers are tuned to record absorption only.

Absorption and dispersion shapes come in pairs connected to each other through the Kramers–Kronig relations. The Lorentzian dispersion signal has the form

$$d = \frac{2[(H - H_0)/\frac{1}{2}\Delta H_{1/2}]}{1 + [(H - H_0)/\frac{1}{2}\Delta H_{1/2}]^2} \qquad (11\text{-}57)$$

and its first derivative d' is

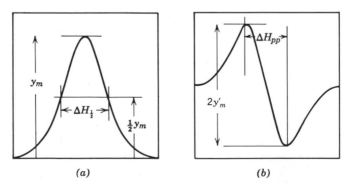

(a)   (b)

Fig. 11-7. (a) Absorption and (b) absorption first derivative lineshapes, showing definitions of amplitude and width parameters.

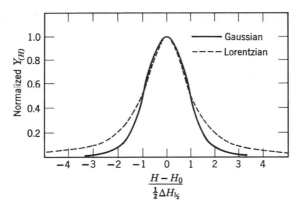

Fig. 11-8. Gaussian and Lorentzian absorption curves with the same half-amplitude linewidth.

$$d' = 3 \; \frac{3 - [(H - H_0)/\frac{1}{2}\Delta H_{pp}]^2}{\{3 + [(H - H_0)/\frac{1}{2}\Delta H_{pp}]^2\}^2} \qquad (11\text{-}58)$$

These two expressions are individually normalized so that each has a maximum amplitude of 1, and as usual $\Delta H_{1/2} = \sqrt{3}\Delta H_{pp}$. The derivative d' vanishes at $H - H_0 = \pm \frac{1}{2}\Delta H_{1/2}$ and reaches extrema at

$$\frac{(H - H_0)}{\frac{1}{2}\Delta H_{pp}} = 0, \pm 3 \qquad (11\text{-}59)$$

The dispersion lineshape resembles that of an absorption derivative, and the dispersion derivative signal looks like an absorption second derivative line.[17]

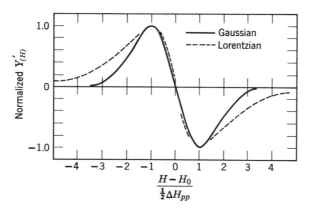

Fig. 11-9. Gaussian and Lorentzian absorption first derivative curves with the same peak-to-peak linewidth.

TABLE 11-1 Comparison of Gaussian and Lorentzian Lineshapes[a]

Parameter	Gaussian Shape	Lorentzian Shape
$\Delta H_{1/2}/\Delta H_{pp}$	$(2\ln 2)^{1/2} = 1.1774$	$3^{1/2} = 1.7321$
$y_m/(y_m'\Delta H_{pp})$	$e^{1/2}/2 = 0.8244$	$4/3 = 1.3333$
$y_m/(y_m''\Delta H_{pp}^2)$	$1/4 = 0.2500$	$3/8 = 0.3750$
$A/(y_m\Delta H_{1/2})$	$\tfrac{1}{2}(\pi/\ln 2)^{1/2} = 1.0645$	$\pi/2 = 1.5708$
$A/(y_m'\Delta H_{pp}^2)$	$\tfrac{1}{2}(\pi e/2)^{1/2} = 1.0332$	$2\pi/3^{1/2} = 3.6276$
$\langle H^2\rangle/(\Delta H_{1/2})^2$	$1/(8\ln 2) = 0.1803$	∞
$\langle H^4\rangle/(\Delta H_{1/2})^4$	$3/[64(\ln 2)^2] = 0.0976$	∞
$\langle H^2\rangle/(\Delta H_{pp})^2$	$1/4 = 0.2500$	∞
$\langle H^4\rangle/(\Delta H_{pp})^4$	$3/16 = 0.1875$	∞
y_1''/y_2''	$\tfrac{1}{2}e^{3/2} = 2.2408$	$64^{1/3} = 4$
$H_1''/\Delta H_{pp}$	0.626	0.567
$H_2''/\Delta H_{pp}$	$3^{1/2} = 1.7321$	$3^{1/2} = 1.7321$
$H_3''/\Delta H_{pp}$	2.52	$81^{1/4} = 3$
$d'(0)/d'(3^{1/2}) = A/B$	$7/2 = 3.50$	$2^3 = 8$

For solid DPPH, $A/[y_m'(\Delta H_{pp})^2] = 2.2$

$e = 2.718282$	$\left.\dfrac{\ln\ 2}{\log_e 2}\right\} = 0.693147$	
$\pi = 3.141593$		
$2^{1/2} = 1.414214$	$\pi^{1/2} = 1.772454$	
$3^{1/2} = 1.732051$	$(\ln 2)^{1/2} = 0.832555$	
$e^{1/2} = 1.648721$		

[a]Compare Ref. 17, Table 12-5, p. 476.

11-8 VOIGT LINESHAPE

A combination of Lorentzian and Gaussian lines can result from several types of interaction. For example, an exchange-narrowed line is Lorentzian in the center and Gaussian in the wings, and an inhomogeneously broadened line is a superposition of many narrow Lorentzians comprising a Gaussian envelope, as shown in Fig. 11-10. This type of combination is called a convolution. The Voigt convolution lineshape obtained for the inhomogeneous case has the following form, using the notation of Section 11-7:

$$Y_{(H-H_0)} = \frac{(\ln 2)^{1/2}}{\pi}\left(\frac{\Delta H_{1/2}^L}{\Delta H_{1/2}^G}\right)$$

$$\times \int_{-\infty}^{\infty} \frac{e^{-x^2}dx}{(\Delta H_{1/2}^L/\Delta H_{1/2}^G)^2 \ln 2 + [2(\ln 2)^{1/2}\{(H-H_0)/\Delta H_{1/2}^G\} - x]^2}$$

$$(11\text{-}60)$$

where the superscript G denotes Gaussian, and L signifies Lorentzian. This equation can not be integrated in closed form, but within certain limits it was evaluated by Zemansky[18] using a combination of series expansions, while Farach and Teitelbaum[19] obtained a graphic solution. Castner[20] used this equation in his treatment of the saturation of inhomogeneously

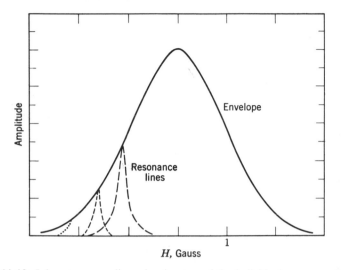

Fig. 11-10. Inhomogeneous line, showing two of the individual component lines.

broadened lines. Homogeneous and inhomogeneous broadening has been reviewed elsewhere.[3,4,17]

11-9 POWDER PATTERN LINESHAPES

When the Hamiltonian is anisotropic the positions of the resonant lines in a single crystal depend on its orientation in the magnetic field. For a powder sample with a very large number of randomly oriented microcrystallites one obtains a powder pattern lineshape arising from the randomly distributed principal directions in the magnetic field.[21-25]

An example of a powder pattern lineshape is that arising from a completely anisotropic g-factor, as shown in Fig. 11-11. For axial symmetry the line assumes the form of Fig. 11-12. If both the g-factor and the hyperfine coupling constant are axially symmetric with the same principal directions, the observed spectrum is the superposition of $(2I+1)$ components, each of which has the shape shown in Fig. 11-13.

We illustrate the general procedure by calculating the powder pattern for the case of an isotropic g-factor and an axially symmetric $I=1$ hyperfine coupling. The Hamiltonian has two terms, an isotropic Zeeman interaction $g\beta H$ and an axially symmetric hyperfine term H_T,

$$H = g\beta H + H_T \tag{11-61}$$

where

$$H_T = \vec{S} \cdot \vec{\vec{T}} \cdot \vec{I} \tag{11-62}$$

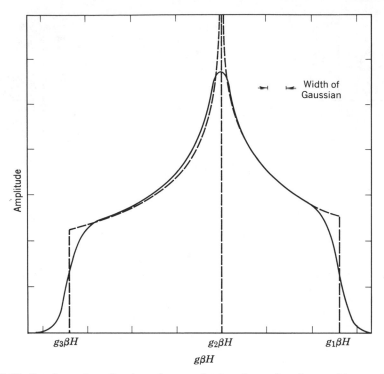

Fig. 11-11. Powder pattern lineshape for completely anisotropic g-factor with zero (---) and finite (—) component linewidth. The infinity at g_2 integrates to a finite area.

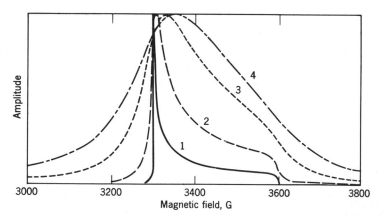

Fig. 11-12. The effect of increasing the component linewidth on the calculated spectrum for an axially symmetric powder pattern. Lorentzian linewidths: (1) 1 G, (2) 10 G, (3) 50 G, (4) 100 G (from Ref. 23).

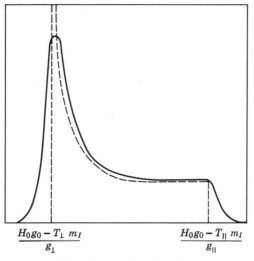

Fig. 11-13. Powder pattern lineshape for one hyperfine component m_I when $g_0 \beta H_0 \gg T_{\parallel}, T_{\perp}$ (from Ref. 24).

It was shown in Chapter 6 that in the principal axis system the hyperfine tensor has the form

$$\vec{\vec{T}} = \begin{pmatrix} T_{\perp} & 0 & 0 \\ 0 & T_{\perp} & 0 \\ 0 & 0 & T_{\parallel} \end{pmatrix} \tag{11-63}$$

with the principal values $T_{\perp}$ and $T_{\parallel}$.

Since the matrix $\vec{\vec{T}}$ is axially symmetric, it is invariant under rotations about the z axis. In addition, a rotation about the x axis is equivalent to one about any axis in the xy plane, so $\vec{\vec{T}}$ is a function of the azimuthal angle θ. Therefore at an arbitrary angle $\vec{\vec{T}}$ becomes

$$\vec{\vec{T}}'(\theta) = \vec{\vec{R}}_{\theta x} \vec{\vec{T}} \vec{\vec{R}}_{\theta x} \tag{11-64}$$

with the rotation matrix $\vec{\vec{R}}_{\theta x}$ given explicitly by

$$\vec{\vec{R}}_{\theta x} = \begin{pmatrix} 1 & 0 & 0 \\ 0 & \cos\theta & -\sin\theta \\ 0 & \sin\theta & \cos\theta \end{pmatrix} \tag{11-65}$$

Carrying out the rotation operations gives (compare Eq. 6-18)

$$\vec{\vec{T}} = \begin{pmatrix} T_{\perp} & 0 & 0 \\ 0 & (T_{\parallel}\sin^2\theta + T_{\perp}\cos^2\theta) & (T_{\perp} - T_{\parallel})\sin\theta\cos\theta \\ 0 & (T_{\perp} - T_{\parallel})\sin\theta\cos\theta & (T_{\parallel}\cos^2\theta + T_{\perp}\sin^2\theta) \end{pmatrix} \tag{11-66}$$

which reduces to the matrix of Eq. 11-63 when $\theta = 0$. The rotated hyperfine Hamiltonian 11-62,

$$H_T = (S_x \quad S_y \quad S_z) \begin{pmatrix} T_{11} & 0 & 0 \\ 0 & T_{22} & T_{23} \\ 0 & T_{32} & T_{33} \end{pmatrix} \begin{pmatrix} I_x \\ I_y \\ I_z \end{pmatrix} \qquad (11\text{-}67)$$

may be expanded and added to the Zeeman term of Eq. 11-61 to give

$$H = g\beta H S_z + T_\perp S_x I_x + (T_\parallel \sin^2 \theta + T_\perp \cos^2 \theta) S_y I_y$$
$$+ (T_\parallel \cos^2 \theta + T_\perp \sin^2 \theta) S_z I_z + [(T_\perp - T_\parallel) \sin \theta \cos \theta](S_y I_z + S_z I_y)$$
$$(11\text{-}68)$$

We assume that the hyperfine term is much smaller than the Zeeman term, and this permits us to write the 6×6 Hamiltonian matrix in the form

$$\begin{pmatrix} 3 \times 3 & \vdots & \text{To be} \\ \text{Submatrix} & \vdots & \text{neglected} \\ \hdashline \text{To be} & \vdots & 3 \times 3 \\ \text{neglected} & \vdots & \text{Submatrix} \end{pmatrix} \qquad (11\text{-}69)$$

where by second-order perturbation theory the neglected terms are of the order $T^2/g\beta H$. This approximation is equivalent to omitting the terms in the Hamiltonian 11-68 which contain the operators S_x and S_y, and so we can write

$$H = g\beta H S_z + (T_\parallel \cos^2 \theta + T_\perp \sin^2 \theta) S_z I_z + (T_\perp - T_\parallel) \sin \theta \cos \theta \, S_z I_y$$
$$(11\text{-}70)$$

Each of the two 3×3 submatrices of Eq. 11-69 which correspond to the two $S_z = \pm \frac{1}{2}$ electronic spin orientations are the negatives of each other, and so it is only necessary to consider one of them. Using the standard direct product expansion procedure the submatrix assumes the form

$$\begin{pmatrix} \frac{1}{2}(T_\parallel \cos^2 \theta + T_\perp \sin^2 \theta) & -\frac{i}{2\sqrt{2}}(T_\perp - T_\parallel)\sin\theta\cos\theta & 0 \\ \frac{i}{2\sqrt{2}}(T_\perp - T_\parallel)\sin\theta\cos\theta & 0 & -\frac{i}{2\sqrt{2}}(T_\perp - T_\parallel)\sin\theta\cos\theta \\ 0 & \frac{i}{2\sqrt{2}}(T_\perp - T_\parallel)\sin\theta\cos\theta & -\frac{1}{2}(T_\parallel \cos^2 \theta + T_\perp \sin^2 \theta) \end{pmatrix}$$
$$(11\text{-}71)$$

The corresponding secular determinant

$$\begin{vmatrix} A + \frac{1}{2}g\beta H - E_+ & -iB & 0 \\ iB & \frac{1}{2}g\beta H - E_+ & -iB \\ 0 & iB & -A + \frac{1}{2}g\beta H - E_+ \end{vmatrix} = 0 \quad (11\text{-}72)$$

where

$$A = \tfrac{1}{2}(T_{\parallel} \cos^2 \theta + T_{\perp} \sin^2 \theta)$$

$$B = \frac{(T_{\perp} - T_{\parallel}) \sin \theta \cos \theta}{2\sqrt{2}}$$

(11-73)

is easily evaluated to give the cubic equation

$$(E_+ - \tfrac{1}{2}g\beta H)[(E_+ - \tfrac{1}{2}g\beta H)^2 - (A^2 + 2B^2)] = 0$$

(11-74)

which has the three roots

$$E_+ = \tfrac{1}{2}g\beta H, \quad \tfrac{1}{2}g\beta H \pm \tfrac{1}{2}[T_{\perp}^2 + (T_{\parallel}^2 - T_{\perp}^2) \cos^2 \theta]^{1/2}$$

(11-75)

The subscript on the energy E indicates that it comes from the upper left submatrix of 11-70. The lower right submatrix gives the analogous energies

$$E_- = -\tfrac{1}{2}g\beta H, \quad -\tfrac{1}{2}g\beta H \mp \tfrac{1}{2}[T_{\perp}^2 + (T_{\parallel}^2 - T_{\perp}^2) \cos^2 \theta]^{1/2}$$

(11-76)

These expressions are valid for the condition

$$|T| \ll g\beta H$$

(11-77)

on the Hamiltonian of Eq. 11-61. The observed energy splittings are the differences $E_+ - E_-$ between the two sets of levels

$$E_+ - E_- = g\beta H, \quad g\beta H \pm [T_{\perp}^2 + (T_{\parallel}^2 - I_{\perp}^2) \cos^2 \theta]^{1/2}$$

(11-78)

corresponding to the selection rules $\Delta m_S = \pm 1$, $\Delta m_I = 0$.

Ordinary ESR experiments are carried out at a constant frequency ω_0

$$E_+ - E_- = \hbar \omega_0$$

(11-79)

and the center or unshifted resonance line occurs at the magnetic field H_0

$$\hbar \omega_0 = g\beta H_0$$

(11-80)

The remaining two hyperfine lines appear at magnetic field values H which satisfy the condition

$$H_0 = H \pm [T_{\perp}^2 + (T_{\parallel}^2 - T_{\perp}^2) \cos^2 \theta]^{1/2}$$

(11-81)

where T is now in the units of gauss instead of energy $(T \rightarrow T/g\beta)$. These hyperfine lines will occur for magnetic field values H within the range

$$H_{min} \leq H \leq H_{max}$$

(11-82)

where H_{min} and H_{max} are the limiting magnetic field values for $\theta = 0$ and $\theta = \tfrac{1}{2}\pi$. Which of the two is the minimum and which is the maximum depends on the signs and relative magnitudes of $T_{\parallel}$ and $T_{\perp}$ and also on the particular hyperfine transition being observed. From Eq. 11-81 the limiting values H_{lim} for these various cases are given explicitly by

$$H_{lim} = \begin{cases} H_0 \pm T_{\parallel} \\ H_0 \pm T_{\perp} \end{cases}$$

(11-83)

To obtain the powder pattern lineshape we assume that the microcrystal-lites have no preferred orientation. The probability of finding a crystallite at an angle θ is proportional to the area of an annular ring of width $r\,d\theta$ and circumference $2\pi r \sin\theta$, as shown in Fig. 11-14. This will cause the number of microcrystallites, dN, at a given orientation θ to vary as

$$dN = \tfrac{1}{2}N_0 \sin\theta \, d\theta \tag{11-84}$$

where

$$0 \le \theta \le \tfrac{1}{2}\pi \tag{11-85}$$

with N_0 denoting the total number of them. This number dN is related to the number of microcrystallites which have resonant frequencies within a magnetic field range dH centered at H through the expression

$$dN = W(H)dH \tag{11-86}$$

where $W(H)$ is the shape function of the powder pattern. This shape function is obtained by solving Eq. 11-81 for $\cos\theta$

$$\cos\theta = \left[\frac{(H_0 - H)^2 - T_{\perp}^2}{T_{\parallel}^2 - T_{\perp}^2} \right]^{1/2} \tag{11-87}$$

differentiating this equation, and substituting the resulting expression for $\sin\theta \, d\theta$ into Eq. 11-84 to give

$$\frac{\partial N}{\partial H} = \frac{\tfrac{1}{2}N_0(H_0 - H)}{\{[(H_0 - H)^2 - T_{\perp}^2](T_{\parallel}^2 - T_{\perp}^2)\}^{1/2}} \tag{11-88}$$

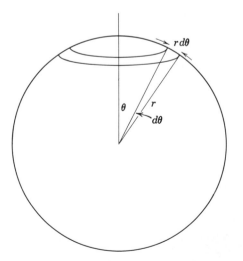

Fig. 11-14. Annular ring between θ and $\theta + d\theta$ with area $2\pi r^2 \sin\theta \, d\theta$.

This shape function increases monotonically from the value $T_{\parallel}/(T_{\parallel}^2 - T_{\perp}^2)$ at $\theta = 0$ to infinity at $\theta = \frac{1}{2}\pi$.

It is more convenient to have a shape function normalized to 1 at its minimum where $\theta = 0$. This is accomplished by multiplying the right-hand side of Eq. 11-88 by the ratio $[T_{\parallel}^2 - T_{\perp}^2]/T_{\parallel}$, and the resulting normalized shape function is

$$W(H) = \left[\frac{T_{\parallel}^2 - T_{\perp}^2}{(H_0 - H)^2 - T_{\perp}^2}\right]^{1/2}\left(\frac{H_0 - H}{T_{\parallel}}\right) \qquad (11\text{-}89)$$

This is plotted in Fig. 11-15.

The function $W(H)$ is the envelope of the powder pattern shape for infinitely narrow lines. The infinity that occurs at the field position $H_0 - H = T_{\perp}$ integrates to a finite area. If the individual lines have a finite width and the lineshape in a single crystal is denoted by $g_{(H)}$, then the overall shape is given by the convolution

$$Y_{(H)} = \int_{-\infty}^{\infty} g_{(H-H')} W_{(H')} dH' \qquad (11\text{-}90)$$

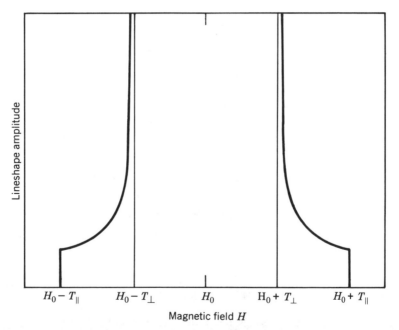

Fig. 11-15. Powder pattern spectrum for an $S = \frac{1}{2}$, $I = 1$ triplet arising from an axially symmetric hyperfine tensor showing the absorption lineshape for infinitely narrow component lines (—). The spectrum is drawn for the condition $T_{\parallel} = 2T_{\perp}$, where both $T_{\parallel}$ and $T_{\perp}$ are positive. When they have opposite signs the hyperfine components overlap. The g-factor is assumed to be isotropic. The infinitely narrow central line at $H = H_0$ is not shown.

where $W_{(H')}$ is given by Eq. 11-89, and in a typical case $g_{(H-H')}$ is Gaussian or Lorentzian.

To explain saturation transfer spectra, which are discussed in Chapter 18, the quantity of interest is the derivative $\partial H/\partial\theta$. This is obtained by combining Eqs. 11-84 and 11-86,

$$\frac{\partial H}{\partial \theta} = \frac{N_0 \sin \theta}{2W(H)} \tag{11-91}$$

and eliminating H with the aid of Eqs. 11-87 and 11-89 to give the equation

$$\frac{\partial H}{\partial \theta} = \frac{(T_\parallel^2 - T_\perp^2)\sin\theta\cos\theta}{(T_\perp^2 \sin^2\theta + T_\parallel^2 \cos^2\theta)^{1/2}} \tag{11-92}$$

which passes through a maximum at the angle θ_{max} given by

$$\tan\theta_{max} = \left(\frac{T_\parallel}{T_\perp}\right)^{1/2} \tag{11-93}$$

This expression is used in Section 18-4 to explain the mechanism of saturation transfer. The quantity $\partial H/\partial\theta$ can also be written as a function of the magnetic field H

$$\frac{\partial H}{\partial \theta} = \frac{[(H_0 - H)^2 - T_\perp^2]^{1/2}[T_\parallel^2 - (H_0 - H)^2]^{1/2}}{2(H_0 - H)} \tag{11-94}$$

The field dependence of $\partial H/\partial\theta$ sketched in Fig. 11-16 for the two outer hyperfine lines of Fig. 11-15 reaches a maximum in the center of each line where $(H_0 - H)^2 = T_\parallel T_\perp$ and vanishes at the edge of each line where $(H_0 - H)^2 = T_\parallel^2$ and $(H_0 - H)^2 = T_\perp^2$.

The treatment given in this section can also be applied to an axially symmetric g-factor, and we list here the equations that are obtained. The resonant magnetic field H for an orientation θ corresponding to Eq. 11-81 is

$$H = \frac{g_0 H_0}{(g_\perp^2 \sin^2\theta + g_\parallel^2 \cos^2\theta)^{1/2}} \tag{11-95}$$

The lineshape expression (11-88) for this case is

$$\frac{\partial N}{\partial H} = \frac{N_0 (g_0 H_0)^2}{2H^2}\left(\frac{g_\parallel^2 - g_\perp^2}{g_0^2 H_0^2 - g_\perp^2 H^2}\right)^{1/2} \tag{11-96}$$

where $g_0 H_0 = g_\perp H_\perp$ so there is an infinity at the field value $H = H_\perp$. The angular dependence of H is

$$\frac{\partial H}{\partial \theta} = \frac{g_0 H_0 (g_\parallel^2 - g_\perp^2)\sin\theta\cos\theta}{(g_\perp^2 \sin^2\theta + g_\parallel^2 \cos^2\theta)^{3/2}} \tag{11-97}$$

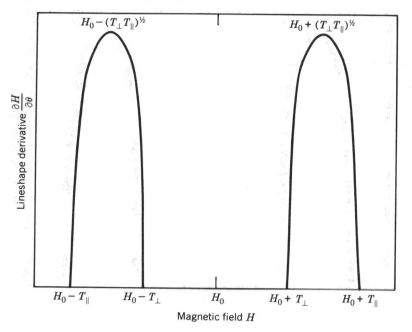

Fig. 11-16. Dependence of the derivative $\partial H / \partial \theta$ on the magnetic field for the two outer hyperfine components illustrated on Fig. 11-15. The peaks occur at $H = H_0 \pm (T_\perp T_\parallel)^{1/2}$.

which is zero for $\theta = 0$ and $\theta = \frac{1}{2}\pi$ and reaches a maximum at the angle

$$\tan \theta_{\text{max}} = \left(\frac{g_\parallel}{g_\perp} \right)^{1/2} \tag{11-98}$$

An alternate expression is

$$\frac{\partial H}{\partial \theta} = \frac{H[(g_0^2 H_0^2 - g_\parallel^2 H^2)(g_\perp^2 H^2 - g_0^2 H_0^2)]^{1/2}}{(g_0 H_0)^2} \tag{11-99}$$

which vanishes for $H = H_\parallel$ and $H = H_\perp$ and reaches a maximum at the field

$$H_{\text{max}} = \frac{g_0 H_0}{(g_\parallel g_\perp)^{1/2}} \tag{11-100}$$

Before closing, several additional references will be mentioned. Kittel and Abrahams[26] calculated the lineshape for magnetically diluted crystals; they found that for a fractional magnetic population $f > 0.1$ the lineshape is approximately Gaussian with a width proportional to $f^{1/2}$, while for $f < 0.01$ the lineshape is approximately Lorentzian with a width proportional to f. A detailed analysis of the change in lineshape with concentration was made by Swarup.[27]

11-10 RELAXATION

The earlier chapters have been devoted mainly to what might be called the static problem in magnetic resonance. In other words, we emphasized the energy levels—the transitions that occur between those levels—and the positions and relative intensities of the resulting spectral lines. Now we say a few words about the dynamic problem in magnetic resonance, which is concerned with the manner in which spins in excited energy states relax back to the ground state by passing their energy on to the lattice or temperature reservoir.

Relaxation processes[2,3,28-40] involve a transverse or spin–spin relaxation time T_2 which is the time constant for the spin system to establish thermal equilibrium within itself. In addition, there is a longitudinal or spin–lattice relaxation time T_1 which is the time constant associated with the time that is required for the spin system to come to thermal equilibrium with the lattice. In low-viscosity liquids or in solids with strong spin exchange interactions the linewidths are quite narrow, and the two relaxation times are almost equal to each other. In solids the spin–lattice relaxation time T_1 becomes quite long and greatly exceeds T_2:

$$T_1 \sim T_2 \quad \text{(Low viscosity liquids)} \tag{11-101}$$

$$T_1 \gg T_2 \quad \text{(Most solids)} \tag{11-102}$$

Thus we see that the relaxation mechanisms in liquids differ from those in solids. We treat each in turn.

The dominant relaxation mechanism in liquids is the rapid Brownian motion of the molecules. This is characterized by a correlation time τ_c and in the simple Debye approximation it is related to the viscosity η and to the molecular radius a through the expression

$$\tau_c = \frac{4\pi\eta a^3}{3kT} \tag{11-103}$$

For water at 20°C we have $\eta = 0.01$ P and $a = 1.5$ Å, to give

$$\tau_c = 3.5 \times 10^{-12} \text{ sec} \tag{11-104}$$

Values of τ_c obtained from NMR data agree well with those determined from dielectric relaxation measurements.

As an example of relaxation in a liquid or solution consider the relaxation as dominated by dipolar interactions from the surrounding solvent molecules. For $I = \frac{1}{2}$ nuclear spins with magnetic moments $g\mu_N I$ separated by the distance r, the relaxation times are

$$\frac{1}{T_1} = \frac{3}{10}\left(\frac{g^4\mu_N^2}{\hbar^2 r^6}\right)\left(\frac{\tau_c}{1 + \omega_0^2\tau_c^2} + \frac{4\tau_c}{1 + 4\omega_0^2\tau_c^2}\right) \tag{11-105}$$

$$\frac{1}{T_2} = \frac{3}{20}\left(\frac{g^4\mu_N^2}{\hbar^2 r^6}\right)\left(3\tau_c + \frac{5\tau_c}{1+\omega_0^2\tau_c^2} + \frac{2\tau_c}{1+4\omega_0^2\tau_c^2}\right) \qquad (11\text{-}106)$$

where the correlation time was given in Eq. 11-103. These expressions were derived under the assumption that rotational effects dominate the relaxation. A more complete calculation would take into account translation. We see from these expressions for T_1 and T_2 that there are two limiting cases, namely,

$$T_1 \sim T_2 \quad \text{for } \omega_0\tau_c \ll 1 \qquad (11\text{-}107)$$
$$T_1 \gg T_2 \quad \text{for } \omega_0\tau_c \gg 1 \qquad (11\text{-}108)$$

Figure 11-17 shows T_1 and T_2 plotted against η/T for two frequencies. The data on this graph are for the NMR for protons in water for which $T_1 = T_2 = 3.6\,\text{sec}$ at room temperature. The addition of paramagnetic impurities of the water considerably reduces T_1. For example, $10^{-2}\,M/L$ of Mn^{2+} dissolved in the water reduces T_1 to $\sim 0.1\,\text{sec}$.

There are two principal types of resonant line in solids: those that are homogeneously broadened and those that are inhomogeneously broadened.[1,3,17] Homogenous broadening arises from dipolar interactions, the spin–lattice interaction, diffusion of spin excitation, and motional narrowing. The Brownian motion relaxation in liquids produces homogeneous broadening. Inhomogeneous broadening arises from anisotropic effects that are only partially averaged, unresolved hyperfine structure, and inhomogeneities in the magnetic field.

The linewidth of a homogeneously broadened line is inversely proportional to the spin–spin relaxation time through the following expressions:

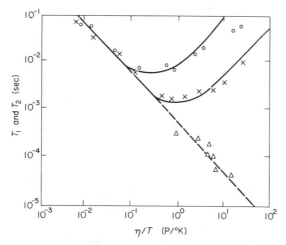

Fig. 11-17. Dependence of the spin–lattice T_1 and spin–spin T_2 relaxation times on the viscosity/temperature ratio η/T for glycerine at 29 and 4.8 MHz. $\bigcirc$, T_1 (29 MHz); $\times$, T_1 (4.8 MHz); $\triangle$, T_2 (29 MHz) (from Ref. 40).

TABLE 11-2 Typical Temperature and Field Dependence Formulas for the Spin–Lattice Relaxation Time T_1^a

Spin System	Crystal Field and Debye Energies	Two phonon or Raman Process Relaxation $(1/T_1)$	One Phonon or Direct Process Relaxation $(1/T_1)$
Kramers salt	$k\Theta_D > E_{CF}$	$a\exp(-E_{CF}/kT) + bT^9$	aH^4T
Kramers salt	$k\Theta_D < E_{CF}$	$\begin{cases} bT^9 \\ bH^2T^7 \end{cases}$	aH^4T
Non-Kramers salt	$k\Theta_D > E_{CF}$	$a\exp(-E_{CF}/kT) + bT^7$	$a\coth(E_{CF}/2kT)$
Non-Kramers salt	$k\Theta_D < E_{CF}$	bT^7	$a\coth(E_{CF}/2kT)$

^{a} Values given at the conditions such as temperature wherein each process is dominant (typically 3 K for direct and 60 K for Raman processes in rare-earth salts). The symbol E_{CF} denotes the crystal field energy. The parameters a and b differ for the various formulas (from Ref. 3).

$$T_2 = \frac{2}{\gamma \Delta H_{1/2}} = \frac{2}{\sqrt{3}\gamma \Delta H_{pp}} \qquad \text{Lorentzian} \qquad (11\text{-}109)$$

$$T_2 = \frac{2(\pi \ln 2)^{1/2}}{\gamma \Delta H_{1/2}} = \frac{(2\pi)^{1/2}}{\gamma \Delta H_{pp}} \qquad \text{Gaussian} \qquad (11\text{-}110)$$

for Lorentzian and Gaussian lines. In frequency units we have

$$\Delta \omega_{1/2} = \gamma \Delta H_{1/2} \qquad (11\text{-}111)$$
$$\Delta \omega_{pp} = \gamma \Delta H_{pp} \qquad (11\text{-}112)$$

As mentioned above, in solids ordinarily T_1 greatly exceeds T_2

$$T_1 \gg T_2 \qquad (11\text{-}113)$$

except when exchange effects predominate.

In many solids the principal mechanism of homogeneous broadening is the dipolar interaction of each spin with all neighboring spins, and this produces a Gaussian lineshape. If exchange interactions are present, they narrow the line in the center and make it close to Lorentzian, while it remains Gaussian in the wings. When exchange narrowing becomes pronounced the spin–lattice and spin–spin relaxation times become equal,

$$T_1 \sim T_2 \quad \text{(Extreme exchange narrowing)} \qquad (11\text{-}114)$$

The spin–lattice relaxation of solids arising from direct processes involves one lattice vibration phonon, and Raman processes involve two such phonons. The relaxation mechanism differs for a Kramers case with a half-integral spin and a non-Kramers case with an integral spin. Table 11-2 presents typical temperature dependencies obtained with both cases. For the first transition series the crystal field energies E_{CF} greatly exceed the Debye energy $k\Theta_D$, where Θ_D is the Debye temperature. For rare earths the Debye energy can often exceed the crystal field energies, and the result is called an Orbach process. Hence we have, in general,

$$k\Theta_D < E_{CF} \quad \text{First transition series} \qquad (11\text{-}115)$$
$$k\Theta_D > E_{CF} \quad \text{Rare earths (Orbach processes)} \qquad (11\text{-}116)$$

Table 11-2 gives the temperature dependence of $1/T_1$ for various typical cases.

11-11 MEASUREMENT OF RELAXATION TIMES

Two widely used methods of determining relaxation times are the saturation method and the pulse method. The saturation method, which is convenient to apply with standard ESR and NMR spectrometers, entails the recording of spectra at several power levels in the neighborhood of the onset of saturation. The pulse method, on the other hand, entails the excitation of the sample with pulses of radiofrequency energy and monitors the recovery or return to equilibrium after the pulse.

The saturation method[3,17] makes use of the following expression for the absorption lineshape $Y(H)$ and its first derivative counterpart $Y'(H)$ [compare Eqs. 11-49 and 11-51]

$$Y(H) = \frac{sH_1 y_m^0}{1 + s(H - H_0)^2 \gamma^2 T_2^2} \tag{11-117}$$

$$Y'(H) = \frac{16s^2(H - H_0)\gamma T_2 H_1 y_m^{0'}}{3^{3/2}[1 + s(H - H_0)^2\gamma^2 T_2^2]^2} \tag{11-118}$$

where s is the saturation factor defined by

$$s = (1 + H_1^2\gamma^2 T_1 T_2)^{-1} \tag{11-119}$$

Here y_m^0 and $y_m^{0'}$ are the respective amplitude factors of Y at $H = H_0$ and of Y' at $(H - H_0) = \pm\frac{1}{2}\Delta H_{pp}$, and the superscript "0" denotes values below saturation where $s \sim 1$.

In a typical case the square root of the power is proportional to H_1,

$$H_1 = k\sqrt{P} \tag{11-120}$$

and the amplitude y_m corresponding to Y at $H = H_0$ is given by

$$y_m = sH_1 y_m^0 = \frac{k\sqrt{P} y_m^0}{1 + kP\gamma^2 T_1 T_2} \tag{11-121}$$

with a similar expression for the amplitude y_m' at $(H - H_0) = \frac{1}{2}\Delta H_{pp}$

$$y_m' = s^{3/2} H_1 y_m^{0'} = \frac{k\sqrt{P} y_m^{0'}}{(1 + kP\gamma^2 T_1 T_2)^{3/2}} \tag{11-122}$$

The linewidth $\Delta H_{1/2}$ is related to its counterpart $\Delta H_{1/2}^0$ below saturation as follows:

$$\Delta H_{1/2} = \frac{\Delta H_{1/2}^0}{\sqrt{s}} = \Delta H_{1/2}^0(1 + kP\gamma^2 T_1 T_2)^{1/2} \tag{11-123}$$

and similarly for the peak-to-peak width

$$\Delta H_{pp} = \Delta H_{pp}^0(1 + kP\gamma^2 T_1 T_2)^{1/2} \tag{11-124}$$

Figure 11-18 shows a plot of y_m' versus $\sqrt{P}$ and Fig. 11-19 presents ΔH_{pp} plotted against $\sqrt{P}$. We see from the first of these figures that y_m' is linear in $\sqrt{P}$ below saturation, it reaches a maximum, and then falls off as $1/P$ at high powers. The maximum occurs at

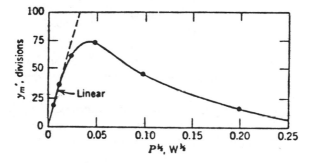

Fig. 11-18. Peak-to-peak first derivative amplitude y'_m plotted as a function of the square root of the microwave power P. The dashed line is an extrapolation of the linear dependence at low powers (from Ref. 17).

$$\frac{dy'_m}{dH_1} = 0 \tag{11-125}$$

corresponding to

$$s = \tfrac{2}{3} \tag{11-126}$$

As a result, we can determine T_2 from the linewidth below saturation,

$$T_2 = \frac{2}{\gamma \Delta H^0_{1/2}} = \frac{2}{\sqrt{3}\gamma \Delta H^0_{pp}} \tag{11-127}$$

and T_1 from the value of H_1 at the peak of the y'_m versus $\sqrt{P}$ plot

$$T_1 = \frac{\sqrt{3}\Delta H^0_{pp}}{\gamma(2H_1)^2} \qquad \text{NMR case} \tag{11-128}$$

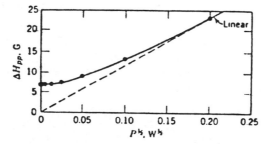

Fig. 11-19. Peak-to-peak linewidth H_{pp} plotted as a function of the square root of the microwave power P. The dashed line gives the linear asymptotic behavior at very high powers (from Ref. 17).

$$T_1 = \frac{1.97 \times 10^{-7} \Delta H_{pp}^0}{g(2H_1)^2} \quad \text{ESR case} \quad (11\text{-}129)$$

The simplest pulse method[4,17,41] of studying relaxation consists in saturating the spin system with a pulse of sufficiently high power that the amplitude y_m' is considerably beyond the maximum on the curve of Fig. 11-18. This pulse tends to equalize the populations of the spin levels, and the saturated spin system will return exponentially to its equilibrium condition with a time constant T_1. To measure this return the amplitude $y_m'(t)$ is measured by scanning through the line at various times during the recovery. The time dependence of this amplitude is given by

$$y_m'(t) = y_m(\text{equil}) - [y_m'(\text{equil}) - y_m'(\text{satur})]e^{-t/T_1} \quad (11\text{-}130)$$

where $y_m'(\text{satur})$ is the amplitude recorded immediately after the pulse $(t \sim 0)$ and $y_m'(\text{equil})$ is the equilibrium amplitude $(t \to \infty)$. A plot of $y_m'(t)$ against time gives the spin–lattice relaxation time T_1.

The pulse used for this type of saturation recovery measurement has a duration longer than the spin–spin relaxation time. This means that the spin system reaches an equilibrium state during the simultaneous application of the strong radiofrequency field H_1 and the applied field H_0.

Another type of pulse experiment called a spin echo experiment is carried out with a pulse width t_w that is short compared to the effective spin–spin relaxation time T_m. For a typical NMR experiment T_m arises from the intrinsic spin–spin relaxation time T_2 and the contribution T_2^* due to magnetic field inhomogeneities in accordance with the expression

$$\frac{1}{T_m} = \frac{1}{T_2} + \frac{1}{T_2^*} \quad (11\text{-}131)$$

The NMR linewidth in a continuous wave experiment is given by

$$\Delta H_{1/2} = \frac{2}{\gamma T_m} \quad (11\text{-}132)$$

and of course in the absence of inhomogeneous broadening this experiment gives T_2 directly.

To perform a spin echo experiment it is important to select the duration of the pulse short compared to the reciprocal of the linewidth $\Delta H_{1/2}$ in frequency units and long enough to exceed several radiofrequency (rf) periods,

$$\frac{2\pi}{\omega_0} < t_w < T_m \quad (11\text{-}133)$$

Pulse echo experiments rely on the selection of rf amplitudes H_1 and pulse widths t_w to produce particular pulse phase lengths such as a 90° pulse corresponding to

$$\gamma H_1 t_w = \tfrac{1}{2}\pi \tag{11-134}$$

and a 180° pulse given by

$$\gamma H_1 t_w = \pi \tag{11-135}$$

A 90° pulse will bend the magnetization from the z direction to the xy plane and a 180° pulse reverses its direction.

If a single 180° pulse is applied to an NMR sample at resonance, it will invert the direction of magnetization so that instead of pointing in the applied magnetic field direction $\vec{H}_0$ it points antiparallel to it. After achieving its alignment opposite to the field, M_z will return to its orientation along the field direction in accordance with the expression

$$M_z(t) = M_0(1 - 2e^{-t/T_1}) \tag{11-136}$$

In a spin echo measurement one uses two or more pulses of width t_w separated by a time interval τ subject to the condition

$$\frac{2\pi}{\omega_0} < t_w < T_m < T_2^* < \tau \ll T_2 \tag{11-137}$$

For example, an initial 90° pulse bends the magnetization into the xy plane, and then a subsequent series of 180° pulses are used to monitor the rate at which it loses coherence in this plane, as will be explained in Section 19-6.

REFERENCES

1. A. Abragam, *Principles of Nuclear Magnetism*, Oxford University Press, London, 1961, Chap. 10.

2. A. Abragam and M. Goldman, *Nuclear Magnetism: Order and Disorder*, Oxford University Press, London, 1982.

3. C. P. Poole, Jr. and H. A. Farach, *Relaxation in Magnetic Resonance*, Academic Press, New York, 1971.

4. C. P. Poole, Jr. and H. A. Farach, *Bull. Magn. Reson.*, **1**, 162 (1980).

5. P. W. Anderson, *J. Phys. Soc. Japan*, **9**, 316 (1954).

6. H. A. Farach, E. F. Strother, and C. Poole, Jr., *J. Phys. Chem. Solids*, **31**, 1491 (1970).

7. H. S. Gutowsky, D. W. McCall, and C. P. Slichter, *J. Chem. Phys.*, **21**, 279 (1953).

8. H. S. Gutowsky and A. Saika, *J. Chem. Phys.*, **21**, 1688 (1953).

9. H. S. Gutowsky and C. H. Holm, *J. Chem. Phys.*, **25**, 1228 (1956).

10. H. M. McConnell, *J. Chem. Phys.*, **28**, 430 (1958).

11. J. I. Kaplan and G. Fraenkel, *NMR in Chemically Exchanging Systems*, Academic Press, New York, 1980.

12. Y. N. Molin, K. M. Salikhov, and K. I. Zamaraev, *Spin Exchange: Principles and Applications in Chemistry & Biology*, Springer-Verlag, Berlin, 1980.

13. L. T. Muus and P. W. Atkins, eds., *Electron Spin Relaxation in Liquids*, Plenum, New York, 1972.

14. J. W. Emsley, J. Feeney, and L. H. Sutcliffe, *High Resolution Nuclear Magnetic Resonance Spectroscopy*, Pergamon, New York, 1965.

15. R. Harris and B. Mann, eds., *NMR and the Periodic Table*, Academic Press, New York, 1978.

16. P. Laszlo, ed., *NMR of Newly Accessible Nuclei*, Vol. 1: *Chemical and Biochemical Applications*, Vol. 2: *Chemically and Biochemically Important Elements*, Academic Press, New York, 1983.

17. C. P. Poole, Jr., *Electron Spin Resonance*, Wiley, New York, 1967, especially Chap. 20; 2nd ed., 1983, Chap. 12.

18. M. W. Zemansky, *Phys. Rev.*, **36**, 219 (1930).

19. H. A. Farach and H. Teitelbaum, *Can. J. Phys.*, **45**, 2913 (1967).

20. T. G. Castner, *Phys. Rev.*, **115**, 1506 (1959); **130**, 58 (1963).

21. B. Bleaney, *Philos. Mag.*, **42**, 441 (1951).

22. R. H. Sands, *Phys. Rev.*, **99**, 1222 (1955).

23. J. A. Ibers and J. D. Swalen, *Phys. Rev.*, **127**, 1914 (1962).

24. D. E. O'Reilly, *J. Chem. Phys.*, **29**, 1188 (1958).

25. T. S. Johnston and H. G. Hecht, *J. Mol. Spectrosc.*, **17**, 98 (1965).

26. C. Kittel and E. Abrahams, *Phys. Rev.*, **90**, 238 (1953).

27. P. Swarup, *Can J. Phys.*, **37**, 848 (1959).

28. D. Wolf, *Spin Temperature and Nuclear Spin Relaxation in Matter: Basic Principles and Applications*, Oxford University Press, London, 1979.

29. L. T. Muus and P. W. Atkins, eds., *Electron Spin Relaxation in Liquids*, Plenum, New York, 1972.

30. L. Banci, I. Bertini, and C. Luchinat, *Magn. Reson. Rev.*, **11**, No. 1, 1986.

31. P. W. Atkins, *Adv. M. Relaxation Processes*, **2**, 121 (1972).

32. M. Kolz and M. D. Zeidler, *Nucl. Magn. Reson. Spec. Perodical Rep.*, **5**, 92 (1977).

33. R. A. Vaughan, *Magn. Reson. Rev.*, **4**, 58 (1975).

34. K. J. Standley and R. A. Vaughan, *Electron Spin Relaxation Phenomena in Solids*, Plenum, New York, 1970.

35. A. G. Redfield, *Adv. Magn. Reson.*, **1**, 1 (1966).

36. R. G. Gordon, *Adv. Magn. Reson.*, **3**, 1 (1968).

37. L. G. Werbelow and D. M. Grant, *Adv. Magn. Reson.*, **9**, 190 (1977).

38. D. Kivelson and K. Ogan, *Adv. Magn. Reson*, **7**, 72 (1974).

39. C. P. Slichter, *Principles of Magnetic Resonance*, Springer-Verlag, Berlin, 1980.

40. N. Bloembergan, E. M. Purcell, and R. V. Pound, *Phys. Rev.*, **73**, 679 (1948).

41. E. Fukurshima and S. B. W. Roeder, *Experimental Pulse NMR: A Nuts and Bolts Approach*, Addison-Wesley, Reading, MA, 1981.

12

DOUBLE RESONANCE

12-1 INTRODUCTION

In Chapter 3 we treated the general spin $(\frac{1}{2}, \frac{1}{2})$ case and in Chapters 4 and 5, respectively, we discussed the NMR and ESR formulations of it. In this and the following four chapters we are concerned with double resonance experiments carried out with the same spin $(\frac{1}{2}, \frac{1}{2})$ system. More specifically, we treat electron–nuclear double resonance (ENDOR) and dynamic nuclear polarization (DNP) which involve one ESR and one NMR transition, electron–electron double resonance (ELDOR) which involves the concurrent excitation of two ESR transitions, and nuclear–nuclear double resonance in which both transitions are of the NMR type. In these experiments one transition is irradiated at a high power level by what is called the pumping power and this causes the populations of the two corresponding energy levels to become closer to equal, a process called saturation. Then another transition is measured at a low power level by what is referred to as the monitoring or observe frequency. The basic principle behind such double resonance experiments is the fact that the high-power pumping of the first pair of levels influences the amplitude of the signal that is observed by the low-power measurement of the second pair. In some double resonance schemes such as DNP the monitored transition is greatly enhanced in amplitude while in others such as ENDOR and spin decoupling the resolution is considerably improved. Double resonance schemes such as ENDOR and spin tickling are often employed to identify the nuclei that are responsible for producing multiplet structure.

In the next two sections we describe the Hamiltonian and the energy level configurations involved in double resonance, then we discuss the relaxation rates that are responsible for the transitions, and finally we proceed to analyze the dynamic aspects of changing energy level populations during varying degrees of saturation which are required for the explanation of the observed spectra.

Most of the double resonance experiments to be described depend on the establishment of non-Boltzmann population differences between the energy levels by the action of the pumping power, and the influence of these differences on the observed transition. However, other schemes such as spin decoupling and spin tickling in NMR do not result from the population changes produced by the pumping power. Instead, they take the pumping power into account by adding to the spin Hamiltonian a time-dependent Zeeman term of the type $g\beta H_1 \cos \omega t$ and transforming to a rotating coordinate system to calculate the eigenvalues and transition probabilities.

In the previous chapters the symbol T was employed for the hyperfine coupling term. In this chapter and the remainder of the book we employ the symbol A for this purpose. This will avoid confusion with the use of T for the temperature and for relaxation times.

12-2 TYPES OF DOUBLE RESONANCE

This section provides a brief description of the principal types of double resonance experiment which we will examine in the next four chapters.

Electron–nuclear double resonance, which is treated in the next chapter, is a method that may be employed to greatly increase the resolution of hyperfine structure. It consists in scanning through the NMR transitions with a high radiofrequency pump power and monitoring the effect of this power on an ESR absorption line. A change occurs in the ESR amplitude when the radiofrequency scan passes through resonance. This technique can be employed to distinguish spectral lines that are unresolved in ordinary ESR spectra.

Electron–electron double resonance, to be discussed in Chapter 14, is a useful technique for studying cross-relaxation and other relaxation processes between the levels of hyperfine multiplets. It consists in pumping one electron spin Zeeman level and observing the effect of this pumping on another ESR transition.

Dynamic nuclear polarization, which is the topic for Chapter 15, is a technique that can provide NMR spectral lines with the intensity of ESR transitions. The result can be an enhancement of the NMR signal by as much as three orders of magnitude. This is accomplished by pumping the ESR transition to establish electron spin Boltzmann population differences between the nuclear spin levels, a process called polarizing the sample. A scan through the NMR transition provides the enhanced signal which can occur either in absorption or in emission. This method can provide NMR data on samples too small to be detected by ordinary NMR.

Nuclear–nuclear double resonance experiments, which are described in Chapter 16, are very helpful in establishing the identity of the nuclei that contribute to complex multiplet structures, and in the determination of spin–spin coupling constants. The technique consists in pumping one NMR frequency and observing the effect on another NMR line. Under nuclear Overhauser conditions the pumped nucleus changes the intensities of the nuclei coupled to it and sometimes the spectra indicate how close together the nuclei are. Under spin decoupling conditions the contribution of the pumped nucleus to the spin–spin multiplet of another nucleus is eliminated and the resulting decoupled multiplet spectrum simplifies. The result can be a very pronounced gain in resolution. Under spin tickling conditions pumping one NMR transition causes the multiplet lines of another nucleus to double, thereby identifying their origin. These double resonance methods are routinely employed in NMR.

12-3 HAMILTONIAN AND ENERGY LEVELS

Since most of the double resonance experiments to be described in this book involve an electron with spin $S = \frac{1}{2}$ interacting with a nucleus of spin $I = \frac{1}{2}$, we describe the techniques in terms of the isotropic spin Hamiltonian.

$$\mathcal{H} = g\beta H S_z - g_N \beta_N H I_z + A\vec{S} \cdot \vec{I} \tag{12-1}$$

which gives the energies listed in Section 5-2. In the high-field approximation where the hyperfine term $A\vec{S} \cdot \vec{I}$ can be approximated by $AS_z I_z$ these energies are all of the form

$$E = g\beta H m_S - g_N \beta_N H m_I + A m_S m_I \tag{12-2}$$

as indicated on Fig. 12-1. This figure is drawn for the case of the hyperfine term $A\vec{S} \cdot \vec{I}$ being larger than the nuclear Zeeman term $g_N \beta_N H m_I$. It shows the m_S and m_I values for each level, the numbering scheme of 1, 2, 3, 4 from top to bottom and the six transitions—two allowed ESR ones, two forbidden ESR ones, and two NMR ones. This energy level diagram forms the basis for describing the various double resonance schemes in this and the next few chapters. In particular, we always associate the level numbers 1, 2, 3, 4 with the m_S, m_I values in the sequence $++$, $+-$, $--$, $-+$ irrespective of the actual ordering of the levels.

12-4 RELAXATION RATES

Most of the double resonance experiments to be described are explained in terms of the six relaxation rates or transition probabilities between the levels shown in Fig. 12-2.[1,2] The transition probability W_i for each level pair shown on the figure is labeled with a parameter M which indicates the change in the total magnetic quantum number $M = m_S + m_I$ which takes place during the transition. There are electronic transition probabilities W_1' which involve changes only in the electron's quantum number m_S, nuclear ones W_1 which

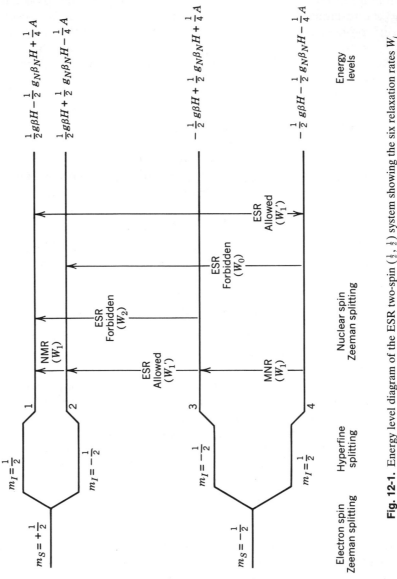

Fig. 12-1. Energy level diagram of the ESR two-spin ($\frac{1}{2}$, $\frac{1}{2}$) system showing the six relaxation rates W_i and the four energies.

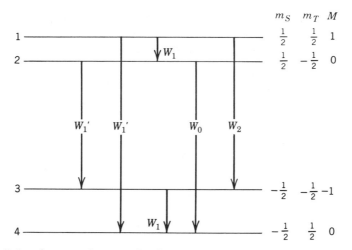

Fig. 12-2. Relaxation rates between the six pairs of levels of the $S = \frac{1}{2}$, $I = \frac{1}{2}$ system. The quantum numbers are given for each level.

involve changes only in the nuclear quantum number m_I, and mixed or forbidden transitions W_0 and W_2 in which both m_S and m_I change.

In Section 3-5 we mentioned that the transition probability, which we denote by the symbol W_{if}, is proportional to the square of the matrix element

$$W_{if} \propto |\langle f|V|i\rangle|^2 \tag{12-3}$$

between the initial and final states of the interaction V that is responsible for the relaxation. For example, V may arise from dipolar interactions, orbit–lattice interactions, Brownian motion, phonon scattering, and so on. In the next section we will evaluate this matrix element for the dipolar interaction.

In the next few chapters we will examine double resonance experiments involving different values of W_i for the various pairs of levels, and in addition we will consider several choices of level pairs for pumping and monitoring, with an emphasis on the spin ($\frac{1}{2}$, $\frac{1}{2}$) system.

12-5 DIPOLAR RELAXATION

Some relaxation processes are mediated by the dipolar Hamiltonian, and in this section we show how relaxation transitions can be induced by this process.

The dipole–dipole interaction between an electronic and a nuclear spin has the form

$$\mathcal{H}_{DD} = \left(\frac{\mu_0}{4\pi}\right)\hbar^2\gamma_S\gamma_I\left[\frac{\vec{S}\cdot\vec{I}}{r^3} - 3\frac{(\vec{S}\cdot\vec{r})(\vec{I}\cdot\vec{r})}{r^5}\right] \tag{12-4}$$

where the factor $\mu_0/4\pi$ which applies for mks units is omitted in the case of cgs units. In spherical coordinates ($x = r \sin\theta \cos\phi$, $y = r \sin\theta \sin\phi$, $z = r \cos\theta$) this may be written as the sum of six terms[1,3]

$$H_{DD} = \frac{K}{r^3}[A + B + C + D + E + F] \qquad (12\text{-}5)$$

where

$$K = \frac{\mu_0}{4\pi} \sqrt{\frac{\pi}{10}}\, \hbar^2 \gamma_S \gamma_I \qquad (12\text{-}6)$$

$$A = 4\sqrt{2}S_z I_z Y_{20}$$

$$B = \sqrt{2}(S^+ I^- + S^- I^+)Y_{20}$$

$$C = -2\sqrt{3}(S^+ I_z + S_z I^+)Y_{2-1}$$

$$D = 2\sqrt{3}(S^- I_z + S_z I^-)Y_{21} \qquad (12\text{-}7)$$

$$E = -2\sqrt{3}S^+ I^+ Y_{2-2}$$

$$F = -2\sqrt{3}S^- I^- Y_{22}$$

using definitions 2-52 and the usual expressions for the second-order ($l = 2$) spherical harmonics.

$$Y_{20}(\theta, \phi) = \frac{1}{4}\sqrt{\frac{5}{\pi}}(3\cos^2\theta - 1)$$

$$Y_{21}(\theta, \phi) = -\frac{1}{2}\sqrt{\frac{15}{2\pi}}\sin\theta \cos\theta\, e^{i\phi} = -Y^*_{2-1}(\theta, \phi) \qquad (12\text{-}8)$$

$$Y_{22}(\theta, \phi) = \frac{1}{4}\sqrt{\frac{15}{2\pi}}\sin^2\theta\, e^{2i\phi} = Y^*_{2-2}(\theta, \phi)$$

The spin part $\langle m_S m_I | H_{DD} | m'_S m'_I \rangle$ of the matrix elements $\langle \psi | H_{DD} | \psi' \rangle$ for the various transitions may be evaluated with aid of Eqs. 2-50 and 2-53, noting that the matrix elements of S_z and I_z add a factor of $\frac{1}{2}$ compared to those of S^+, S^-, I^+, and I^-. For example, when $\Delta M = +1$, term C has the only nonzero matrix elements as follows:

$$\langle \tfrac{1}{2}\,\tfrac{1}{2} | \mathscr{H}_{DD} | \tfrac{1}{2} -\tfrac{1}{2} \rangle = \langle \tfrac{1}{2}\,\tfrac{1}{2} | \mathscr{H}_{DD} | -\tfrac{1}{2}\,\tfrac{1}{2} \rangle = -\langle \tfrac{1}{2} -\tfrac{1}{2} | \mathscr{H}_{DD} | -\tfrac{1}{2} -\tfrac{1}{2} \rangle$$

$$= -\langle -\tfrac{1}{2}\,\tfrac{1}{2} | \mathscr{H}_{DD} | -\tfrac{1}{2} -\tfrac{1}{2} \rangle$$

$$= -\sqrt{3}K\,\frac{Y_{2-1}(\theta_{(t)}, \phi_{(t)})}{r^3} \qquad (12\text{-}9)$$

After evaluating all the matrix elements they may be gathered together in the following transition probability matrix:

	$\lvert + +\rangle$	$\lvert + -\rangle$	$\lvert - +\rangle$	$\lvert - -\rangle$
$\langle + +\rvert$	—	$-\dfrac{\sqrt{3}KY_{2-1}}{r^3}$	$-\dfrac{\sqrt{3}KY_{2-1}}{r^3}$	$-\dfrac{2\sqrt{3}KY_{2-2}}{r^3}$
$\langle + -\rvert$	$\dfrac{\sqrt{3}KY_{21}}{r^2}$	—	$\dfrac{\sqrt{2}KY_{20}}{r^3}$	$\dfrac{\sqrt{3}KY_{2-1}}{r^3}$
$\langle - +\rvert$	$\dfrac{\sqrt{3}KY_{21}}{r^2}$	$\dfrac{\sqrt{2}KY_{20}}{r^3}$	—	$\dfrac{\sqrt{3}KY_{2-1}}{r^3}$
$\langle - -\rvert$	$-\dfrac{2\sqrt{3}KY_{22}}{r^3}$	$-\dfrac{\sqrt{3}KY_{21}}{r^3}$	$-\dfrac{\sqrt{3}KY_{21}}{r^3}$	—

$$(12\text{-}10)$$

In this matrix the term B of Eq. 12-7 induces the $\Delta M = 0$ transitions, C and D induce $\Delta M = +1$ and -1, respectively, and E and F induce $\Delta M = \pm 2$ transitions. The first term A is diagonal with $\Delta M = 0$ and so it does not induce any transitions.

For rotational relaxation θ and ϕ are time dependent as the molecules tumble and for translational relaxation r is also time dependent. For the rotational case the functions $Y_{2m}(t)$ vary randomly with the time and may be treated in terms of correlation function[1] $G(\tau)$

$$G(\tau) = \overline{Y_{2m}(t) Y^*_{2m}(t + \tau)} \tag{12-11}$$

with the value

$$G(0) = \overline{Y_{2m}(t)^2} \tag{12-12}$$

for $\tau = 0$, where the average is taken over the time. The relaxation rate is expressed in terms of the spectral density

$$J(\omega) = \int_{-\infty}^{\infty} G(\tau) e^{i\omega\tau}\, d\tau \tag{12-13}$$

which is the Fourier transform of the correlation function.

One often assumes an exponentially decaying correlation function

$$G(\tau) = G(0) e^{-\tau/\tau_c} \tag{12-14}$$

where τ_c is the correlation time. Inserting this expression into Eq. 12-13 we obtain

$$J(\omega) = G(0) \int_{-\infty}^{\infty} e^{-\tau/\tau_c} e^{i\omega\tau} \, d\tau \tag{12-15}$$

$$= 2G(0) \int_{0}^{\infty} e^{-\tau/\tau_c} \cos \omega\tau \, d\tau$$

$$= \frac{2\tau_c G(0)}{1 + \omega^2 \tau_c^2} \tag{12-16}$$

which has the approximate frequency-independent value

$$J(\omega) = 2\tau_c G(0) \tag{12-17}$$

in the high-temperature limit $\omega\tau_c \ll 1$ of short correlation times. This approximation also applies to other spectral density functions such as $J(2\omega)$ which contains the factor $(1 + 4\omega^2 \tau_c^2)^{-1}$.

Dipolar-induced relaxation rates W_M for particular ΔM transitions are proportional to the time average of the matrix element squared,

$$W_M = \overline{|\langle m_S m_I | H_{DD} | m_S' m_I' \rangle|^2}$$

$$= \alpha^2 \overline{\langle |Y_{LM}| \rangle^2} \left\langle \frac{1}{r^3} \right\rangle \tag{12-18}$$

where α is the coefficient that appears in the transition probability matrix 12-10 for each transition. In the high-temperature limit for rotational relaxation arising from random tumbling of the molecules r does not change its length with time and all spherical harmonics average to unity, which gives

$$W_0 = 2\alpha^2 \left\langle \frac{1}{r^3} \right\rangle \qquad \Delta M = 0$$

$$W_1 = 3\alpha^2 \left\langle \frac{1}{r^3} \right\rangle \qquad \Delta M = \pm 1$$

$$\tag{12-19}$$

$$W_1' = 3\alpha^2 \left\langle \frac{1}{r^3} \right\rangle \qquad \Delta M = \pm 1$$

$$W_2 = 12\alpha^2 \left\langle \frac{1}{r^3} \right\rangle \qquad \Delta M = \pm 2$$

We have seen that the dipolar interaction is capable of inducing transitions involving all six of these relaxation paths, as Eqs. 12-19 indicate. The scalar spin–spin interaction $J\vec{S} \cdot \vec{I}$, which can be written in the form

$$J\vec{S} \cdot \vec{I} = \tfrac{1}{2} J[S^+ I^- + S^- I^+] + J S_z I_z \tag{12-20}$$

has the A and B terms of Eq. 12-7 and hence can only induce transitions of the type W_0, as explained in the previous section.

12-6 THERMAL EQUILIBRIUM POPULATIONS

In this section we consider many identical atoms or nuclei with spin $\frac{1}{2}$ located in a strong magnetic field of strength H. We assume that the $m = +\frac{1}{2}$ level lies above the $m = -\frac{1}{2}$ level as indicated on Fig. 12-3. If we wait long enough for thermal equilibrium to be achieved then the number of atoms N_{0+} in the upper energy level is related to the number N_{0-} in the lower level through the Boltzmann factor[1,3]

$$N_{0+} = N_{0-} \exp\left(-\frac{g\beta H}{kT}\right) \tag{12-21}$$

where k is Boltzmann's constant and T is the absolute temperature. It is helpful to write this expression in the form

$$N_{0+} = N_{0-} \exp(-2\epsilon) \tag{12-22}$$

where

$$\epsilon = \frac{g\beta H}{2kT} \tag{12-23}$$

The average population N_{AV} of the two levels is

$$N_{AV} = \tfrac{1}{2}(N_{0+} + N_{0-}) \tag{12-24}$$

At typical experimental temperatures the energy level splitting $g\beta H$ is very small compared to the thermal energy kT so the populations of the two energy levels are quite close to each other. As a result, we find it convenient to talk in terms of populations n_{0j} normalized relative to the average population N_{AV}, and accordingly we write

$$n_{0+} = \frac{N_{0+}}{N_{AV}} \approx 1 - \epsilon \tag{12-25}$$

where use is made of the approximation

$$\exp(\pm\epsilon) \approx 1 \pm \epsilon \tag{12-26}$$

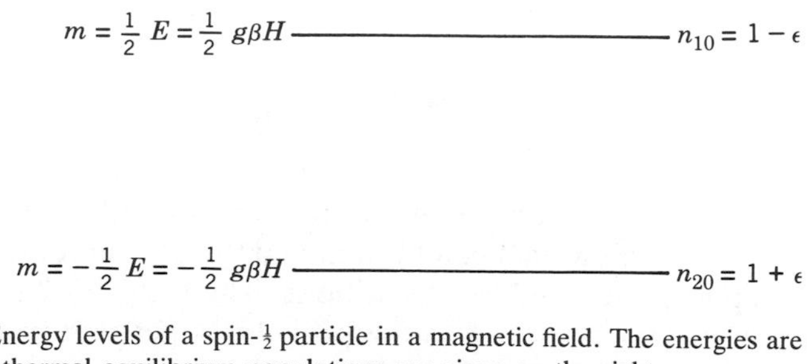

$$m = \tfrac{1}{2} \quad E = \tfrac{1}{2} g\beta H \text{————————————} n_{10} = 1 - \epsilon$$

$$m = -\tfrac{1}{2} \quad E = -\tfrac{1}{2} g\beta H \text{————————————} n_{20} = 1 + \epsilon$$

Fig. 12-3. Energy levels of a spin-$\frac{1}{2}$ particle in a magnetic field. The energies are shown on the left and the thermal equilibrium populations are given on the right.

that is valid for small ϵ. In like manner, we can write

$$n_{0-} \approx 1 + \epsilon \tag{12-27}$$

The difference in population n_0 between the two levels at thermal equilibrium is

$$n_0 = n_{0-} - n_{0+} \approx 2\epsilon \tag{12-28}$$

The population difference n when the two populations n_+ and n_- do not have their thermal equilibrium values is

$$n = n_- - n_+ \tag{12-29}$$

and in typical cases $n(t)$ is a function of the time. We are concerned with the manner in which applied radiofrequency power causes $n(t)$ to deviate from n_0 while relaxation processes act to restore the thermal equilibrium condition, as is explained in the next section.

12-7 DYNAMIC EQUILIBRIUM POPULATIONS

In the previous section we discussed the populations of a pair of energy levels in thermal equilibrium. In this section we examine these populations in the presence of applied pumping power.

When a magnetic resonance experiment is performed there are four main factors that determine the population difference n at any particular time: (1) the Boltzmann factor $\exp(-g\beta H/kT)$, (2) the magnitude of the applied radiofrequency power P, (3) the relaxation rate W, and (4) the initial populations. For the two-level system the relaxation rate is inversely related to the spin–lattice relaxation time T_1 as follows:

$$T_1 = \frac{1}{2W} \tag{12-30}$$

This relaxation time is the time constant associated with the return of the spin system to equilibrium after a pulse causes it to deviate therefrom.

The response of the spin system to the application and removal of applied power can be expressed in terms of the following phenomenological differential equation:[1,3]

$$\frac{dn}{dt} = 2W(n_0 - n) - 2nP \tag{12-31}$$

The first term in this expression gives the rate of inducing relaxation transitions to reestablish thermal equilibrium and the second term is the rate at which the applied power induces transitions. The factor of 2 is placed in front of the power term because P is the probability per unit time that the radiofrequency field will induce a transition, and every transition changes n

by 2. The solution of this equation for the initial population difference equal to n_i at the time $t = 0$ is

$$n(t) = n_i - [n_i - n_0(1 + 2PT_1)^{-1}]\{1 - \exp[-2(W + P)t]\} \quad (12\text{-}32)$$

When the system is initially at equilibrium then $n_i = n_0$ and Eq. 12-32 simplifies to the form

$$n(t) = n_0(1 + 2PT_1)^{-1}\{1 + 2PT_1 \exp[-2W(1 + 2PT_1)t]\} \quad (12\text{-}33)$$

where, with the aid of Eq. 12-30, we replaced $(W + P)$ by $W(1 + 2PT_1)$. From this expression we see that in the presence of a constant power the population difference relaxes in an exponential manner to a final dynamic equilibrium or steady-state value n_s

$$n_s = \frac{n_0}{1 + 2PT_1} \quad (12\text{-}34)$$

This expression can be obtained by either letting time go to infinity in Eq. 12-33 or setting dn/dt equal to zero in Eq. 12-31. In later chapters we will find it convenient to adopt the notation

$$n_{s+} = 1 - \delta$$
$$n_{s-} = 1 + \delta \quad (12\text{-}35)$$
$$n_s = 2\delta$$

where δ plays a role analogous to that of ϵ for the case of pumped levels. Inserting Eqs. 12-28 and 12-35 into Eq. 12-34 gives

$$\delta = \frac{\epsilon}{1 + 2PT_1} \quad (12\text{-}36)$$

which shows that δ has the range

$$0 \le \delta \le \epsilon \quad (12\text{-}37)$$

The quantity δ is close to ϵ for a mild amount of saturation and it approaches zero for strong saturation. Equations 12-34 through 12-37 will be found useful in later chapters. Equation 12-49 gives a more general definition of δ.

If the spin system is in a nonequilibrium state with the initial population n_i and there is no applied power then Eq. 12-32 reduces to the form

$$n(t) = n_0 + (n_i - n_0) \exp\left(-\frac{t}{T_1}\right) \quad (12\text{-}38)$$

corresponding to an exponential return to equilibrium with the time constant T_1. A typical time dependence of $n(t)$ is illustrated in Fig. 12-4a for the case $n_i < n_0$ and in Fig. 12-4b for the case $n_i > n_0$. In both cases the system returns to equilibrium with the time constant T_1.

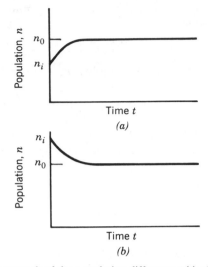

Fig. 12-4. Exponential approach of the population difference $n(t)$ of a spin level from an initial nonequilibrium value n_i to a final equilibrium value n_0 (a) when $n_i < n_0$ and (b) when $n_i > n_0$.

The time constant T_1^P associated with a spin system in the presence of pumping power as it relaxes toward a dynamic equilibrium state is obtained from the exponential term of Eq. 12-33:

$$T_1^P = \frac{1}{2(W + P)}$$
$$= \frac{T_1}{1 + 2PT_1}$$

(12-39)

This is shorter than the relaxation time T_1 in the absence of any applied radiofrequency power. These expressions are illustrated in Fig. 12-5 where we show how the application of radiofrequency power P at the time t_1 causes the population difference $n(t)$ to decrease from its initial equilibrium value n_0 to the value n_s with the shorter time constant T_1^P of Eq. 12-39. This figure shows how the removal of the power at the later time t_2 causes the spin system to relax back to its initial equilibrium value n_0 with the longer time constant T_1.

We see from Eq. 12-34 that the final steady-state difference in population n_s can become quite small at high power levels, and when the product PT_1 is much greater than 1 then n_s is inversely proportional to the power P. If the spin system is pumped to produce saturation and then the power is turned off the population difference $n(t)$ will return exponentially to its equilibrium value n_0 with the time constant T_1 in accordance with Eq. 12-38. This can be monitored by scanning through the resonance at various times using a low power. The ratio $Y(t)/Y_0$ of the amplitude $Y(t)$ of the resulting spectrum to the final equilibrium amplitude Y_0 is equal to the population ratio $n(t)/n_0$.

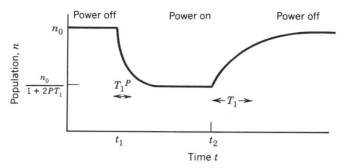

Fig. 12-5. Population n of a spin-$\frac{1}{2}$ energy level which starts at its equilibrium value n_0, and at the time t_1 is subjected to a radiofrequency power which reduces the population to the value $n_0/(1+2PT_1)$ with the short time constant T_1^P. At the later time t_2 the power is removed and the spin system returns to its equilibrium population n_0 with the longer time constant T_1.

12-8 DYNAMIC EQUILIBRIUM OF A FOUR-LEVEL SYSTEM

The previous two sections dealt with the dynamics of the populations of a pair of energy levels both in the absence and in the presence of pumping power. Since in the subsequent chapters we will deal mainly with the spin $(\frac{1}{2}, \frac{1}{2})$ four-level system presented in Fig. 12-1, it is instructive to examine the dynamics of its level populations both in the absence and in the presence of pumping power. This system has the six relaxation paths shown in Fig. 12-6 that were described in Section 12-4. Each level in the figure is labeled with its thermal equilibrium population in terms of the three Boltzmann factors for the electronic g-factor, for the nuclear g-factor, and for the hyperfine interaction, respectively:

$$\epsilon = \frac{g\beta H}{2kT}$$

$$\epsilon_N = \frac{g_N\beta_N H}{2kT} \tag{12-40}$$

$$\epsilon_A = \frac{A}{2kT}$$

When a pump power is applied the dynamics of the changing level populations are governed by the master equation,[2,4] and the four such equations for the system depicted in Fig. 12-6 in which the $2\rightarrow 1$ transition is pumped with high-power radiowaves are as follows:[4]

$$\frac{dn_1}{dt} = W_1[(n_2 - n_{20}) - (n_1 - n_{10})] + W_2[(n_3 - n_{30}) - (n_1 - n_{10})]$$

$$+ W_1'[(n_4 - n_{40}) - (n_1 - n_{10})] - P(n_1 - n_2) \tag{12-41a}$$

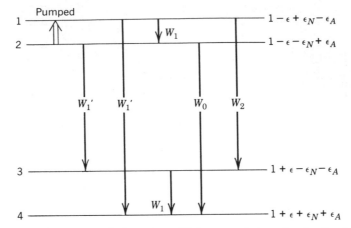

Fig. 12-6. Relaxation rates and thermal equilibrium populations for the $S = \frac{1}{2}$, $I = \frac{1}{2}$ spin system. Applying the pump power to the $2 \rightarrow 1$ transition alters these populations.

$$\frac{dn_2}{dt} = W_1[(n_1 - n_{10}) - (n_2 - n_{20})] + W_1'[(n_3 - n_{30}) - (n_2 - n_{20})]$$
$$+ W_0[(n_4 - n_{40}) - (n_2 - n_{20})] - P(n_2 - n_1) \tag{12-41b}$$

$$\frac{dn_3}{dt} = W_2[(n_1 - n_{10}) - (n_3 - n_{30})] + W_1'[(n_2 - n_{20}) - (n_3 - n_{30})]$$
$$+ W_1[(n_4 - n_{40}) - (n_3 - n_{30})] \tag{12-41c}$$

$$\frac{dn_4}{dt} = W_1'[(n_1 - n_{10}) - (n_4 - n_{40})] + W_0[(n_2 - n_{20}) - (n_4 - n_{40})]$$
$$+ W_1[(n_3 - n_{30}) - (n_4 - n_{40})] \tag{12-41d}$$

These equations can be solved to determine how the populations change with time. The corresponding equations for other pumping conditions are easily written down by analogy.

Ordinarily, we are interested in the steady-state solution to the above set of equations, and this is obtained by setting the time derivatives of Eqs. 12-41 equal to zero. The result is the following expressions:

$$(W_1 + P)(n_2 - n_1) + W_2(n_3 - n_1) + W_1'(n_4 - n_1)$$
$$= 2\epsilon(W_1' + W_2) + 2\epsilon_A(W_1 + W_1') - 2\epsilon_N(W_1 + W_2)$$

$$- (W_1 + P)(n_2 - n_1) + W_1'(n_3 - n_2) + W_0(n_4 - n_2) \tag{12-42}$$
$$= 2\epsilon(W_0 + W_1') - 2\epsilon_A(W_1 + W_1') + 2\epsilon_N(W_0 + W_1)$$

$$W_2(n_3 - n_1) + W_1'(n_3 - n_2) - W_1(n_4 - n_3)$$
$$= 2\epsilon(W_1' + W_2) - 2\epsilon_A(W_1 + W_1') - 2\epsilon_N(W_1 + W_2)$$

$$W_1'(n_4 - n_1) + W_0(n_4 - n_2) + W_1(n_4 - n_3)$$
$$= 2\epsilon(W_0 + W_1') + 2\epsilon_A(W_1 + W_1') + 2\epsilon_N(W_0 + W_1) \quad \text{(12-42 cont.)}$$

where we have inserted the thermal equilibrium populations from Fig. 12-7. The normalization condition for a four-level system

$$n_1 + n_2 + n_3 + n_4 = 4 \qquad (12\text{-}43)$$

may be added to this set of equations.

To gain some insight into the nature of the solutions to Eqs. 12-42 we consider the simpler, somewhat unrealistic case in which the same level pair is pumped but only the two W_1' and the upper W_1 relaxation paths are operative, as indicated in Fig. 12-7. The corresponding master equations are

$$(W_1 + P)(n_2 - n_1) + W_1'(n_4 - n_1)$$
$$= 2\epsilon W_1' + 2\epsilon_A(W_1 + W_1') - 2\epsilon_N W_1 \qquad (12\text{-}44a)$$

$$-(W_1 + P)(n_2 - n_1) + W_1'(n_3 - n_2)$$
$$= 2\epsilon W_1' - 2\epsilon_A(W_1 \mid W_1') + 2\epsilon_N W_1 \qquad (12\text{-}44b)$$

$$n_3 - n_2 = 2(\epsilon - \epsilon_A) \qquad (12\text{-}44c)$$
$$n_4 - n_1 = 2(\epsilon + \epsilon_A) \qquad (12\text{-}44d)$$

The third and fourth equations have immediate solutions which can be inserted into the first or second equation to give the expression

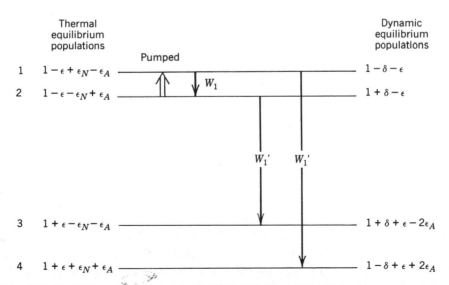

Fig. 12.7. Thermal equilibrium populations (left) and dynamic equilibrium populations (right) when the 2→1 transition is pumped and three relaxation processes are operative.

$$n_2 - n_1 = \frac{2(\epsilon_A - \epsilon_N)}{1 + 2PT_R} \tag{12-45}$$

$$= 2\delta \tag{12-46}$$

where

$$\delta = \frac{\epsilon_A - \epsilon_N}{1 + 2PT_R} \tag{12-47}$$

and we have employed the notation

$$W_1 = \frac{1}{2T_R} \tag{12-48}$$

in which T_R denotes the relaxation time of the pumped transition alone. It is analogous to the quantity T_1 of Eq. 12-30. This provides us with three equations (12-44c, 12-44d, and 12-46), and if we add the normalization condition (12-43) we have four equations for the four unknown populations n_1, n_2, n_3, and n_4. These are easily solved to give the dynamic steady-state populations listed in Fig. 12-7. Figure 12-8 gives the population differences for the five nonpumped transitions both before and during the pumping. We see from these population differences, which are proportional to the intensities of the corresponding observed spectral lines, that the two allowed transitons $3 \rightarrow 2$ and $4 \rightarrow 1$ are unchanged by the pumping. The forbidden transitions $3 \rightarrow 1$ and $4 \rightarrow 2$, on the other hand, have intensities that are functions of the pumping factor δ and hence are changed by the pumping.

12-9 STEADY-STATE POPULATIONS

The example that has been discussed at the end of the last section provided us with a set of equations and solutions from which we can infer some general conclusions about the steady-state populations of pairs of energy levels of systems being pumped. These conclusions are as follows:

(*a*) If one level of a pumped pair has no other connection to another level then its population difference is the pumped factor 2δ. An example of this is the $7 \rightarrow 1$ pumped pair A of Fig. 12-9 in which level 1 has no other connection to another level.

(*b*) If one level of a relaxed pair has no connection with another level then that pair will have a Boltzmann difference $2\epsilon_{BOLT}$ in population. An example is relaxation path B of Fig. 12-9 in which level 8 is unconnected.

(*c*) When sequences of levels are connected together with individual pumped or relaxed paths which terminate at an unconnected level then each pair in the sequence acquires its appropriate 2δ or $2\epsilon_{BOLT}$ population ratio. The sequence $B - C - D$ of Fig. 12-9 has this property.

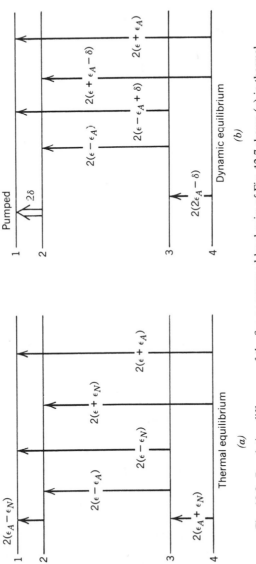

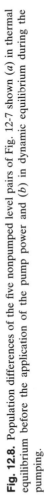

Fig. 12.8. Population differences of the five nonpumped level pairs of Fig. 12-7 shown (*a*) in thermal equilibrium before the application of the pump power and (*b*) in dynamic equilibrium during the pumping.

223

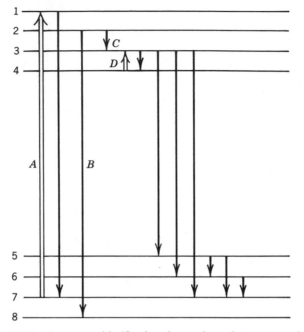

Fig. 12-9. An eight-level system with 10 relaxation paths and two pumped transitions. The relaxation transitions B and C produce Boltzmann distributions and the pumping powers A and D produce population distributions determined by the pumping factor δ between their respective level pairs, as explained in the text.

(d) Multiply connected pairs of levels such as the group $3 \rightarrow 5$, $3 \rightarrow 6$, $3 \rightarrow 7$, $5 \rightarrow 6$, $5 \rightarrow 7$, and $6 \rightarrow 7$ of Fig. 12-9 can not be treated in this simple fashion.

(e) The sum of the relative populations Σn_i in all of the levels equals the number of levels in the system (cf. Eq. 12-43).

The five rules that we have just given suggest a more general definition of δ than that given by Eq. 12-47

$$\delta = \frac{\epsilon_{\text{BOLT}}}{1 + 2PT_R} \tag{12-49}$$

where $2\epsilon_{\text{BOLT}}$ is the Boltzmann population difference that would exist in thermal equilibrium, in the absence of the pump. The values of ϵ_{BOLT} for the various relaxation transitions of the spin $(\frac{1}{2}, \frac{1}{2})$ case may be read off Fig. 12-7, and they are as follows:

$$
\begin{aligned}
\epsilon_{\text{BOLT}} &= \epsilon_A - \epsilon_N & 1 &\rightarrow 2 \\
\epsilon_{\text{BOLT}} &= \epsilon - \epsilon_N & 1 &\rightarrow 3 \\
\epsilon_{\text{BOLT}} &= \epsilon + \epsilon_A & 1 &\rightarrow 4 \\
\epsilon_{\text{BOLT}} &= \epsilon - \epsilon_A & 2 &\rightarrow 3
\end{aligned}
\tag{12-50}
$$

$$\epsilon_{BOLT} = \epsilon + \epsilon_N \qquad 2 \to 4$$
$$\epsilon_{BOLT} = \epsilon_N + \epsilon_A \qquad 3 \to 4$$

(12-50 cont.)

These expressions will be found useful for comparing δ when different levels are pumped. They can be found on the right-hand side of Eqs. 12-42. In some cases the nuclear and hyperfine Boltzmann factors ϵ_N and ϵ_A, respectively, can be neglected relative to the electronic Boltzmann factor ϵ.

12-10 ABSORPTION AND EMISSION

Ordinarily, double resonance responses are observed as absorption signals because the lower of the two energy levels is more populated than the upper one. This is the case for the five transitions which have their population differences presented in Fig. 12-8. There are some combinations of pumping powers and relaxation paths which distribute the populations among the levels in such a way that one level acquires a larger number of spins than another level below it. When this occurs a transition induced between that pair of levels will appear as an emission signal which can be recognized by the opposite phase of its recorder trace, as illustrated in Fig. 12-10. Figures 15-6 and 15-9 provide additional examples of emission signals.

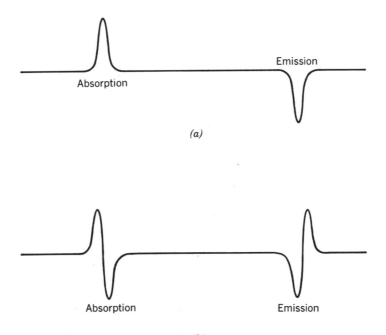

(a)

(b)

Fig. 12-10. Spectral lines for (a) direct absorption and emission and (b) first derivative absorption and emission.

REFERENCES

1. C. P. Poole, Jr. and H. A. Farach, *Relaxation in Magnetic Resonance*, Academic Press, New York, 1971.
2. R. H. Webb, *Am. J. Phys.*, **29**, 429 (1961).
3. E. R. Andrew, *Nuclear Magnetic Resonance*, Cambridge University Press, Cambridge, 1956.
4. A. Abragam, *Principles of Nuclear Magnetism*, Oxford University Press, London, 1961.

13

ELECTRON–NUCLEAR
DOUBLE RESONANCE

13-1 INTRODUCTION

The previous chapter introduced the basic principles behind double resonance experiments and described how most of them depend on the establishment of non-Boltzmann distributions between energy levels by the action of high pumping powers. In this chapter we discuss the type of double resonance called electron–nuclear double resonance or ENDOR. This technique, which was originally devised by Feher,[1] provides a large increase in resolution over ordinary ESR, particularly when the number of hyperfine lines is quite large, as occurs frequently with organic free radicals.

In a typical ENDOR experiment a saturating radiofrequency field is scanned through the region where it induces transitions directly between hyperfine levels, and the effect of inducing these hyperfine transitions is observed by detecting the change in amplitude of a partly saturated ESR line. When the radiofrequency passes through regions of resonance it alters the populations of the various energy levels, and these changes in population are reflected in the population difference of the levels being monitored by the ESR signal, hence its change in absorption. Thus ENDOR depends on the manner in which the population differences are affected by the applied pumping power, and on the relaxation paths whereby they return to equilibrium. When studies are made of liquid and solid samples with short relaxation times (<0.1 msec) then radiofrequency power levels of 10–1000 W are required and the technique is referred to as high-power ENDOR. When the relaxation times are between 0.1 and 10 sec, on the other hand, then low-power ENDOR is used with pumping powers of 0.1–10 W.

In this chapter we treat the two-spin $S = \frac{1}{2}$, $I = \frac{1}{2}$ system whose four

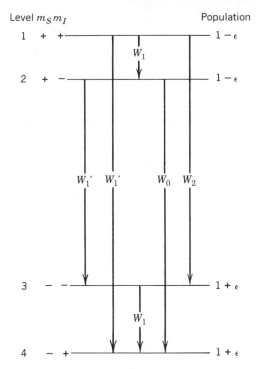

Fig. 13-1. Energy level diagram of the $S = \frac{1}{2}$, $I = \frac{1}{2}$ spin system showing the six relaxation transitions and the level populations in thermal equilibrium. The hyperfine and nuclear Zeeman terms are neglected in writing these populations [compare with Fig. 12-6].

energy levels have, before the application of the pumping power, the thermal equilibrium populations given in Fig. 13-1. This equilibrium is maintained by some or all of the relaxation rates indicated in the figure. The influence of the nuclear Zeeman and the hyperfine terms on the level populations given in Fig. 12-6 can be neglected in the treatment of ENDOR, as is done for the present figure. For simplicity we assume complete saturation of the scanned NMR transition and partial saturation of the monitored ESR transition. The monograph of Kevan and Kispert[2] provides a more extensive treatment of ENDOR, Wertz and Bolton[3] describe the method, and recent reviews[4,5] describe experimental methods employed in ENDOR.

13-2 HAMILTONIAN AND ENERGY LEVELS

The spin Hamiltonian for the $S = \frac{1}{2}$, $I = \frac{1}{2}$ system is given by

$$\mathcal{H} = g\beta\vec{\mathbf{S}}\cdot\vec{\mathbf{H}} - g_N\beta_N\vec{\mathbf{H}}\cdot\vec{\mathbf{I}} + \hbar A\vec{\mathbf{S}}\cdot\vec{\mathbf{I}} \tag{13-1}$$

ENDOR experiments are ordinarily carried out under conditions where the hyperfine term $A\vec{S}\cdot\vec{I}$ is much smaller than the electronic Zeeman term $g\beta\vec{S}\cdot\vec{H}$ so that it can be approximated by AS_zI_z, and therefore the six energies are given by Eq. 12-2:

$$E = g\beta H m_S - g_N\beta_N H m_I + \hbar A m_S m_I \qquad (13\text{-}2)$$

If we adopt the notation

$$\hbar\omega_S = g\beta H$$
$$\hbar\omega_N = g_N\beta_N H \qquad (13\text{-}3)$$

and select the symbols g, g_N, H, and A as positive then we obtain the energies, in frequency units, of the two allowed ESR transitions

$$\omega_{\text{ESR}} = \omega_S + \tfrac{1}{2}A$$
$$\omega_{\text{ESR}} = \omega_S - \tfrac{1}{2}A \qquad (13\text{-}4)$$

which obey the standard ESR selection rules

$$\Delta m_S = \pm 1$$
$$\Delta m_I = 0 \qquad (13\text{-}5)$$

the two forbidden ESR transitions

$$\omega_{\text{FORB}} = \omega_S + \omega_N$$
$$\omega_{\text{FORB}} = \omega_S - \omega_N \qquad (13\text{-}6)$$

with the selection rules

$$\Delta m_S = \pm 1$$
$$\Delta m_I = \mp 1 \qquad (13\text{-}7)$$

and the two ENDOR transitions

$$\omega_{\text{ENDOR}} = \tfrac{1}{2}A + \omega_N$$
$$\omega_{\text{ENDOR}} = |\tfrac{1}{2}A - \omega_N| \qquad (13\text{-}8)$$

which satisfy the NMR selection rules

$$\Delta m_S = 0$$
$$\Delta m_I = \pm 1 \qquad (13\text{-}9)$$

as indicated in Fig. 13-2. The ordering of the energy levels depends on the signs of the g-factors and hyperfine coupling constant, and for simplicity we assume that all three are positive unless otherwise stated, corresponding to the figure. Ordinarily, ENDOR experiments only provide the absolute value of the hyperfine term A.

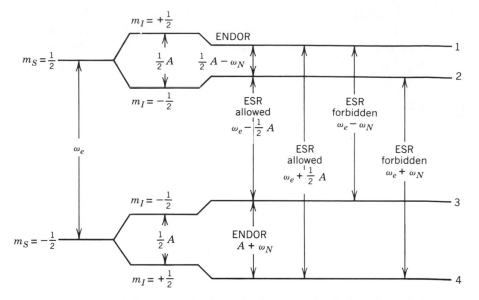

Fig. 13-2. Energy level diagram for the $S = \frac{1}{2}$, $I = \frac{1}{2}$ case showing the hyperfine splitting alone (center) and the hyperfine and nuclear Zeeman splittings together (right). The ESR and ENDOR transition energies are indicated.

In solids anisotropic hyperfine couplings can be determined by studying the angular dependence of the ENDOR spectra,[6] and when $I > \frac{1}{2}$ quadrupole couplings can also be evaluated by this method. The present chapter treats only the isotropic case with $I = \frac{1}{2}$. The signs of the hyperfine coupling constant[7] and of the quadrupole moment[8] have been determined in liquid crystals.

13-3 MASTER EQUATIONS

We mentioned in the introduction that the nuclear Zeeman and hyperfine contributions to the populations can be neglected, and that we assume complete saturation of the NMR transition and partial saturation of the ESR transition. We select the $4 \rightarrow 1$ ESR transition for partial saturation, as indicated in Fig. 13-3, and as in Section 12-8 we write down the master equations for this case, taking into account only the ESR Zeeman contribution ϵ to the equilibrium populations

$$\frac{dn_1}{dt} = W_1(n_2 - n_1) + W_2(n_3 - n_1) + (W_1' + P)(n_4 - n_1) - 2\epsilon(W_1' + W_2)$$

$$(13\text{-}10a)$$

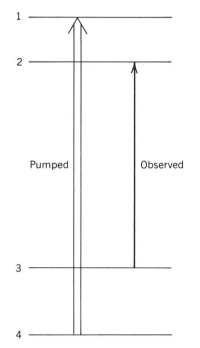

Fig. 13-3. ENDOR experiment in which the $4 \to 1$ level is pumped and the $3 \to 2$ level is observed.

$$\frac{dn_2}{dt} = -W_1(n_2 - n_1) + W_1'(n_3 - n_2) + W_0(n_4 - n_2) - 2\epsilon(W_0 + W_1')$$

(13-10b)

$$-\frac{dn_3}{dt} = W_2(n_3 - n_1) + W_1'(n_3 - n_2) - W_1(n_4 - n_3) - 2\epsilon(W_1' + W_2)$$

(13-10c)

$$-\frac{dn_4}{dt} = (W_1' + P)(n_4 - n_1) + W_0(n_4 - n_2) + W_1(n_4 - n_3) - 2\epsilon(W_0 + W_1')$$

(13-10d)

The corresponding master equations for the partial saturation of a different ESR transition are easily written down by analogy.

To carry out the ENDOR experiment the saturating radiofrequency power scans through the upper and then the lower NMR level pair in turn, equalizing the respective populations as it does so. The steady-state ENDOR response is the difference between the $4 \to 1$ ESR level population difference before and during this equalizing. We can take into account the complete saturation of the upper level pair by combining Eqs. 13-10a and 13-10b and setting $n_1 = n_2$ in the remaining two equations to give

$$n_1 = n_2 \tag{13-11a}$$

$$\frac{d(n_1 + n_2)}{dt} = (W_1' + W_2)(n_3 - n_1) + (W_0 + W_1' + P)(n_4 - n_1)$$
$$- 2\epsilon(W_0 + 2W_1' + W_2) \tag{13-11b}$$

$$-\frac{dn_3}{dt} = (W_1' + W_2)(n_3 - n_1) - W_1(n_4 - n_3) - 2\epsilon(W_1' + W_2) \tag{13-11c}$$

$$-\frac{dn_4}{dt} = (W_0 + W_1' + P)(n_4 - n_1) + W_1(n_4 - n_3) - 2\epsilon(W_0 + W_1') \tag{13-11d}$$

In like manner we can write down the equations for the second case of $n_3 = n_4$ by combining the last two equations of the set 13-10 as follows:

$$n_3 = n_4 \tag{13-12a}$$

$$-\frac{d(n_3 + n_4)}{dt} = (W_1' + W_2 + P)(n_4 - n_1) + (W_0 + W_1')(n_4 - n_2)$$
$$- 2\epsilon(W_0 + 2W_1' + W_2) \tag{13-12b}$$

$$\frac{dn_1}{dt} = W_1(n_2 - n_1) + (W_1' + W_2 + P)(n_4 - n_1) - 2\epsilon(W_1' + W_2) \tag{13-12c}$$

$$\frac{dn_2}{dt} = -W_1(n_2 - n_1) + (W_0 + W_1')(n_4 - n_2) - 2\epsilon(W_0 + W_1') \tag{13-12d}$$

The ENDOR response can be found by solving these sets of coupled equations for $n_3 - n_2$ and comparing the results in the absence and in the presence of the saturating NMR power. In the following sections we will give the results of this procedure for some particular cases.

13-4 THE ENDOR EXPERIMENT

In an ENDOR experiment the magnetic field is set at the value corresponding to resonance for one of the ESR transitions of Eq. 13-4, and a radiofrequency ω is swept through the region corresponding to the ENDOR frequencies of Eq. 13-8. When ω passes through the values $\frac{1}{2}A + \omega_N$ and $|\frac{1}{2}A - \omega_N|$, the populations of the corresponding pairs of levels are altered, and a change occurs in the magnitude of the detected ESR absorption signal

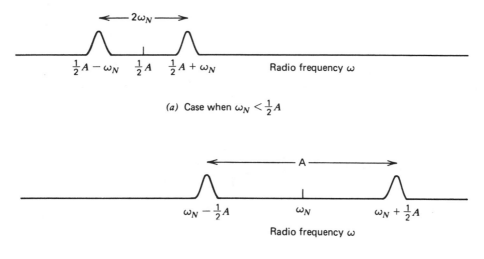

(a) Case when $\omega_N < \frac{1}{2}A$

(b) Case when $\frac{1}{2}A < \omega_N$

Fig. 13-4. ENDOR spectra shown for the two conditions (a) $\omega_N < \frac{1}{2}A$ and (b) $\frac{1}{2}A < \omega_N$. For some systems case (a) occurs at X band (9 GHz) and case (b) at Q band (35 GHz) (from Ref. 4).

that is plotted by a chart recorder, as indicated in Fig. 13-4. When measurements are made at different microwave frequency bands, then ω_N is proportional to ω_e, and hence $|\frac{1}{2}A|$ will be greater than ω_N for low microwave frequencies and less than ω_N for high microwave frequencies. Spectra for the two cases of ω_N less than and greater than $\frac{1}{2}A$ are illustrated in Fig. 13-4.

For an ENDOR signal to be observable both the ESR and NMR transitions must be saturable at available powers. Partial saturation is sufficient for the ESR transition, and if this can not be readily achieved at room temperature, then it can generally be accomplished at low temperatures such as 4 K. More complete saturation is required for the NMR transition. The ENDOR spectral lines are comparable in width to ESR lines, typically 0.1 MHz for free radicals in liquids. The number of lines in the ENDOR spectrum is considerably less, however, so the effective resolution is much greater, and the spectrum is easier to interpret. This is illustrated in Fig. 13-5 for the case of the diphenylanthracene (DPA) negative ion.[9]

All lines on this spectrum correspond to the case in which $\frac{1}{2}A < \omega_N$ so they are equally spaced on each side of the proton NMR frequency (~14.6 MHz). The values of three hyperfine couplings (~0.6 MHz, ~4 MHz, and ~7 MHz) are estimated from spectral line separations in the figure.

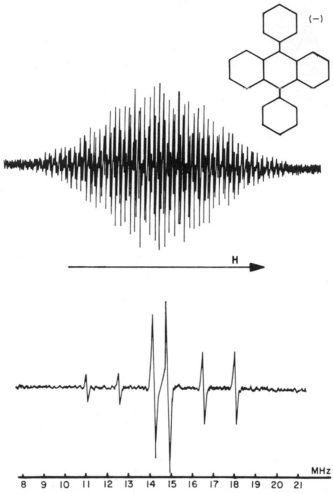

Fig. 13-5. Comparison of the ESR (top) and ENDOR (bottom) spectra of the diphenyl-anthracene negative ion (from Ref. 9).

13-5 STEADY-STATE ENDOR

When the relaxation times of the spin system are short compared to the time required to scan through a resonant line then the pumping radiofrequency produces level population rearrangements which have effectively attained their final steady-state values in terms of their influence on the intensity of the monitored ESR transition. An ENDOR experiment in which this is the case is referred to as steady-state ENDOR. The condition for steady-state ENDOR is the ability of the pumping radiofrequency to produce a dynamic steady-state change in the population difference of the monitored level pair. We now consider several schemes for accomplishing this.

We begin by considering the case illustrated in Fig. 13-6 in which the electronic relaxation rate W_1' is comparable to the forbidden transition relaxation rate W_0 between levels 2 $(+ -)$ and 4 $(- +)$ which involve no change in the total magnetic quantum number $M = m_S + m_I$. All other relaxation rates are assumed to be much slower and hence negligible. When the ESR transition is pumped we see from Fig. 13-6 that the relaxation rates W_1' and W_0 are able to maintain a thermal equilibrium population distribution between the level pairs $3 \to 2$ and $4 \to 2$, respectively, in accordance with rule 3 of Section 12-9, and the population difference 2δ is established between the two pumped levels. The dynamic equilbrium populations shown in Fig. 13-6 are easily calculated by the methods described in Section 12-9. These give the $4 \to 1$ level population difference which is monitored during the ENDOR experiment:

$$n_4 - n_1 = \frac{2\epsilon}{1 + 2PT_p} = 2\epsilon \frac{W_1'}{W_1' + P} = 2\delta \tag{13-13}$$

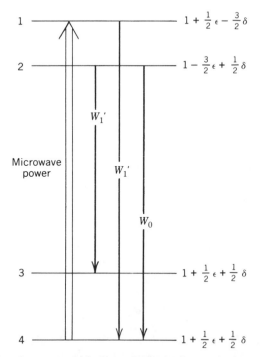

Fig. 13-6. The relaxation rates which are operative in the steady-state ENDOR experiments described on Figs. 13-7 and 13-8. The right-hand side shows the level populations in the presence of the $4 \to 1$ microwave pumping power with the radiofrequency power turned off.

where $W_1' = 1/2T_p$. This population difference will be compared with that calculated during saturation of the NMR level.

We see from the figure that the populations of the lower two levels are already equalized so that no ENDOR transition will be observed by saturating them. This is not the case for the upper NMR level pair so this is favorable for a steady-state ENDOR sweep, and we calculate the enhancement that is attained.

When the upper level pair of Fig. 13-6 is pumped the various levels become connected by the relaxation paths and the elementary methods of Section 12-9 can not be employed to solve the problem. The dynamic steady-state master equations (13-11) with the $(n_4 - n_3)$ terms neglected must be solved for this case

$$W_1'(n_3 - n_1) + (W_0 + W_1' + P)(n_4 - n_1) = 2\epsilon(W_0 + 2W_1') \qquad (13\text{-}14a)$$
$$W_1'(n_3 - n_1) = 2\epsilon W_1' \qquad (13\text{-}14b)$$
$$(W_0 + W_1' + P)(n_4 - n_1) = 2\epsilon(W_0 + W_1') \qquad (13\text{-}14c)$$

The last of these three equations gives the $n_4 - n_1$ population difference immediately:

$$n_4 - n_1 = 2\epsilon \frac{W_0 + W_1'}{W_0 + W_1' + P} \qquad (13\text{-}15)$$

If we divide Eq. 13-15 by 13-13 we obtain the ENDOR enhancement which is the ratio of the ENDOR signal to that of the ESR signal without the NMR pump:

$$\text{Enhancement} = \frac{W_0 + W_1'}{W_1'} \frac{W_1' + P}{W_0 + W_1' + P} \qquad (13\text{-}16)$$

We see that the enhancement depends on the relative values of the two relaxation rates W_0 and W_1', and it also depends on the power level P. The greatest enhancement $(W_0 + W_1')/W_1'$ occurs when the power factor P becomes much larger than the ralaxation rates, but this is not a practical operating microwave power level because the ESR signal is too weak to be detected. The greatest sensitivity occurs for moderate levels of saturation where the ESR signal is still strong and at the same time appreciable enhancement occurs. The condition

$$P = W_0 = W_1' \qquad (13\text{-}17)$$

provides an enhancement of $\frac{4}{3}$ and the higher power level

$$P = 2W_0 = 2W_1' \qquad (13\text{-}18)$$

corresponds to an enhancement of $\frac{3}{2}$.

13-6 DYNAMICS OF ENDOR RESPONSE

The dynamic processes involved in the rearrangement of the level populations by the action of the pumping power may be monitored by setting the microwave frequency on top of the ESR line, turning on the radiofrequency pump power and tracing out on the recorder the time evolution of the ENDOR response. A fast response spectrometer is needed for this purpose.

This experiment is examined for the case illustrated in Fig. 13-6 with the assumptions

$$P = W_0 = W_1'$$ (13-19)

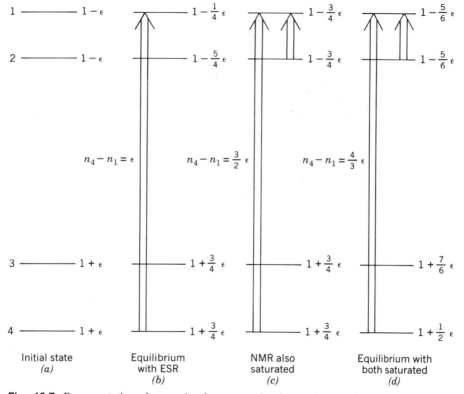

Initial state   Equilibrium   NMR also   Equilibrium with
(a)   with ESR   saturated   both saturated
    (b)   (c)   (d)

Fig. 13-7. Representative changes in the energy level populations during a steady-state ENDOR experiment in which the $4 \rightarrow 1$ ESR and the $2 \rightarrow 1$ NMR transitions are saturated. The figure shows from left to right: (a) the initial equilibrium state of the system, (b) the partial equalization of the populations of levels 1 and 4 by the application of the high microwave power at the time t_E, (c) the equalization of the populations of levels 1 and 2 by the application of the radiofrequency power at the time t_N, and (d) the final dynamic equilibrium state a relatively long time later. The population difference in this final state provides the steady-state ENDOR signal. The time sequence of the events of this figure is presented in Fig. 13-8.

corresponding to $\delta = \frac{1}{2}\epsilon$ from Eq. 13-13 and assuming complete saturation of the $4 \rightarrow 3$ NMR transition which equalizes the associated populations

$$n_3 = n_4 \tag{13-20}$$

We treat this as a three-step process which begins with the thermal equilibrium condition of Fig. 13-7a and follows the time evolution sketched in Fig. 13-8. The first step is the pumping of the $4 \rightarrow 1$ levels and the subsequent rearrangement of all the populations to their dynamic equilibrium values, as indicated in Fig. 13-7b. The second step is the application of the high-power radiowaves to produce a very rapid equalization of the populations of levels 1 and 2 as indicated in Fig. 13-7c. A subsequent somewhat slower redistribution of the populations brings them to their final steady-state values determined in the last section which are indicated in Fig. 13-7d for the condition 13-19.

The explicit time dependence of the population difference $n_4 - n_1$ can be calculated with the aid of the master equations 13-10 and it is sketched in Fig. 13-8 for the case $\delta = \frac{1}{2}$. The figure shows the turning on of the microwave power at the time t_E to saturate the ESR line. Then at the time t_N the radiowave pump is turned on and the ESR response begins to rise, passes through a maximum, and then settles into its final dynamic equilibrium value. This rise and passage through a maximum constitutes the transient ENDOR response. The final population values result from the competition between the two applied powers acting to equalize the level pairs 1, 2 and 1, 4 and the relaxation mechanisms trying to establish Boltzmann ratios between the level pairs 2, 3 and 2, 4.

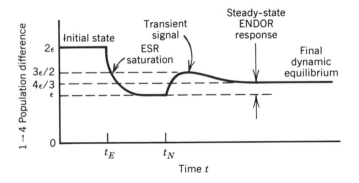

Fig. 13-8. Population difference of the 1, 4 level pair during a steady-state ENDOR experiment showing, for increasing times, the initial equilibrium state, the turning on of the microwave power at the time t_E followed by the decay of the population difference from 2ϵ to ϵ and the turning on of the radiowave power at the time t_N followed by a rapid rise in population difference to the value $3\epsilon/2$ and then a slower decay to the dynamic final equilibrium value $4\epsilon/3$. This final value provides the steady-state ENDOR signal. This figure is drawn for the condition $\delta = \frac{1}{2}\epsilon$. The populations and energy level diagrams for this case are presented in Fig. 13-7.

The dynamic processes described here permit ENDOR to be employed for the study of relaxation effects[10-15] and tunneling.[16-20]

13-7 TRANSIENT ENDOR

In this section we consider the situation when the only operant relaxation paths are W_1' corresponding to the two allowed ESR transitions, as illustrated in Fig. 13-9. For this case it is not possible to detect ENDOR under steady-state conditions, and a transient experiment must be carried out.

The experiment begins by turning on the high microwave power at the time t_E to saturate the $3 \rightarrow 2$ transition, as indicated in Figs. 13-10b and 13-11. Then high radiofrequency power is applied to saturate the $2 \rightarrow 1$ transition at the time t_N indicated in Fig. 13-11. It is assumed that the radiofrequency signal is sufficiently high in power so that it completely saturates the $2 \rightarrow 1$ levels in a time that is short compared to the electronic relaxation time $1/W_1'$ of the system. After levels 1 and 2 become equalized the system begins to relax, and this brings about a change in the relative populations of levels 2 and 3, which is plotted in Fig. 13-11. The result is a change in the ESR absorption which is monitored and constitutes the

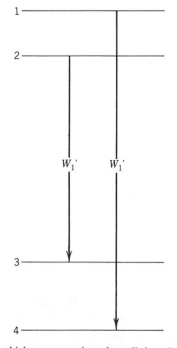

Fig. 13-9. Relaxation rates which are operative when all the relaxation takes place via allowed ESR transitions, as in the transient ENDOR case described by Figs. 13-10 and 13-11.

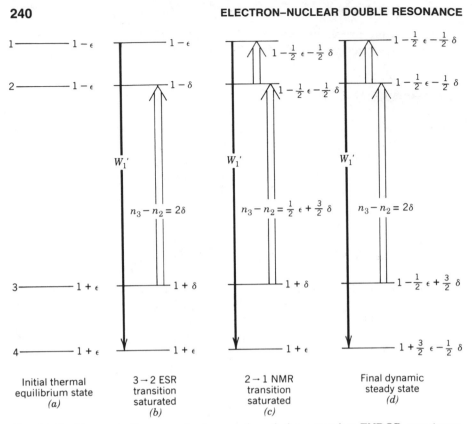

Fig. 13-10. Changes in the energy level populations during a transient ENDOR experiment, showing from left to right: (*a*) the initial equilibrium state of the system, (*b*) the partial equalization of the populations of levels 2 and 3 by the application of the high microwave power, (*c*) the equalization of the populations of levels 1 and 2 before the electron spins can respond, and (*d*) the final dynamic steady state resulting from the action of the two saturating powers and the one relaxation path as shown. The transient ENDOR signal arises from the readjustment of the populations between steps (*c*) and (*d*).

transient ENDOR signal. After an interval lasting several relaxation times the system begins to reach its final steady-state configuration shown in Figs. 13-10*d* and 13-11 with no more changes occurring in the ESR absorption. There is no steady-state ENDOR because levels 2 and 3 have the same degree of saturation before and a long time after the application of the NMR pumping power.

For a transient experiment of this type to be practical it is necessary for the relaxation time $1/W_1'$ to be sufficiently long so that the spectrometer is able to follow the decay of the ENDOR signal on the right side of Fig. 13-11. For typical spectrometers several seconds are required to monitor this response, and when the relaxation time is very short then more sophisticated techniques must be used. In Feher's original ENDOR experiment,[1] carried out with phosphorus-doped silicon at 1.25 K, the electronic relaxation time

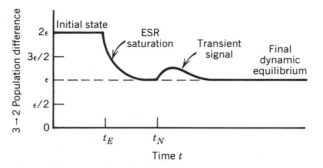

Fig. 13-11. Population difference of the 2, 3 level pair during a transient ENDOR experiment showing, for increasing times, the initial equilibrium state, the turning on of the microwave power at the time t_E followed by the decay of the population difference to $2\delta = \epsilon$, and the turning on of the radiowave power at the time t_N followed by a rapid rise in population difference with a slower decay to 2δ. This rise and subsequent decay provides the transient ENDOR signal. The sketch is drawn using the population values given in Fig. 13-10 with the assumption $\delta = \epsilon/2$.

was about 40 min. Free radicals in solution and paramagnetic solids above liquid nitrogen temperature ordinarily have electronic relaxation times in the range between a microsecond and several milliseconds, and for these cases transient experiments are not practical to carry out.

13-8 DOUBLE ENDOR

ESR and ENDOR generally only provide the magnitudes of hyperfine coupling constants. To determine the relative signs of the couplings in a hyperfine multiplet, a double ENDOR or triple resonance method may be employed.[21-26] This is illustrated for the three-spin case $S = I_1 = I_2 = \frac{1}{2}$.

The energy level diagram for these three spins when one nucleus has a positive hyperfine coupling constant A_1 is presented in Fig. 13-12 for the two cases in which the other coupling constant A_2 is positive and negative. In constructing this diagram it was assumed that both nuclei have the same gyromagnetic ratio γ_N. The four nuclear transitions ω_i that are scanned in ENDOR have the frequencies

$$\omega_1 = \omega_N - \tfrac{1}{2}A_1$$
$$\omega_2 = \omega_N - \tfrac{1}{2}|A_2|$$
$$\omega_3 = \omega_N + \tfrac{1}{2}|A_2|$$
$$\omega_4 = \omega_N + \tfrac{1}{2}A_1$$

(13-21)

where Fig. 13-12 is drawn for the condition $2\omega_N > A_1 > |A_2|$. We note from the figure that the frequencies ω_2 and ω_3 shift between the upper and lower sets of energy levels when the sign of A_2 changes.

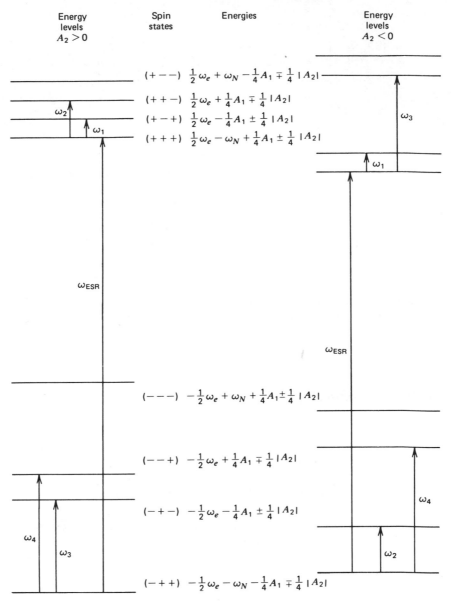

Fig. 13-12. Energy level diagrams for a three-spin system with $S = I_1 = I_2 = \frac{1}{2}$ and a positive hyperfine coupling constant A_1 drawn on the left for the coupling constant A_2 positive and on the right for A_2 negative. The spin states (m_S, m_{I1}, m_{I2}) and the level energies are given, and the ESR transition ω_{ESR} and four ENDOR transitions ω_1, ω_2, ω_3, and ω_4 are indicated. Note that the ω_2 and ω_3 transitions defined by Eqs. 13-21 shift between the upper and lower sets of energy levels when the sign of A_2 changes. Figure 13-13 gives the energy level populations for particular conditions (from Refs. 4 and 5).

To carry out a double ENDOR experiment, one ESR and one NMR transition are saturated, and a separate rf oscillator is employed to scan through other NMR transitions while monitoring the ESR absorption. Figure 13-13 shows the relative populations of the eight levels in the $S = I_1 = I_2 = \frac{1}{2}$ system for the three situations: (1) static thermal equilibrium, (2) dynamic thermal equilibrium in the presence of saturating microwave

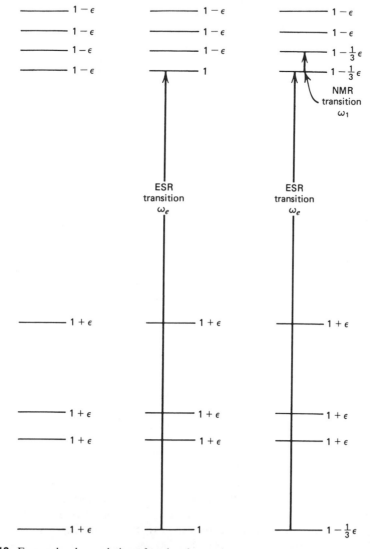

Fig. 13-13. Energy level populations for the three-spin system of Fig. 13-12 shown in the absence of saturation (left), when the ESR transition is saturated (center), and when the ESR and ω_1 ENDOR transitions are saturated (right) (from Refs. 4 and 5). The level populations are shown for complete saturation ($\delta = 0$) before the approach to dynamic equilibrium.

TABLE 13-1 Relative Intensities of the Double ENDOR Transitions ω_2, ω_3, and ω_4. When the ESR and ω_1 Transitions Are Saturated for the Cases When the Hyperfine Coupling Constants A_1 and A_2 Have the Same and Opposite Signs. Ordinary ENDOR Transitions Have the Relative Intensity ϵ (from Ref. 5)

Signs of A_1 and A_2	Relative Intensities		
	ω_2	ω_3	ω_4
Same	$2\epsilon/3$	$4\epsilon/3$	$4\epsilon/3$
Opposite	$4\epsilon/3$	$2\epsilon/3$	$4\epsilon/3$

power at the ESR frequency ω_{ESR}, and (3) dynamic thermal equilibrium in the presence of saturating power from the ESR frequency ω_{ESR} and one of the ENDOR transition frequencies ω_1. This figure is drawn for the case of positive A_2, but the level populations also apply to the negative A_2 case. Here we are assuming the simple model of transient ENDOR in which relaxation mechanisms do not complicate the picture. If a scan is made through another ENDOR line, then the saturated ESR signal will be affected to an extent that is proportional to the population difference between the corresponding ENDOR levels. We see from Table 13-1 that in the case of double ENDOR the ω_2 and ω_3 population differences differ for the two coupling constant cases. Thus the experimental determination of the relative intensities of the ω_2 and ω_3 lines by double ENDOR provides the relative signs of the hyperfine coupling constants.[21–26]

REFERENCES

1. G. Feher, *Phys. Rev.*, **103**, 834 (1956).
2. L. Kevan and L. D. Kispert, *Electron Spin Double Resonance Spectroscopy*, Wiley, New York, 1976.
3. J. Wertz and J. R. Bolton, *Electron Spin Resonance, Elementary Theory and Applications*, Chapman & Hall, New York, 1986, Chap. 13.
4. C. P. Poole, Jr. and H. A. Farach, *Appl. Spectrosc. Rev.*, **19**, 167 (1983).
5. C. P. Poole, Jr., *Electron Spin Resonance*, 2nd ed., Wiley-Interscience, New York, 1983.
6. K. P. Dinse, R. Biehl, K. Möbius, and H. Haustein, *Chem. Phys. Lett.*, **12**, 399 (1971).
7. H. Dinse, K. Möbius, M. Plato, R. Biehl, and H. Haustein, *Chem. Phys. Lett.*, **14**, 196 (1972).
8. R. Biehl, M. Plato, and H. Möbius, *J. Chem. Phys.*, **63**, 3515 (1975).
9. R. Biehl, private communication.
10. A. B. Brik and S. S. Ishchenki, *Fiz. Tverd. Tela*, **18**, 2442 (1976).

11. V. L. Gokhman and B. D. Shanina, *Fiz. Tverd. Tela*, **17**, 1408 (1975).

12. H. Hoostraate, J. Poot, W. Th. Wenckebach, and N. J. Poulis, *Physica B and C*, **79B** and **79C**, 499 (1975).

13. E. C. McIrvine, J. Lambe, and N. Lawrence, *Phys. Rev.*, **A136**, 467 (1964).

14. L. F. Mollenauer and S. Pan, *Phys. Rev. B*, **6**, 772 (1972).

15. V. Ya Zevin and A. B. Brik, *Ukr. Fiz. Zh.*, **17**, 1688 (1972).

16. E. G. Derouane and J. C. Vedrine, *Chem. Phys. Lett.*, **29**, 222 (1974).

17. J. Helbert, L. Kevan, and B. L. Bales, *J. Chem. Phys.*, **57**, 723 (1972).

18. L. Kevan, P. A. Narayana, K. Toriyama, and M. Iwasaki, *J. Chem. Phys.*, **70**, 5006 (1979).

19. S. Clough, J. R. Hill, and M. Punkkinen, *J. Phys. C*, **7**, 3413, 3779 (1974).

20. S. Clough and J. Hobson, *J. Phys. C*, **8**, 1745 (1975).

21. J. M. Baker and W. B. J. Blake, *Phys. Lett. A*, **31**, 61 (1970); *Phys. Rev. C*, **6**, 3501 (1973).

22. R. J. Cook and D. H. Whiffen, *Proc. Phys. Soc.*, **84**, 845 (1964); *J. Chem. Phys.*, **43**, 2908 (1965).

23. J. A. R. Coope, N. S. Dalal, C. A. McDowell, and R. Spinivasan, *Mol. Phys.*, **24**, 403 (1972).

24. N. S. Dalal, C. A. McDowell, and R. Srinivasan, *Mol. Phys.*, **24**, 417 (1972).

25. D. A. Hampton and G. C. Moulton, *J. Chem. Phys.*, **63**, 1078 (1975).

26. I. Ursu, *Rev. Roum. Phys.*, **17**, 955 (1972).

ELECTRON–ELECTRON DOUBLE RESONANCE

14-1 INTRODUCTION

In the previous chapter we described electron–nuclear double resonance, a double resonance technique involving one electronic and one nuclear transition. The present chapter is concerned with electron–electron double resonance or ELDOR which was introduced by Hyde et al.[1] and involves two electronic transitions. In a typical experiment a saturating microwave frequency is scanned through the region where it induces transitions, and this alters the populations of the energy levels of the monitored transition thereby producing a change in the detected absorption signal. The result can be either an enhancement or a reduction in the intensity of the monitored transition. The extent to which the intensity changes depends on the various relaxation paths which are operative, and this makes ELDOR a convenient way to sort out the effect of these relaxation rates. For example, spin exchange processes and spin diffusion alter the ELDOR intensity and can be elucidated by studies of this type. Furthermore, ELDOR does not involve the decrease in sensitivity relative to ESR that is characteristic of ENDOR.

In this chapter we treat the two-spin $S = \frac{1}{2}$, $I = \frac{1}{2}$ system whose four energy levels have, before the application of the pumping power, the thermal equilibrium populations given in Fig. 14-1. This equilibrium is maintained by some or all of the relaxation rates indicated in the figure. The influence of the nuclear Zeeman and the hyperfine terms on the level populations is neglected in the treatment of ELDOR, as it was in the case of ENDOR. For simplicity we generally assume complete saturation of the pumped transition. Our approach is similar to that of Chapters 12 and 13 with an emphasis on such factors as the population differences, the effect of the pumping and monitoring signals, and the changes that occur in the

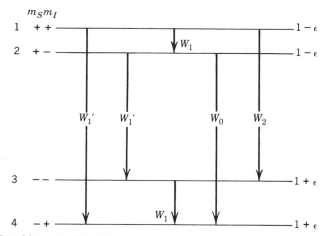

Fig. 14-1. Transition rates W_j of the six relaxation paths of the $S = \frac{1}{2}$, $I = \frac{1}{2}$ spin system. The m_S, m_I values for each level are listed on the left and the thermal equilibrium populations are indicated on the right. The nuclear Zeeman and hyperfine Boltzmann factors are neglected.

observed spectra. The monograph of Kevan and Kispert[2] provides a more extensive treatment of ELDOR, recent reviews[3,4] describe experimental methods employed in this type of investigation and other reviews[5,6] report experimental data on various systems.

14-2 TRANSITION ENERGIES

The ENDOR spin Hamiltonian (13-1) for the $S = \frac{1}{2}$, $I = \frac{1}{2}$ spin system is also used in ELDOR, and it provides the six energies that are given by Eq. 13-2

$$E = g\beta H m_S - g_N \beta_N H m_I + A m_S m_I \qquad (14\text{-}1)$$

The energies, in frequency units, of the two allowed ESR transitions

$$\omega_{\text{ESR}} = \frac{g\beta H}{\hbar} + \frac{1}{2} A \qquad (14\text{-}2)$$

$$\omega_{\text{ESR}} = \frac{g\beta H}{\hbar} - \frac{1}{2} A \qquad (14\text{-}3)$$

obey the standard ESR selection rules of Eqs. 13-5, and the two forbidden ESR transitions

$$\omega_{\text{FORB}} = \frac{g\beta H}{\hbar} + \omega_N \qquad (14\text{-}4)$$

$$\omega_{\text{FORB}} = \frac{g\beta H}{\hbar} - \omega_N \qquad (14\text{-}5)$$

satisfy the NMR selection rules of Eqs. 13-7 as indicated in Fig. 13-2. The ordering of the energy levels depends on the signs of the g-factors and hyperfine coupling constant which, following Section 13-2, we assume are all positive.

14-3 MASTER EQUATIONS

We mentioned in the introduction that the nuclear Zeeman and hyperfine contributions to the populations can be neglected, and that we assume complete saturation of the pumped ESR transition. We select levels $4 \rightarrow 1$ for pumping, as indicated in Fig. 14-2, and by analogy with Eqs. 13-11 and 13-12 we write the master equations for this case in which $n_4 \cong n_1$, taking into account only the ESR Zeeman contribution ϵ to the equilibrium populations and setting all time derivatives equal to zero

$$(W_1 + W_0)(n_2 - n_1) + (W_1 + W_2)(n_3 - n_1) = 2\epsilon(W_2 - W_0) \qquad (14\text{-}6a)$$

$$-(W_1 + W_0)(n_2 - n_1) + (W_1' + P)(n_3 - n_2) = 2\epsilon(W_1' + W_0) \qquad (14\text{-}6b)$$

$$(W_1 + W_2)(n_3 - n_1) + (W_1' + P)(n_3 - n_2) = 2\epsilon(W_1' + W_2) \qquad (14\text{-}6c)$$

Another useful form for these equations is

$$-(W_0 + 2W_1 + W_2)n_1 + (W_1 + W_0)n_2 + (W_1 + W_2)n_3 = 2\epsilon(W_2 - W_0)$$
$$(14\text{-}7a)$$

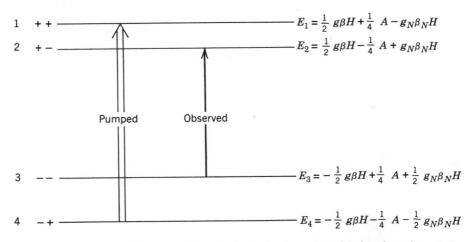

Fig. 14-2. Energy level diagram of the $S = \frac{1}{2}$, $I = \frac{1}{2}$ spin system showing the values of the energies on the right.

$$(W_0 + W_1)n_1 - (W_0 + W_1 + W_1' + P)n_2 + (W_1' + P)n_3 = 2\epsilon(W_1' + W_0)$$

$$(14\text{-}7b)$$

$$(W_1 + W_2)n_1 + (W_1' + P)n_2 - (W_1 + W_1' + W_2 + P)n_3 = -2\epsilon(W_1' + W_2)$$

$$(14\text{-}7c)$$

where the determinant of the coefficients is

$$\text{Det} = 2W_1[W_1'(W_0 + 2W_1 + W_2) + (W_0 + W_1)(W_1 + W_2)] \quad (14\text{-}8)$$

Here P is the power level factor of the monitored transition, and ordinarily it is set equal to zero corresponding to the approximation of monitoring the observed transition at a power level considerably below saturation ($P \ll W_1'$). The corresponding master equations for the partial saturation of another ESR transition are easily written by analogy.

A considerable amount of algebra is involved in solving either the master equations (14-6) or the equivalent set (14-7) to find the population difference $n_3 - n_2$ of the monitored transition when $P = 0$

$$n_3 - n_2 = 2\epsilon \, \frac{W_1'(2W_1 + W_0 + W_2) + W_1(W_0 + W_2) + 2W_0W_2}{W_1'(2W_1 + W_0 + W_2) + (W_1 + W_0)(W_1 + W_2)} \quad (14\text{-}9)$$

where the denominator is the part of the determinant (14-8) in square brackets. The ELDOR response can be found by comparing this result with the population difference

$$n_{30} - n_{20} = 2\epsilon \quad (14\text{-}10)$$

in the absence of the pumping power.

In the next section we will make use of these procedures and results to write expressions for the ELDOR reduction factor, and in the following sections we will apply these results to various cases.

14-4 ELDOR REDUCTION FACTOR

It is customary to express the influence of the pumping microwave power on the amplitude Y_P of the monitored ESR line relative to its amplitude Y_0 in the absence of the pumping power in terms of the ELDOR reduction factor R which is defined as follows:

$$R = \frac{Y_0 - Y_P}{Y_0} \quad (14\text{-}11)$$

The amplitude Y_P is proportional to the population difference $n_L - n_U$ between the lower (L) and the upper (U) monitored levels during pumping. Y_0 is proportional to the thermal equilibrium value $n_{0L} - n_{0U} \approx 2\epsilon$ of this population difference, and when these factors are inserted into Eq. 14-11 we obtain the value

$$R = 1 - \frac{n_L - n_U}{n_{0L} - n_{0U}} \tag{14-12}$$

for the ELDOR reduction factor. When the population difference $n_3 - n_2$ from Eq. 14-9 is inserted into Eq. 14-12 we obtain[1,2,7]

$$R = \frac{(W_1)^2 - W_0 W_2}{W_1'(2W_1 + W_0 + W_2) + (W_1 + W_0)(W_1 + W_2)} \tag{14-13}$$

This expression applies when allowed transitions are pumped and monitored. A negative sign for R corresponds to an enhancement rather than a reduction in amplitude. Ordinarily, the sign is positive indicating a reduction in amplitude, hence the name reduction factor. Equation 14-13 assumes complete saturation ($\delta = 0$) of the pumped ESR transition and it neglects the effect of the nuclear (ϵ_N) and hyperfine (ϵ_A) Boltzmann factors on the level populations. Some of the reduction factors listed in column 3 of Table 14-1 take these additional factors into account.

The procedures described in the previous section can be employed to calculate the ELDOR population difference $n_U - n_L$ for other cases of pumped and observed transitions, and we quote several results with δ, ϵ_A

TABLE 14-1 ELDOR Reduction Factor; for Several Pumped and Monitored Transitions

Pumped Transition	Monitored Transition	ELDOR Reduction Factor	Relaxation Rates	Level Population Conditions	Figure
$4 \rightarrow 1$	$3 \rightarrow 2^a$	$-1 + \dfrac{\delta}{\epsilon}$	Others $\ll W_0, W_2$	$n_4 - n_1 = 2\delta$ $n_3 - n_1 = 2\epsilon$ $n_4 - n_2 = 2\epsilon$	14-4
$3 \rightarrow 2$	$4 \rightarrow 1^a$	0	Others $\ll W_1'$	$n_3 - n_2 = 2\delta$ $n_4 - n_1 = 2\epsilon$	
$4 \rightarrow 1$	$3 \rightarrow 2$	$1 - \dfrac{\delta - 2\epsilon_A}{\epsilon - \epsilon_A}$	Others $\ll W_1$	$n_4 - n_1 = 2\delta$ $n_2 - n_1 = 2(\epsilon_A - \epsilon_N)$ $n_4 - n_3 = 2(\epsilon_A + \epsilon_N)$	14-5
$3 \rightarrow 2$	$4 \rightarrow 1$	$1 - \dfrac{\delta + 2\epsilon_A}{\epsilon + \epsilon_A}$	Others $\ll W_1$	$n_3 - n_2 = 2\delta$ $n_2 - n_1 = 2(\epsilon_A - \epsilon_N)$ $n_4 - n_3 = 2(\epsilon_A + \epsilon_N)$	
$4 \rightarrow 1$	$3 \rightarrow 2$	0	Others $\ll W_1', W_0$	$n_4 - n_1 = 2\delta$ $n_3 - n_2 = 2(\epsilon - \epsilon_A)$ $n_4 - n_2 = 2(\epsilon + \epsilon_N)$	14-6
$3 \rightarrow 1$	$4 \rightarrow 2^a$	$-1 + \dfrac{\delta}{\epsilon}$	Others $\ll W_1'$	$n_3 - n_1 = 2\delta$ $n_4 - n_1 = 2\epsilon$ $n_3 - n_2 = 2\epsilon$	14-7

^{a} The nuclear Boltzmann factors are not taken into account.

and ϵ_N neglected. If the forbidden $3 \to 1$ transition is pumped and the forbidden $4 \to 2$ transition is observed then the reduction factor is given by

$$R = \frac{W_1 - W_1'}{W_1 + W_1' + 2W_2} \qquad (14\text{-}14)$$

This case is included at the end of Table 14-1 and discussed later in Section 14-6. When the allowed $3 \to 2$ transition is pumped and the forbidden $4 \to 2$ one is observed we have

$$R = \frac{W_1 + W_1'}{W_1 + 2W_1' + W_2} \qquad (14\text{-}15)$$

Finally, when the allowed transition $3 \to 2$ is pumped and the forbidden transition $3 \to 1$ is monitored R becomes

$$R = \frac{(W_1 + W_1')(W_1 + W_2)}{(W_1 + 2W_1' + W_0)(W_1 + W_0)} \qquad (14\text{-}16)$$

Kevan and Kispert[2] discuss these expressions and give a table with values of R for various combinations of relaxation rates with $\delta = \epsilon_A = \epsilon_N = 0$.

14-5 THE ELDOR EXPERIMENT

To carry out an ELDOR experiment the magnetic field is initially set at the value H_0 that satisfies the resonant condition of Eq. 14-3

$$\omega_0 = \frac{g\beta H_0}{\hbar} - \tfrac{1}{2}A \qquad (14\text{-}17)$$

for the observed frequency ω_0. In Fig. 14-3 we show the magnetic field scanned at time t_1 through the hyperfine line at the position $\omega_0 + \tfrac{1}{2}A$ and then stopped at time t_2 on the hyperfine line at $\omega_0 - \tfrac{1}{2}A$ corresponding to the $3 \to 2$ transition of Fig. 14-2. The magnet setting is maintained at this value where Eq. 14-17 is satisfied. The pumping klystron is then turned on and its frequency ω_p is scanned. At the time t_3 when ω_p passes through the value that satisfies the resonant condition of Eq. 14-2

$$\omega_p = \frac{g\beta H_0}{\hbar} + \tfrac{1}{2}A \qquad (14\text{-}18)$$

the $4 \to 1$ hyperfine transition of Fig. 14-2 becomes saturated. This produces nuclear spin flips that disturb the populations of levels 2 and 3 and thereby change the intensity of the $3 \to 2$ transition being monitored by the observing klystron. The result is a decrease in detected signal as the pumping frequency ω_p passes through its resonance value at the time t_3, as shown in

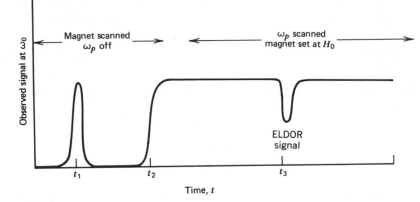

Fig. 14-3. ELDOR experiment showing the amplitude of the signal detected at the observed klystron frequency ω_0. At first the magnet is scanned with the pumping power turned off as shown on the left. At the time t_1 the magnetic field passes through the value $H = \hbar(\omega_0 + \frac{1}{2}A)/g\beta$ corresponding to the $4 \rightarrow 1$ transition of Fig. 14-2, and the scan is stopped at time t_2 when it reaches the value $H_0 = \hbar(\omega_0 - \frac{1}{2}A)/g\beta$ at the center of the $3 \rightarrow 2$ transition. At this point the pumping klystron is turned on and ω_p is scanned for times $t > t_2$ while H is maintained at the value H_0. At time t_3 the pumping frequency ω_p passes through the value that satisfies the resonance condition $\omega_p = g\beta H/\hbar + \frac{1}{2}A$ of the $4 \rightarrow 1$ transition and there is a decrease in the microwave power detected at the frequency ω_0. This decrease constitutes the ELDOR signal (from Refs. 3 and 4).

Fig. 14-3. If the term $g\beta H/\hbar$ is eliminated from Eqs. 14-17 and 14-18 by subtraction, then we obtain the relation

$$\omega_p - \omega_0 = A \tag{14-19}$$

which provides the hyperfine coupling constant A from the frequency difference $\omega_p - \omega_0$ when the pumping and observing klystrons are on resonance.

The technique just described is known as frequency swept ELDOR. Other possibilities exist such as scanning the observed frequency or the magnetic field, the latter being referred to as field swept ELDOR.[1]

Most ELDOR studies are carried out with the two microwave frequencies close together, within 10% of each other at X band (~9 GHz). In contrast to this, Cox et al.[8] used an X-band pumped frequency and a K-band observed frequency, Tanaka et al.[9] employed 22 and 9 GHz, respectively, for the pumped and observed frequencies, and Bowers and Mims[10] selected 3.9 and 4.1 GHz for these purposes.

14-6 STEADY-STATE ELDOR

In a steady-state ELDOR experiment a low-power microwave monitoring frequency detects the effect of the pumping microwave power on the energy level populations in dynamic equilibrium with it. This assumes that the rate of scanning through the resonant line is slow compared to the relaxation times of the system so that a state of dynamic equilibrium is maintained throughout the experiment.

To gain more insight into the ELDOR process we examine several simple cases wherein the populations of the dynamic equilibrium state can be determined by the elementary methods of Section 12-9 without solving the master equations. In addition, we take into account the effect of the applied pumping power through the factor δ that was defined by Eq. 12-49. The expressions for the ELDOR reduction factor that are given in Section 14-4 all assume that the pumping power is sufficiently high so that δ can be neglected.

We begin by examining the case in which the $4 \rightarrow 1$ allowed level is pumped and partly saturated, the $3 \rightarrow 2$ level is observed, and the relaxation takes place through the W_0 and W_2 forbidden transitions from level 2 to 4 and 1 to 3, respectively, as indicated in Fig. 14-4. Using the rules of Section 12-9 we have for the population differences of the levels

$$
\begin{aligned}
n_4 - n_1 &= 2\delta && (4 \rightarrow 1 \text{ pumped transition}) \\
n_3 - n_1 &= 2\epsilon && (1 \rightarrow 3 \text{ relaxation transition}) \qquad (14\text{-}20) \\
n_4 - n_2 &= 2\epsilon && (2 \rightarrow 4 \text{ relaxation transition})
\end{aligned}
$$

and the sum (12-43) over the levels provides a fourth condition

$$
n_1 + n_2 + n_3 + n_4 = 4 \qquad (14\text{-}21)
$$

These equations are easily solved to give the dynamic equilibrium populations indicated in Fig. 14-4. The monitored population difference

$$
n_3 - n_2 = 4\epsilon - 2\delta \qquad (14\text{-}22)
$$

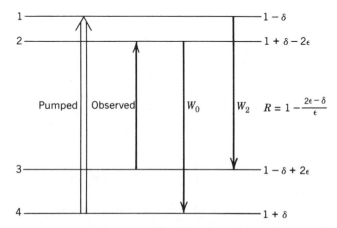

Fig. 14-4. Energy level populations during ELDOR when the forbidden transitions W_0 and W_2 dominate the relaxation. The nuclear Boltzmann factors are not taken into account.

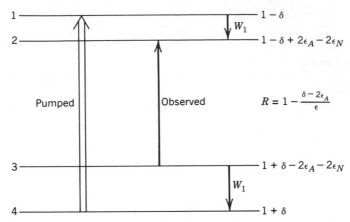

Fig. 14-5. Energy level populations during ELDOR when the nuclear transitions W_1 dominate the relaxation.

inserted into Eq. 14-12 gives the ELDOR reduction factor

$$R = 1 - \frac{2\epsilon - \delta}{\epsilon} \tag{14-23}$$

which corresponds to $R = -1$ for complete saturation ($\delta = 0$), in agreement with Eq. 14-13.

Figures 14-5 through 14-7 give three more examples of ELDOR in which very few relaxation paths are involved and the level populations are readily calculated. In the case presented in Fig. 14-5 the ELDOR reduction factor deviates from 1 only through the pumping factor δ and the hyperfine–

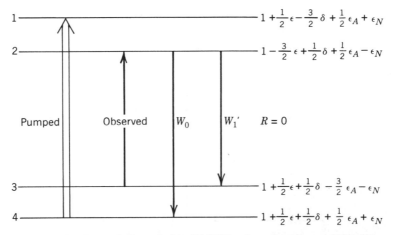

Fig. 14-6. Energy level populations during ELDOR when the allowed ESR W_1' and the $\Delta M = 0$ forbidden ESR W_0 transitions dominate the relaxation.

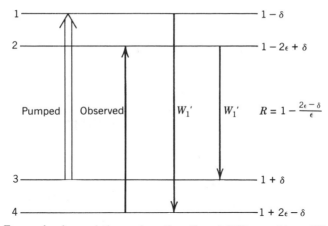

Fig. 14-7. Energy level populations when the allowed ESR transitions W_1' dominate the relaxation and the forbidden transitions $3 \to 1$ and $4 \to 2$, respectively, are pumped and observed. The nuclear Boltzmann factors are not taken into account.

Boltzmann factor ϵ_A. The situation depicted in Fig. 14-6 is one in which the relaxation rate W_1' maintains the Boltzmann equilibrium between levels 2 and 3 so that the pumping has no effect on the monitored intensity and $R = 0$. Figure 14-7 depicts a case in which both the pumped and observed transitions are of the forbidden type, and the reduction factor

$$R = 1 - \frac{2\epsilon - \delta}{\epsilon} \qquad (14\text{-}24)$$

depends on δ and ϵ as in Fig. 14-4. ELDOR can also be observed when either the pumped or the observed transition is forbidden and the other one is allowed. Several cases are summarized in Table 14-1.

14-7 TRANSIENT ELDOR

This chapter has emphasized steady-state ELDOR where dynamic equilibrium has been established. We see from the case that is illustrated in Fig. 14-6 that sometimes the ELDOR reduction factor R is zero so that no ELDOR effect is observed. Under these conditions it is possible to carry out a transient experiment in which the observed intensity is monitored during the period from the onset of the pumping power to the final attainment of dynamic equilibrium. To observe this transient ELDOR signal it is necessary to detect it over time intervals that are comparable to the relaxation times, which can be quite short. This experiment is analogous to the transient ENDOR one described in Section 13-7.

REFERENCES

1. J. S. Hyde, J. C. W. Chien, and J. H. Freed, *J. Chem. Phys.*, **48**, 4211 (1968).
2. L. Kevan and L. D. Kispert, *Electron Spin Double Resonance Spectroscopy*, Wiley, New York, 1976.
3. C. P. Poole, Jr. and H. A. Farach, *Appl. Spectrosc. Rev.*, **19**, 167 (1983).
4. C. P. Poole, Jr., *Electron Spin Resonance*, Wiley-Interscience, New York, 1983, Chap. 14.
5. L. R. Dalton and L. A. Dalton, *Magn. Reson. Rev.*, **1**, 301 (1972); **2**, 361 (1973).
6. M. M. Dorio, *Magn. Reson. Rev.*, **4**, 105 (1977).
7. J. H. Freed, *Electron Spin Resonance in Liquids*, L. T. Muus and P. W. Atkins, eds., Plenum, New York, 1972, pp. 503–530.
8. P. Cox, T. W. Flynn, and E. B. Wilson., *J. Chem. Phys.*, **42**, 3094 (1965).
9. S. Tanaka, A. Koma, and M. Kobayashi, *J. Phys. Soc. Japan*, **22**, 127 (1967).
10. K. D. Bowers and W. B. Mims, *Phys. Rev.*, **155**, 285 (1959).

DYNAMIC POLARIZATION

15-1 INTRODUCTION

In Chapter 13 we discussed electron–nuclear double resonance (ENDOR) in which a saturated ESR transition is monitored while a sweep is made at a high rf power level through one of the NMR transitions. The saturation of the NMR transition equalizes the populations of the corresponding nuclear spin levels, and this alters the electron spin populations of the pair of levels being monitored by ESR. As a result, the latter signal undergoes a detectable change which we refer to as an ENDOR signal. Thus ENDOR measures the influence of the saturating NMR power on the ESR transition.

The inverse experiment called dynamic nuclear polarization (DNP) was first proposed by Overhauser[1] and is sometimes referred to as the Over-hauser effect. In this case the NMR transition is monitored while the ESR transition is saturated, and the altering of the electron spin populations produces a pronounced change in the detected NMR signal. Under favorable conditions the population difference of the two NMR levels can increase to the point where it is determined by an ESR Boltzmann factor 2ϵ rather than an NMR one $2\epsilon_N$, and when this occurs the NMR signal can be enhanced by the ratio ϵ/ϵ_N which is a factor of 10^3. This process of dynamic nuclear polarization has been reviewed by several authors.[2-7]

15-2 THE HAMILTONIAN

In this chapter we treat the two-spin $S = \frac{1}{2}$, $I = \frac{1}{2}$ system in terms of the Hamiltonian

$$\mathcal{H} = g\beta HS_z - g_N\beta_N HI_z + A\vec{\mathbf{S}} \cdot \vec{\mathbf{I}} + \left(\frac{gg_N\beta\beta_N}{r^3}\right)[3(\vec{\mathbf{S}} \cdot \hat{r})(\vec{\mathbf{I}} \cdot \hat{r}) - \vec{\mathbf{S}} \cdot \vec{\mathbf{I}}] \qquad (15-1)$$

where for convenience ESR notation is used for both Zeeman terms. The energy level splittings shown in Fig. 15-1 are produced by a truncated Hamiltonian containing the Zeeman terms alone

$$\mathcal{H} = g\beta H S_z - g_N \beta_N H I_z \tag{15-2}$$

and this means that the two ESR transitions $3 \rightarrow 2$ and $4 \rightarrow 1$ are equal in energy, as are the two NMR transitions $1 \rightarrow 2$ and $4 \rightarrow 3$. The hyperfine term $A\vec{S} \cdot \vec{I}$ does not contribute to splitting the observed DNP signal into a doublet because the electron spins flip rapidly on the time scale of the NMR measurement and average to zero.[8,9] The effect of the hyperfine interaction is to shift the nuclear resonance frequency from the value ω_0 in the absence of this interaction to a new value ω given by[9,10]

$$\omega = \omega_0 \left(1 + \frac{A}{4kT} \frac{g\beta}{g_N \beta_N} \right) \tag{15-3}$$

This shift, not shown in Fig. 15-1, is small but observable. In a typical case $A/4kT \sim 10^{-6}$ and $g\beta/g_N\beta_N \sim 10^3$. The principal effect of the hyperfine term, however, is its contribution to the relaxation processes of the system, as will be explained in Section 15-6. The dipole–dipole interaction between the electron and nuclear spins, which appears at the end of the Hamiltonian (15-1), contributes to the relaxation without shifting the levels, as will also be explained in Section 15-6. In Section 12-5 we showed how these two Hamiltonian terms can be written in forms that are convenient for describing relaxation phenomena.

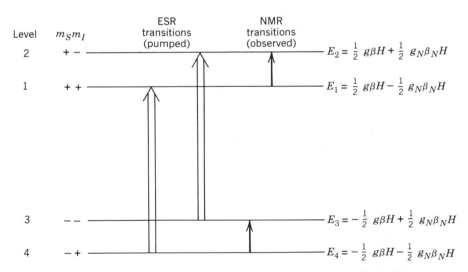

Fig. 15-1. Energy level diagram for an $S = \frac{1}{2}$, $I = \frac{1}{2}$ dynamic nuclear polarization experiment in which the allowed ESR $3 \rightarrow 2$ and $4 \rightarrow 1$ transitions are pumped and the NMR $1 \rightarrow 2$ and $4 \rightarrow 3$ transitions are observed. The level labels are given on the left and the energies on the right.

The practical result of the averaging of the hyperfine interaction is the effective equality of the two ESR transition frequencies and also of the two NMR frequencies. This causes the microwave pumping power to saturate both ESR level pairs simultaneously and the monitoring radiofrequency receiver to detect both NMR transitions at the same time. In addition, the analysis of the relaxation processes is somewhat simplified by this situation. One should also note that the ordering of the energy levels in Fig. 15-1 differs from that in Figs. 12-2, 12-6, 13-1, 13-2, 14-1, etc. which are drawn for the case of a positive hyperfine interaction that is much larger than the nuclear Zeeman term. The level ordering in the present case arises from the Zeeman interactions only. The values given in the figure of the m_S, m_I labels are not exact because of the mixing of the wavefunctions in the manner described in Sections 3-4 and 5-3, but we will find them a helpful guide during the discussions of the population changes.

15-3 MASTER EQUATIONS

We mentioned in the previous section that the hyperfine coupling does not contribute to the energy level splittings in a dynamic polarization experiment. This has two effects: (1) the $4 \rightarrow 1$ and $3 \rightarrow 2$ allowed ESR transitions become degenerate so the microwave pumping power saturates them simultaneously as shown in Fig. 15-2, and (2) only the Zeeman–Boltzmann factors ϵ and ϵ_N, determine the thermal equilibrium population distribution in the spin levels. The master equations (12-41) for these conditions are

$$\frac{dn_1}{dt} = W_1(n_2 - n_1) + W_2(n_3 - n_1) + (W_1' + P)(n_4 - n_1)$$
$$- 2\epsilon(W_1' + W_2) + 2\epsilon_N(W_1 + W_2) \tag{15-4a}$$

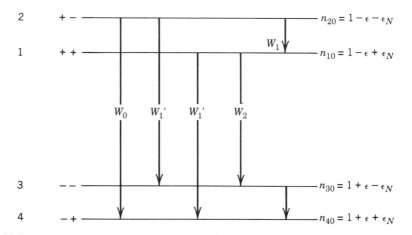

Fig. 15-2. Boltzmann populations and relaxation mechanisms for the $S = \frac{1}{2}$, $I = \frac{1}{2}$ spin system.

$$\frac{dn_2}{dt} = -W_1(n_2 - n_1) + (W_1' + P)(n_3 - n_2) + W_0(n_4 - n_2)$$
$$- 2\epsilon(W_0 + W_1') - 2\epsilon_N(W_0 + W_1) \tag{15-4b}$$

$$\frac{dn_3}{dt} = -W_2(n_3 - n_1) - (W_1' + P)(n_3 - n_2) + W_1(n_4 - n_3)$$
$$+ 2\epsilon(W_1' + W_2) - 2\epsilon_N(W_1 + W_2) \tag{15-4c}$$

$$\frac{dn_4}{dt} = -(W_1' + P)(n_4 - n_1) - W_0(n_4 - n_2) - W_1(n_4 - n_3)$$
$$+ 2\epsilon(W_0 + W_1') + 2\epsilon_N(W_0 + W_1) \tag{15-4d}$$

It is convenient to make the following substitutions in Eqs. 15-3a to 15-3d, respectively:

$$\begin{aligned} n_3 - n_1 &= (n_2 - n_1) + (n_3 - n_2) \\ n_4 - n_2 &= (n_1 - n_2) + (n_4 - n_1) \\ n_3 - n_1 &= (n_3 - n_4) + (n_4 - n_1) \\ n_4 - n_2 &= (n_4 - n_3) + (n_3 - n_2) \end{aligned} \tag{15-5}$$

which give

$$\frac{dn_1}{dt} = (W_1 + W_2)(n_2 - n_1) + W_2(n_3 - n_2) + (W_1' + P)(n_4 - n_1)$$
$$- 2\epsilon(W_1' + W_2) + 2\epsilon_N(W_1 + W_2) \tag{15-6a}$$

$$\frac{dn_2}{dt} = -(W_1 + W_0)(n_2 - n_1) + (W_1' + P)(n_3 - n_2) + W_0(n_4 - n_1)$$
$$- 2\epsilon(W_0 + W_1') - 2\epsilon_N(W_0 + W_1) \tag{15-6b}$$

$$\frac{dn_3}{dt} = (W_1 + W_2)(n_4 - n_3) - (W_1' + P)(n_3 - n_2) - W_2(n_4 - n_1)$$
$$+ 2\epsilon(W_1' + W_2) - 2\epsilon_N(W_1 + W_2) \tag{15-6c}$$

$$\frac{dn_4}{dt} = -(W_0 + W_1)(n_4 - n_3) - W_0(n_3 - n_2) - (W_1' + P)(n_4 - n_1)$$
$$+ 2\epsilon(W_0 + W_1') + 2\epsilon_N(W_0 + W_1) \tag{15-6d}$$

If we add the first and fourth equations from this group and subtract from this sum the second and third equations we obtain the expression

$$\frac{d}{dt}[n_1 + n_4 - n_2 - n_3] = -W[n_1 + n_4 - n_2 - n_3] + 4W\epsilon_N + (W_0 - W_2)$$
$$\times [4\epsilon - (n_3 + n_4 - n_1 - n_2)] \tag{15-7}$$

where W is the reciprocal of the spin–lattice relaxation time

$$W = \frac{1}{T_1} = W_0 + 2W_1 + W_2 \tag{15-8}$$

Our main interest is the dynamic steady-state solution to Eq. 15-7 which is obtained by setting the time derivative equal to zero:

$$W[n_1 + n_4 - n_2 - n_3] = 4W\epsilon_N + (W_0 - W_2)[4\epsilon - (n_3 + n_4 - n_1 - n_2)] \tag{15-9}$$

This relation will be used in the next section to write an expression for the nuclear polarization P_N. The time-dependent solution to Eq. 15-7 will be given in Section 15-8.

15-4 POLARIZATION AND ENHANCEMENT

The enhancement of the NMR signal can be described in terms of the nuclear polarization P_N which is defined as the number of nuclear spins in energy levels where they point upward ($m_I = +\frac{1}{2}$) minus the number in levels where they point downward ($m_I = -\frac{1}{2}$). With the aid of Fig. 15-1 we can write

$$P_N = (n_1 + n_4) - (n_2 + n_3) \tag{15-10}$$

The thermal equilibrium polarization P_{N0} is obtained by writing Eq. 15-10 for the corresponding equilibrium populations

$$P_{N0} = (n_{10} + n_{40}) - (n_{20} + n_{30}) \tag{15-11}$$

and using the values in Fig. 15-2 gives

$$P_{N0} = 4\epsilon_N \tag{15-12}$$

We can write an expression analogous to Eq. 15-10 for the polarization P_S of the electronic spins by defining P_S as the number of electron spins in energy levels where they point downward ($m_S = -\frac{1}{2}$) minus the number in levels where they point upward ($m_S = +\frac{1}{2}$). The interchange of the words "upward" and "downward" in this definition relative to the nuclear case results from the difference in the signs of the two g-factors in Eq. 15-1. With the help of Fig. 15-1 we write

$$P_S = (n_3 + n_4) - (n_1 + n_2) \tag{15-13}$$

and the thermal equilibrium value

$$P_{S0} = (n_{30} + n_{40}) - (n_{10} + n_{20}) \tag{15-14}$$
$$= 4\epsilon \tag{15-15}$$

is easily written from the level populations in Fig. 15-2.

A DNP experiment measures the enhancement E which is the ratio of the nuclear polarization P_N to its value P_{N0} in thermal equilibrium

$$E = \frac{P_N}{P_{N0}}$$
$$= \frac{P_N}{4\epsilon_N} \tag{15-16}$$

The quantity in square brackets on the left-hand side of Eq. 15-9 is the nuclear polarization (15-10), and this may be substituted into Eq. 15-16 to give

$$E = 1 + \frac{W_0 - W_2}{4\epsilon_N W}(4\epsilon - P_S) \tag{15-17}$$

where we have used the fact that the square brackets on the right-hand side of Eq. 15-9 enclose the populations that constitute the electronic polarization (15-13).

We see from this equation that the DNP signal is proportional to the extent to which the electron spin polarization P_S deviates from its thermal equilibrium value $P_{S0} = 4\epsilon$, so it is convenient to define the saturation factor S as follows:

$$S = \frac{P_{S0} - P_S}{P_{S0}}$$
$$= 1 - \frac{P_S}{4\epsilon} \tag{15-18}$$

In addition, the ratio of the Boltzmann factors ϵ/ϵ_N is the same as the ratio of the electron spin gyromagnetic ratio γ_S to the nuclear spin gyromagnetic ratio γ_N

$$\frac{\epsilon}{\epsilon_N} = \frac{\gamma_S}{\gamma_N} = \frac{g\beta}{g_N\beta_N} \tag{15-19}$$

which has a value of about 10^3, and for protons on free radicals with $g = 2.00$ it equals 657.5. Using this ratio and the saturation factor S permits Eq. 15-17 to be written in the convenient form

$$E = 1 + \frac{\gamma_S}{\gamma_N}\left(\frac{W_0 - W_2}{W}\right)S \tag{15-20}$$

If additional relaxation processes W_A contribute to the spin–lattice relaxation time (15-8)

$$\frac{1}{T_1} = W + W_A \tag{15-21}$$

but not to the enhancement, we can take this into account by a correction factor called the leakage factor

$$f = \frac{W}{W + W_A} \qquad (15\text{-}22)$$

which represents energy which "leaks" to the lattice without contributing to the enhancement. Some authors define the quantity

$$\xi = \frac{W_0 - W_2}{W} \qquad (15\text{-}23)$$

called the coupling factor. Using this plus the leakage factor permits us to write the enhancement in the final form[4,11,12]

$$E = 1 + \frac{\gamma_S}{\gamma_N} \xi f S \qquad (15\text{-}24)$$

which can be compared with experiment.

Since the amplitude Y of the DNP signal is proportional to the population difference and hence to the polarization, it follows that the enhancement can be written as the ratio of the signal amplitude Y_P detected from a polarized sample to that Y_0 measured under thermal equilibrium conditions

$$E = \frac{Y_P}{Y_0} \qquad (15\text{-}25)$$

15-5 INDIVIDUAL RELAXATION PATHS

It is of interest to write the nuclear polarization and the enhancement for situations in which one of the relaxation paths, W_0, W_1, or W_2, is operative. This is easily handled by the methods of Section 12-9 and the results are illustrated in Figs. 15-3 to 15-5. We see from these figures that in each case the electron spin polarization P_S has the value

$$P_S = 4\delta \qquad (15\text{-}26)$$

which gives for the saturation factor

$$S = 1 - \frac{\delta}{\epsilon} \qquad (15\text{-}27)$$

where δ is defined by Eq. 12-49.

From an inspection of Fig. 15-3 we can write the polarizations and enhancement for the W_0 case as follows:

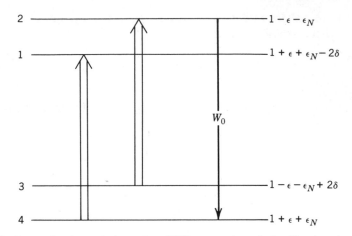

Fig. 15-3. Energy level populations when DNP occurs through the W_0 relaxation process.

$$P_N = 4(\epsilon + \epsilon_N - \delta)$$

$$E = 1 + \frac{\gamma_S}{\gamma_N} S \tag{15-28}$$

where use was made of Eqs. 15-10, 15-16, 15-19 and 15-27. This positive enhancement corresponds to the true Overhauser effect. When W_1 acts alone we have from Fig. 15-4

$$P_N = 4\epsilon_N$$

$$E = 1 \tag{15.29}$$

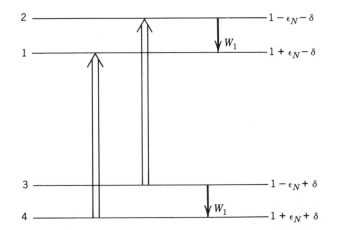

Fig. 15-4. Energy level populations when DNP occurs through the W_1 relaxation processes.

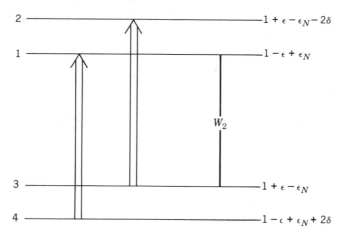

Fig. 15-5. Energy level populations when DNP occurs through the W_2 relaxation process.

which means that no enhancement occurs, while for W_2 the enhancement has a negative component, see Fig. 15-5,

$$P_N = 4(\epsilon_N - \epsilon + \delta)$$

$$E = 1 - \frac{\gamma_S}{\gamma_N} S$$

(15-30)

commonly referred to as the underhauser effect.[13]

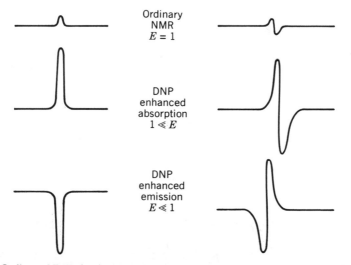

Fig. 15-6. Ordinary NMR (top), enhanced absorption (center), and enhanced emission (bottom) signals recorded in the absorption mode (left) and in the absorption first derivative mode (right).

A positive enhancement corresponds to a stronger absorption signal while a negative enhancement corresponds to a stronger emission signal, as was explained in Section 12-10 and as is illustrated in Fig. 15-6. We see from this figure that an emission signal is recorded opposite in phase to an absorption one.

Equations 15-28 to 15-30 are, of course, special cases of Eq. 15-20. They are written separately with the aid of Figs. 15-3 to 15-5 to give more insight into the dynamic polarization process.

15-6 SCALAR AND DIPOLAR RELAXATION

There are two cases of particular importance to which we will apply the general formula 15-20 for the nuclear polarization, namely, the scalar and dipolar cases that were examined in Section 12-5.

The scalar mechanism results from the operation of the hyperfine interaction $A\vec{S} \cdot \vec{I}$, and we saw in Section 12-5 that this interaction induces only transitions with $M = 0$ corresponding to the transition rate W_0. This is the case of Eq. 15-28 which gives the positive enhancement

$$E = 1 + \frac{\gamma_S}{\gamma_N} fS \qquad (15\text{-}31)$$

of the Overhauser effect where we have included the leakage factor f of Eq. 15-22. This enhancement was originally reported in metals where the coupling to the electrons is scalar.[14,15]

Dipolar relaxation is more complicated because the five terms B, C, D, E, and F of Eq. 12-7 all contribute to the mechanism. We showed in Section 12-5 that in the high-temperature limit corresponding to the inequality $\omega_0 \tau_c \ll 1$ the molecular tumbling averages all the spherical harmonics of Eq. 12-8 to the same value, and Eqs. 12-19 provide the following ratios of the relaxation rates:

$$\frac{W_0}{2} = \frac{W_1}{3} = \frac{W_2}{12} \qquad (15\text{-}32)$$

Inserting these into Eq. 15-23 provides the coupling factor

$$\xi = -\tfrac{1}{2} \qquad (15\text{-}33)$$

which gives a negative enhancement

$$E = 1 - \frac{1}{2} \frac{\gamma_S}{\gamma_N} fS \qquad (15\text{-}34)$$

or underhauser effect.

15-7 DYNAMIC NUCLEAR POLARIZATION VIA FORBIDDEN TRANSITIONS

The polarization schemes that we have discussed in the previous sections all involved saturating either the $4 \to 1$ or the $3 \to 2$ allowed ESR transitions which satisfy the condition $\Delta M = \pm 1$ and are degenerate in energy. Nuclear polarizations can also be achieved by saturating the forbidden transitions $4 \to 2$ with $\Delta M = 0$ and $3 \to 1$ with $\Delta M = \pm 2$. When the coupling is scalar so that $W_2 = 0$ and the $4 \to 2$ transition is pumped with the radiofrequency magnetic field H_1 aligned along the static field H_0 direction the experiment is referred to as the Jeffries effect.[3,16] This effect is ordinarily studied for spins $I > \frac{1}{2}$. Another phenomenon called the solid effect[17] occurs with tensor coupling when the two relaxation rates W_0 and W_2 are comparable to each other and either the $4 \to 2$ or $3 \to 1$ forbidden transition is saturated. Explicit expressions of the polarization involved in these effects can be derived from the master equations introduced in Section 12-8.

15-8 TRANSIENT NUCLEAR POLARIZATION

The discussion until now has concerned the dynamic equilibrium state that results from setting all time derivatives in Eqs. 15-4, 15-6, and 15-7 equal to zero. When these time derivatives are not set equal to zero, we obtain the differential equation 15-7 which can be written in the following form:

$$\frac{dP_N}{dt} = -W(P_N - 4\epsilon_N) + 4\epsilon S(W_0 - W_2) \tag{15-35}$$

with the aid of Eqs. 15-10, 15-13, and 15-18.

To solve this equation for the time evolution of the nuclear polarization from the initial turning on of the ESR saturating power to the final establishment of the dynamic equilibrium state we make the change of variable

$$x = \frac{P_N - P_{N0}}{P_{N0}} \tag{15-36}$$

to obtain the simpler equation

$$\frac{dx}{dt} = -Wx + 4\epsilon S(W_0 - W_2) \tag{15-37}$$

If the system starts with thermal equilibrium of the nuclear polarization $(x = 0)$ at the time $t = 0$ then the solution is

$$\frac{P_N}{P_{N0}} = 1 + \frac{\gamma_S}{\gamma_N} \frac{W_0 - W_2}{W} S(1 - e^{-Wt}) \tag{15-38}$$

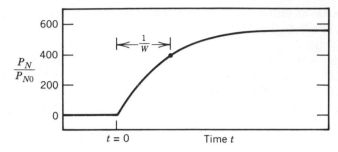

Fig. 15-7. Time evolution of the nuclear polarization when the pumping microwave power is turned on at time $t = 0$. The time constant $T_1 = 1/W$ of the process is indicated.

which satisfies the initial condition of $P_N = P_{N0}$ at $t = 0$ and gives the equilibrium result (15-20) when the time t goes to infinity.

Figure 15-7 shows a graph of $P_N(t)/P_{N0}$ plotted against time. The time constant for this process is the reciprocal $1/W$ of the overall relaxation rate from Eq. 15-8, as indicated in the figure. For times short compared to this time constant the nuclear polarization approximated by expanding the exponential e^{-Wt} in a power series and retaining only the first two terms $1 - Wt$

$$\frac{P_N}{P_{N0}} = 1 + \frac{\gamma_S}{\gamma_N}(W_0 - W_2)St \quad \text{for } t \ll \frac{1}{W} \tag{15-39}$$

increases linearly with the time.

15-9 CHEMICALLY INDUCED DYNAMIC NUCLEAR POLARIZATION

Until now we have discussed the case of a two-spin $(\frac{1}{2}, \frac{1}{2})$ system which starts in thermal equilibrium and is disturbed therefrom by saturating microwave power to produce strongly polarized nuclear levels. When a free radical is formed during a chemical reaction its initial spin level populations, in general, are not in thermal equilibrium with each other. After formation the various relaxation mechanisms rearrange the populations and return them to their equilibrium distributions. This causes the polarizations to change with time, hence the name chemically induced dynamic nuclear polarization (CIDNP), and a measurement of these changes provides important information needed for understanding the kinetics of the radical formation and recombination processes.[18, 19]

Glarum[20] has analyzed a simple model in which the radical is formed with all four of its spin levels equally populated and relaxation occurs through rotational modulation of the dipolar couplings between the spins. The relaxation times between the various pairs of energy levels sketched in Fig.

15-2 are related through Eq. 15-32. The initial polarization of the equally populated levels is, of course, zero. It grows with time, passes through a maximum, and then decays to its thermal equilibrium value.

Glarum[20] evaluated the overall time dependence with the aid of the master equations and he obtained for the enhancement

$$E = 1 - \frac{\epsilon}{2\epsilon_N} \exp(-Wt)[1 - \exp(-2Wt)] \tag{15-40}$$

where the negative sign indicates an NMR emission signal, and

$$W = \frac{10W_1'}{3} \tag{15-41}$$

Equation 15-40 is plotted as a function of Wt in Fig. 15-8. We see from the figure that the magnitude of the enhancement passes through a maximum before decaying exponentially to its final equilibrium value. If E is differentiated with respect to time and the derivative is set equal to zero we obtain the condition for the maximum

$$t_{MAX} = \frac{\ln 3}{2W} \tag{15-42}$$

The curve of Fig. 15-8 is what would be measured experimentally if NMR could be used to follow the course of an individual radical's formation and decay. If the radical lifetime t_{RAD} is much longer than T_{MAX} then the ratio of the maximum polarization to the final product signal from an ensemble of radicals will decrease by the ratio t_{MAX}/t_{RAD}. Other factors such as viscous effects, distributions of radical lifetimes, additional relaxation mechanisms, and so on cause this simple model to break down.

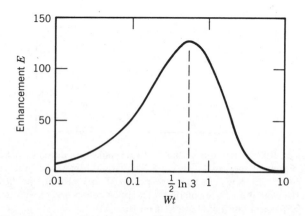

Fig. 15-8. Time evolution of a CIDNP process (from Ref. 20).

Most CIDNP results have been attributed to specific mechanisms. One of them involves the interactions between radical pairs that can be in either singlet or triplet states. The transitions between the singlet and triplet states involve transition probabilities that depend on the nuclear spin states, and hence the product radicals are created with nonequilibrium population distributions. Some photochemically produced radicals have a particular initial polarization while others evolve through the radical pair mechanism.[21-29]

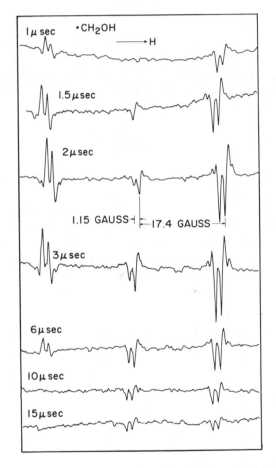

Fig. 15-9. Polarized ESR spectrum from the ·CH_2OH radical produced by pulse radiolysis in a 10% aqueous solution of methanol (pH = 2.6) at various times after the cessation of the electron beam pulse. The low-field lines are polarized in emission, the high-field ones in absorption, and the central line is unpolarized. The polarization decays with time to produce a final symmetric ESR spectrum (not shown) (from Ref. 27).

15-10 CHEMICALLY INDUCED DYNAMIC ELECTRON POLARIZATION

In the previous section we discussed the measurement of the nuclear polarization of radicals produced in nonequilibrium nuclear spin states. Free radicals or triplet states can also be produced with a nonequilibrium population distribution in different electron spin m_S states, and this chemically induced dynamic electron polarization (CIDEP) causes the relative intensities of the hyperfine components to differ from the usual binomial or equal intensity type distributions.[19,21]

The spin state populations often deviate so much from a Boltzmann distribution that some lines appear in emission,[27] as shown in Fig. 15-9. This is explained by postulating that a molecule S called a precursor splits to form a pair of radicals R_A^* and R_B^* in excited electronic states denoted by superscripted asterisks ($*$)

$$S \rightarrow R_A^* + R_B^* \tag{15-43}$$

One member of the pair R_A^* is formed in inverted spin states so that it produces emission lines and the other member R_B^* is polarized in the opposite sense so that it produces enhanced absorption spectra, as illustrated in Fig. 15-9. Such an intense asymmetric spectrum decays with time to a symmetric absorption pattern of ordinary intensity.

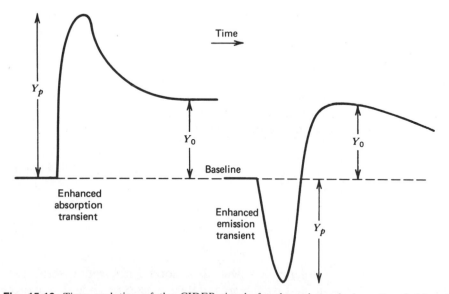

Fig. 15-10. Time evolution of the CIDEP signals for the enhanced absorption (left) and enhanced emission (right) cases. Figure 15-11 presents the enhanced emission on a broader time scale to show the subsequent decay process (from Refs. 28, 29).

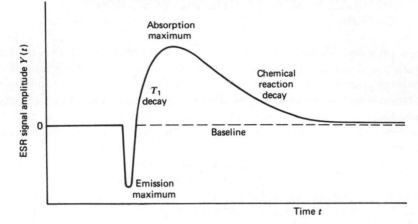

Fig. 15-11. Time evolution of the ESR absorption $Y(t)$ for the CIDEP enhanced emission case showing the initial rapid rise of the number of emitting radicals, followed by their rapid decay to the normal absorption condition with the time constant T_1, and the subsequent slower decay through radical recombination or chemical reactions. The emission profile is sketched on the right side of Fig. 15-10 on a narrower time scale (from Refs. 28, 29).

The time evolution of the electron spin polarization can be studied by producing the radicals in a liquid placed in a resonant cavity using a pulse of radiation shorter than the spin–lattice relaxation time T_1, typically between 2 nsec and 20 μsec in length. The spectrometer monitors the radical population with time, and Fig. 15-10 shows the time evolution of the ESR signal amplitude in a CIDEP experiment that exhibits transient absorption and another that corresponds to transient emission.[26,27] In both cases the signal $Y(t)$ passes through an extremum polarization amplitude Y_p and then decays to an ordinary absorption signal Y_0. In each case we can define the polarization enhancement factor E':

$$E' = \frac{|Y_p| - |Y_0|}{|Y_0|} \quad \text{Enhanced absorption} \qquad (15\text{-}44)$$

$$E' = \frac{|Y_p| + |Y_0|}{|Y_0|} \quad \text{Emission} \qquad (15\text{-}45)$$

The decay from the polarization peak to a normal absorption state takes place at a rate determined by the spin–lattice relaxation time T_1, and the radicals themselves typically decay at a much slower rate via chemical reactions in the manner illustrated in Fig. 15-11.[28,29]

REFERENCES

1. A. W. Overhauser, *Phys. Rev.*, **92**, 411 (1953).
2. R. H. Webb, *Am. J. Phys.*, **29**, 428 (1961).
3. C. D. Jeffries, *Dynamic Nuclear Orientation*, Wiley-Interscience, New York, 1963.
4. K. H. Hausser and D. Stehlik, *Adv. Magn. Reson.*, **3**, 79 (1968).
5. L. M. Jackman and F. A. Cotton, eds., *Dynamic Nuclear Magnetic Resonance Polarization*, Academic Press, New York, 1975.
6. L. L. Buishvili, T. G. Vardosanidze, and A. I. Ugulava, *Fiz. Tverd. Tela*, **18**, 558 (1976).
7. W. Müller-Warmuth and K. Meise-Gresch, *Adv. Magn. Reson.*, **11**, 1 (1983).
8. I. Solomon and N. Bloembergen, *J. Chem. Phys.*, **25**, 261 (1956).
9. T. H. Brown, D. H. Anderson, and H. S. Gutowsky, *J. Chem. Phys.*, **33**, 720 (1960).
10. K. H. Hausser, H. Brunner, and J. C. Jochims, *Mol. Phys.*, **10**, 253 (1966).
11. A. Abragam, *Phys. Rev.*, **98**, 1729 (1955).
12. A. Abragam, *The Principles of Nuclear Magnetism*, Oxford University Press, London, 1961.
13. L. H. Bennett and H. C. Torrey, *Phys. Rev.*, **108**, 449 (1957).
14. T. R. Carver and C. P. Slichter, *Phys. Rev.*, **92**, 212 (1953).
15. T. R. Carver and C. P. Slichter, *Phys. Rev.*, **102**, 975 (1956).
16. C. D. Jeffries, *Phys. Rev.*, **117**, 1056 (1960).
17. A. Abragam and W. G. Proctor, *Compt. Rend.*, **246**, 2253 (1958).
18. L. T. Muus, P. W. Atkins, K. A. McLauchlan, and J. B. Peterson, eds., *Chemically Induced Magnetic Polarization*, Reidel, Dordrecht, Holland, 1977.
19. A. R. Lepley and G. L. Closs, eds., *Chemically Induced Magnetic Polarization*, Wiley, New York, 1973.
20. S. H. Glarum, in *Chemically Induced Magnetic Polarization*, A. R. Lepley and G. L. Closs, eds., Wiley, New York, 1973, pp. 1–38.
21. J. H. Freed and J. B. Pederson, *Adv. Magn. Reson.*, **8**, 1 (1976).
22. J. K. S. Wan, *Adv. Photochem.*, **12**, 283 (1980).
23. A. D. Trifunac, *Magn. Reson. Rev.*, **7**, 147 (1982).
24. S. I. Weissman, *Ann. Rev. Phys. Chem.*, **33**, 101 (1982).
25. M. C. Depew and J. K. S. Wan, *Magn. Reson. Rev.*, **8**, 85 (1983).
26. G. L. Closs, *Adv. Magn. Reson.*, **7**, 157 (1974).
27. A. D. Trifunac and M. C. Thurnauer, *J. Chem. Phys.*, **62**, 4889 (1975).
28. C. P. Poole, Jr., and H. A. Farach, *Appl. Spectrosc. Rev.*, **19**, 167 (1983).
29. C. P. Poole, Jr., *Electron Spin Resonance*, 2nd ed., Wiley-Interscience, New York, 1983.

NUCLEAR–NUCLEAR
DOUBLE RESONANCE

16-1 INTRODUCTION

The previous three chapters have treated double resonance experiments involving one electronic spin $S = \frac{1}{2}$ and one nuclear spin $I = \frac{1}{2}$; this chapter is concerned with nuclear spins only.[1-4] The three static Hamiltonian terms are the Zeeman interactions of each spin with the external magnetic field and the hyperfine coupling of the two spins with each other. Most of the double resonance experiments that were described previously involved the saturation of a particular transition to change the populations of the energy levels and the observation of the effect of this population redistribution on the intensities of the spectral lines arising from other transitions. When the same type of double resonance is carried out with nuclear spin multiplet systems it is referred to as the nuclear Overhauser effect.[5-8] However, some of the most common nuclear–nuclear double resonance experiments such as spin tickling and spin decoupling[8-11] do not arise from redistributions of level populations but rather from the presence of the pumping rf field term in the spin Hamiltonian which is diagonalized to provide the energies. In Section 16-5 we will solve the spin Hamiltonian of the AX spin system containing a time-varying Zeeman term in the x direction arising from the pumping field. This solution will provide us with the proper conditions for carrying out spin tickling and decoupling experiments.

In Sections 16-3 and 16-4 we examine the nuclear Overhauser effect and then proceed to the discussion of spin tickling and spin decoupling. The Overhauser effect is the enhancement or reduction in the intensity of the lines from one nuclear species when another is selectively saturated. It can provide the distances between the two interacting nuclei. Spin tickling is a method for identifying the particular spin–spin coupling interactions that contribute to individual lines in multiplets by saturating the transition frequency of one line and observing the doubling of another. Spin decoupling provides an increase in the resolution of the multiplets of one nuclear spin by pumping the NMR frequency of another spin that is coupled to it. The result is to eliminate the influence of the pumped nucleus on the spectral lines of the other, thereby reducing the number of lines in the latter's spectrum. For example, one can decouple protons and eliminate their influence on ^{13}C spectra.

16-2 PRODUCING NUCLEAR DOUBLE RESONANCE

Nuclear double resonance experiments are carried out over a wide range of pumping powers, and different effects are observed for different ranges of power. This occurs because the rf field term H_{rf} given by

$$\mathcal{H}_{rf} = \gamma H_1 I_z \cos \omega t \tag{16-1}$$

can play two distinct roles in producing the spectra. Its first is that of inducing the transition via the matrix element $\langle i|H_{rf}|j\rangle$, and this is its usual role at low power levels where the magnitude of H_{rf} is small compared to the other terms in the Hamiltonian. When H_{rf} is comparable to other Hamiltonian terms it also contributes to the spin Hamiltonian which is diagonalized to provide the energy levels. In this role H_{rf} splits and shifts the energy levels and spectral lines, and the result is the observation of phenomena such as spin tickling and decoupling. Population changes can not shift or split energy levels.

Nuclear double resonance methods are especially useful when working with spin systems in which the multiplet structure of each nucleus is well resolved and widely separated from the multiplets of other nuclear spins. For such a spectrum the chemical shift $\omega_0 \delta_{ij} = (\gamma_j - \gamma_i)H$ between each pair of nuclei is much greater than the corresponding spin–spin coupling constant J_{ij}. In this chapter we investigate the AX system which produces a well-resolved quartet spectrum of the type presented in Fig. 16-1. Such a spectrum has parameters which satisfy the following inequalities:

$$\frac{1}{(T_1 T_2)^{1/2}} \le \frac{1}{T_2} \ll J \ll |\omega_A - \omega_X| \tag{16-2}$$

The various types of double resonance phenomenon occur when γH_1 assumes different values in this range of inequalities. Most high-resolution NMR experiments are carried out at low power where γH_1 is much less than $1/(T_1 T_2)^{1/2}$ since this is the condition for obtaining narrow lines. If the rf field H_1 is made strong enough so that saturation effects begin to be detected, corresponding to the condition (see Eq. 11-119)

$$1 < \gamma^2 H_1^2 T_1 T_2 \tag{16-3}$$

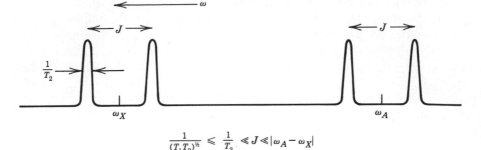

$$\frac{1}{(T_1 T_2)^{1/2}} \leqslant \frac{1}{T_2} \ll J \ll |\omega_A - \omega_X|$$

Fig. 16-1. Conditions for a well-resolved AX NMR spectrum.

which may also be written in the form

$$\frac{1}{(T_1 T_2)^{1/2}} < \gamma H_1 \tag{16-4}$$

then the lines broaden and the populations of the energy levels become redistributed. This latter result produces double resonance effects of the type discussed in Chapters 12–15 in which the relative intensities of the various spectral lines are altered. The observance of these intensity changes in NMR spin systems is referred to as the nuclear Overhauser effect, and this will be discussed in the next two sections. The Overhauser effect is dominant in the following range of γH_1 values,

$$\frac{1}{(T_1 T_2)^{1/2}} \leq \gamma H_1 \ll J \tag{16-5}$$

but continues to occur for higher power levels where new phenomena begin to appear.

The use of pumping powers in excess of the amount needed to produce Overhauser enhancements is found to shift and split the observed spectral lines, and this indicates that the radiofrequency magnetic field H_1 is now strong enough to contribute to the Zeeman term in the spin Hamiltonian in the manner mentioned above. Higher values of the radiofrequency magnetic field corresponding to the range

$$\frac{1}{(T_1 T_2)^{1/2}} \ll \gamma H_1 < J \tag{16-6}$$

produce satellite lines and the phenomenon called spin tickling, while even stronger radiofrequency fields corresponding to

$$J \leq \gamma H_1 \ll |\omega_A - \omega_X| \tag{16-7}$$

produce spin decoupling. The next sections will discuss these various effects.

16-3 NUCLEAR OVERHAUSER EFFECT

The nuclear Overhauser effect (NOE) is a dynamic polarization experiment involving two nuclear species[5-8] which is analogous to the dynamic nuclear polarization (DNP) experiment treated in Chapter 15 which involved an electron spin and a nuclear spin. In an NOE experiment one nuclear species I_X is pumped and this polarizes the spins of another nuclear species I_A so that a scan through the resonant frequency of the latter will produce an enhanced or reduced signal, depending on the dominant relaxation mechanisms and the relative signs of the gyromagnetic ratios. Our procedure is to start with the Hamiltonian of the $I_X = I_A = \frac{1}{2}$ spin system written in NMR notation and to adopt the relevant master equations and subsidiary relations from Chapter 15 to the present case as appropriate.

The spin Hamiltonian for the AX spin system is

$$\mathcal{H} = \hbar H_0 [\gamma_A I_{Az} + \gamma_X I_{Xz}] \tag{16-8}$$

where only Zeeman terms contribute to the energies. For simplicity we assign positive signs to the two terms of Eq. 16-8. The four energies of this system are listed in Fig. 16-2. The numbering of the levels on the left side of this figure uses the convention of Section 12-3 that is illustrated in Fig. 12-2. This makes it easier to apply the results obtained for the electron–nucleus DNP case of Chapter 15 to the NMR cases of the present chapter. In Chapter 4 (e.g., in Fig. 4-1) we followed the usual NMR convention which interchanges the labels 3 and 4 of the lowest two levels.

We see from Fig. 16-2 that the two pumped transitions, namely, the $3 \rightarrow 2$ and $4 \rightarrow 1$ transitions, are both equal in energy

$$E_{\text{PUMP}} = \hbar \gamma_X \tag{16-9}$$

as are the two observed transitions $2 \rightarrow 1$ and $3 \rightarrow 4$

$$E_{\text{OBS}} = \hbar \gamma_A \tag{16-10}$$

The double degeneracy of the pumped transition energy given by Eq. 16-9 causes all the spins of the observed or A nucleus to become polarized, and the extent of this polarization is easily deduced from the master equations. There is a formal identity between the spin Hamiltonian (16-8) introduced here and the truncated spin Hamiltonian (15-2) which was responsible for the dynamic nuclear polarization described in Chapter 15. The only difference between the Hamiltonians, aside from the notation, is the negative sign that appears on one of the terms in Eq. 15-2. This means that all the results of Sections 15-3 and 15-4 can be taken over here directly if we make sure to include the requisite sign change in γ_A. For the equations that we quote here we will give the source equations from Chapters 12 and 15 in square brackets.

The A and X nuclei have the following polarizations [(15-10), (15-13)]

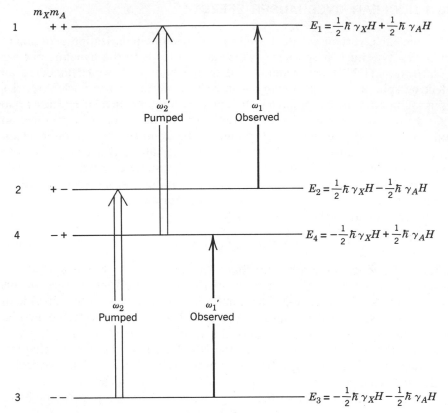

Fig. 16-2. Energy levels for the $I_A = I_X = \frac{1}{2}$, AX spin system showing the pumped $3 \to 2$ and $4 \to 1$ transitions, the observed $2 \to 1$ and $3 \to 4$ transitions, the level labels on the left, and the energies on the right.

$$P_A = (n_1 + n_4) - (n_2 + n_3)$$
$$P_X = (n_1 + n_2) - (n_3 + n_4) \tag{16-11}$$

and thermal equilibrium polarizations [(15-12), (15-15)]

$$P_{A0} = 4\epsilon_A$$
$$P_{X0} = 4\epsilon_X \tag{16-12}$$

where the Boltzmann factors given in Fig. 16-3 for the various energy levels have the usual definitions [(12-23), (12-40)]

$$\epsilon_A = \frac{\hbar \gamma_A H}{2kT}$$

$$\epsilon_X = \frac{\hbar \gamma_X H}{2kT} \tag{16-13}$$

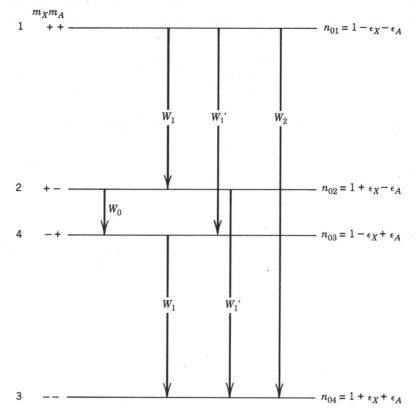

Fig. 16-3. Energy levels, transition rates, and thermal equilibrium populations of an AX spin system.

The saturation factor for the pumped X nucleus is [(15-18)]

$$S_X = \frac{4\epsilon_X - P_X}{4\epsilon_X} \qquad (16\text{-}14)$$

The enhancement of the I_A resonance caused by pumping the I_X transitions is [(15-20)]

$$E = 1 + \frac{\gamma_X}{\gamma_A} \left(\frac{W_2 - W_0}{W} \right) S_X \qquad (16\text{-}15)$$

The leakage factor correction term f of Eq. 15-22 may be included in this expression if desired. In writing this equation we reversed the sign before the last term of Eq. 15-20 to take into account the presence of the two positive signs in the spin Hamiltonian 16-8. Table 16-1 gives values for the ratios γ_X/γ_A for several important γ_A nuclei when γ_X corresponds to protons.

TABLE 16-1 Ratio of Gyromagnetic Ratios γ_X/γ_A for Several $I_A = \frac{1}{2}$ Nuclei When the X Nucleus Is a Proton

Nucleus	$\gamma_A (\text{rad/sec} \cdot \text{G})^a$	γ_X/γ_A	Abundance (%)
^{1}H	26,751	1.00	99.985
^{13}C	6,726	3.98	1.108
^{15}N	−2,711	−9.87	0.37
^{19}F	25,167	1.06	100.
^{29}Si	−5,314	−5.03	4.70
^{31}P	10,829	2.47	100.
^{77}Se	5,101	5.24	7.58

^{a} Note that the gyromagnetic ratios of ^{15}N and ^{29}Si are negative.

Section 15-6 gives the values of the negative coupling factor $(W_2 - W_0)/W$ equal to $+\frac{1}{2}$ and -1 for dipolar[12] and scalar relaxation, respectively, which gives for the enhancements

$$E = 1 + \frac{1}{2}\frac{\gamma_X}{\gamma_A} S_X \quad \text{(Dipolar)}$$

$$E = 1 - \frac{\gamma_X}{\gamma_A} S_X \quad \text{(Scalar)}$$

(16-16)

Here use was made of Eqs. 15-34 and 15-31, respectively, taking into account the change in sign of the gyromagnetic ratio for the nuclear versus the electronic case.

Figure 16-4 illustrates how a nuclear Overhauser experiment can be carried out. The first spectrum shown is an ordinary NMR one with a single resonance line of amplitude Y_0 at the frequency ω_X of nucleus X and a second line also of amplitude Y_0 at the position ω_A of nucleus A. The second spectrum is a scan made with I_X saturated so its resonance is missing, and the resonance of I_A has the reduced amplitude Y_P due to the NOE. When these two spectra are subtracted from each other we obtain the NOE difference spectrum shown at the bottom of Fig. 16-4. From Eq. 15-25 the enhancement E given by

$$E = \frac{Y_P}{Y_0}$$

permits us to write for Eq. 16-15

$$\frac{\gamma_X}{\gamma_A}\left(\frac{W_2 - W_0}{W}\right) S_X = -\frac{Y_0 - Y_P}{Y_0}$$

(16-17)

The enhancement reflects a reduction from scalar coupling since $Y_P < Y_0$.

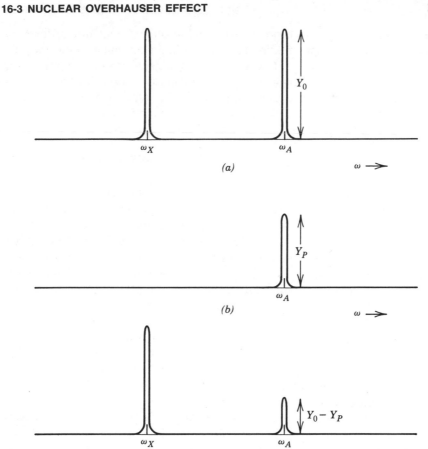

Fig. 16-4. AX NMR spectrum shown (a) before pumping, (b) while pumping the I_X line, and (c) NOE difference spectrum obtained by subtracting (b) from (a). The amplitudes of the I_A lines are indicated.

The time evolution of the NOE signal after the onset of the saturating pulse is found by combining Eqs. 15-16, 15-25, and 15-38

$$\frac{Y_P}{Y_0} = 1 + \frac{\gamma_X}{\gamma_A}\left(\frac{W_2 - W_0}{W}\right)S_X(1 - e^{-Wt}) \tag{16-18}$$

Radiofrequency power levels corresponding to the following range of γH_1 values [Eq. 16-5]

$$\frac{1}{(T_1 T_2)^{1/2}} \leq \gamma H_1 \ll J \tag{16-19}$$

are in what may be called the nuclear Overhauser range because the power is strong enough to produce polarization enhancements but not sufficiently strong to produce the spin tickling and decoupling phenomena which will be described in Section 16-7. However, at the higher powers required for tickling and decoupling the nuclear levels still become polarized and Overhauser enhancements still occur. This should be kept in mind in reading those sections, although it will not be explicitly mentioned there.

16-4 TWO-DIMENSIONAL NUCLEAR OVERHAUSER EFFECT

The previous section discussed the ordinary or one-dimensional NOE. To avoid ambiguous results when several nuclear species are present with closely spaced resonant lines, this technique requires the selective pumping of one nuclear species without irradiating nearby ones, and in addition it has a low sensitivity. These limitations can be overcome by the use of two-dimensional techniques[7,13,14] which are explained in Section 19-8. Two-dimensional nuclear Overhauser effect spectroscopy (2D NOESY) is carried out using the three 90° pulse sequence shown in Fig. 16-5.[15,16] The spins are frequency labeled during the evolution time t_1, nearby protons exchange magnetization during the mixing time t_M, and detection takes place a time t_2 after the pulse following the mixing. The acquired data undergo a double Fourier transform which involves integration over t_1 and t_2 to provide spectral line amplitudes which are functions of the two variables ω_1 and ω_2 which are conjugate to t_1 and t_2, respectively, as explained in Section 19-8.

The amplitude data for a particular mixing time are plotted on a two-dimensional graph in which the variables ω_1 and ω_2 constitute the ordinate and abscissa, respectively, as indicated in Fig. 16-6. The peaks along the diagonal of this figure correspond to the ordinary one-dimensional NOE that was discussed in the previous section. The symmetrically located

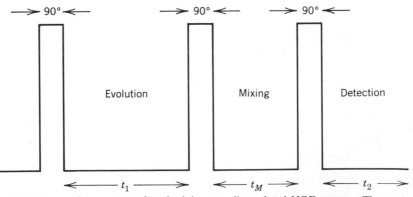

Fig. 16-5. Three-pulse sequence for obtaining two-dimensional NOE spectra. The evolution time t_1, the mixing time t_M, and the detection time t_2 are shown between the 90° pulses.

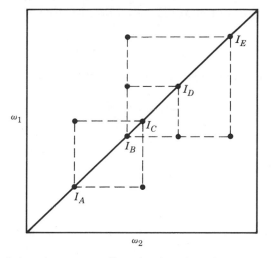

Fig. 16-6. Plot of data from a two-dimensional nuclear Overhauser effect spectroscopy experiment showing the points of the ordinary NOE spectrum along the diagonal and the cross-peaks at symmetric off-diagonal positions. NOE lines occur between the A, C pair, the B, D pair, and the B, E pair.

off-diagonal peaks indicate selective NOE polarizations between pairs of individual lines. This is shown in the figure for the pair A, C, the pair B, D, and the pair B, E. When data are acquired for successively increasing mixing times t_M they represent buildup of the NOEs of the correlated pairs of lines, and these buildup rates are related to the inverse sixth powers (r^{-6}) of the respective proton–proton distances.[17-20] For short mixing times only the closest protons provide off-diagonal peaks. NOE spectra from large molecules sometimes exhibit two-dimensional plots with large numbers of cross-peaks.

Thus we see that the transient nuclear Overhauser effect provides information on proton–proton distances. Information of this type can not be obtained from a steady-state NOE enhancement.

A simpler two-dimensional technique which employs two 90° pulses without the mixing time provides ω_1, ω_2 plots with the ordinary one-dimensional NMR spectrum appearing along the diagonal and cross-peaks arising from J coupling connectivities present at symmetric off-diagonal locations.[12] This is referred to as correlated spectroscopy (COSY).

16-5 HAMILTONIAN WITH HIGH RADIOFREQUENCY POWER

In this section we treat the case of two half-integer nuclear spins I_A and I_X of the AX system in the presence of a strong radiofrequency field H_1 which satisfies the inequalities of either Eq. 16-5 or 16-6. The Hamiltonian for H_1 in the x direction is

$$\mathcal{H} = \hbar H[\gamma_A I_{Az} + \gamma_X I_{Xz}] + \hbar J \vec{I}_A \cdot \vec{I}_X + \hbar \gamma_X H_1 I_{Xx} \cos \omega_P t \quad (16\text{-}20)$$

where the static magnetic field H is along z. We adopt the notation

$$\begin{aligned} \omega_A &= \gamma_A H \\ \omega_X &= \gamma_X H \end{aligned} \quad (16\text{-}21)$$

and select the pumping frequency ω_P close to the resonant frequency ω_X of the X spins. In a frame of reference rotating at the pump or stirring frequency ω_P we have

$$\mathcal{H} = (\omega_A - \omega_P) I_{Az} - (\omega_P - \omega_X) I_{Xz} + \gamma_X H_1 I_{Xx} + J \vec{I}_A \cdot \vec{I}_X \quad (16\text{-}22)$$

where all terms are in frequency units. We have omitted from this expression other time-dependent terms which average to zero and so do not contribute to the spectra.[21]

Substituting the appropriate 2×2 matrices for the various Hamiltonian terms and carrying out the direct product expansion gives the following 4×4 Hamiltonian matrix:

$$\begin{pmatrix} (\omega_A - \omega_P) - (\omega_P - \omega_X) + \tfrac{1}{2}J & \gamma_X H_1 & 0 & 0 \\ \gamma_X H_1 & (\omega_A - \omega_X) - \tfrac{1}{2}J & J & 0 \\ 0 & J & -(\omega_A - \omega_X) - \tfrac{1}{2}J & \gamma_X H_1 \\ 0 & 0 & \gamma_X H_1 & -(\omega_A - \omega_P) + (\omega_P - \omega_X) + \tfrac{1}{2}J \end{pmatrix}$$

$$(16\text{-}23)$$

If we recall that for an AX spin system

$$J \ll |\omega_A - \omega_X| \quad (16\text{-}24)$$

we see that the off-diagonal term J can be neglected. This is equivalent to omitting the terms $JI_{Ax}I_{Xx} + JI_{Ay}I_{Xy}$ and only retaining $JI_{Az}I_{Xz}$ in the original Hamiltonian. This omission causes the Hamiltonian to reduce to two 2×2 matrices which are easily solved in closed form to give the four energies[9,10]

$$E_j = \pm \omega_A + a_\pm \quad (16\text{-}25)$$

where

$$a_\pm = \tfrac{1}{2}[(\omega_P - \omega_X \pm \tfrac{1}{2}J)^2 + (\gamma_X H_1)^2]^{1/2} \quad (16\text{-}26)$$

and as before the energies are in frequency units. The following four transitions with the observed frequencies ω_0 appear close to the Larmor frequency ω_A of the A nucleus:

$$\omega_0 = \begin{cases} \omega_A + a_+ + a_- \\ \omega_A + a_+ - a_- \\ \omega_A - a_+ + a_- \\ \omega_A - a_+ - a_- \end{cases} \quad (16\text{-}27)$$

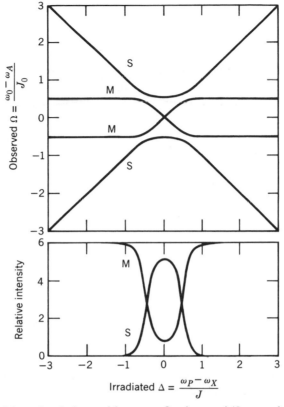

Fig. 16-7. Plot of the reduced observed frequency $\Omega = (\omega_0 - \omega_A)/J$ versus the reduced pumped frequency $\Delta = (\omega_P - \omega_X)/J$ for a reduced radiofrequency amplitude $\gamma_X H_1/J = 0.2$. The positions of the main lines (M) and satellite lines (S) are shown at the top and the relative intensities are presented at the bottom. Spin tickling occurs for an offset of $\Delta = \pm \frac{1}{2}$ (from Ref. 9).

Anderson and Freeman[9] gave a graphic presentation of these solutions by plotting the reduced observed frequency Ω,

$$\Omega = \frac{\omega_0 - \omega_A}{J} \qquad (16\text{-}28)$$

versus the reduced pumped frequency Δ

$$\Delta = \frac{\omega_P - \omega_X}{J} \qquad (16\text{-}29)$$

for various ratios $\gamma_X H_1/J$ of the pumping amplitude to the spin–spin coupling constant. Figures 16-7 and 16-8 show these plots for $\gamma_X H_1/J = 0.2$ and 1, respectively. In addition, Anderson and Freeman calculated the wavefunctions and the transition probabilities for the various transitions, and these are also presented in Figs. 16-7 and 16-8.

We see from these plots that there are four transitions everywhere except in the center ($\Omega = 0$ and $\Delta = 0$) where two lines cross. Most of the time the

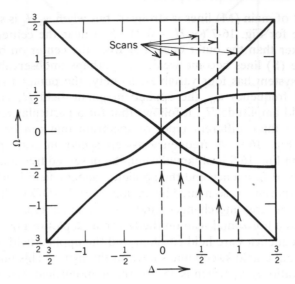

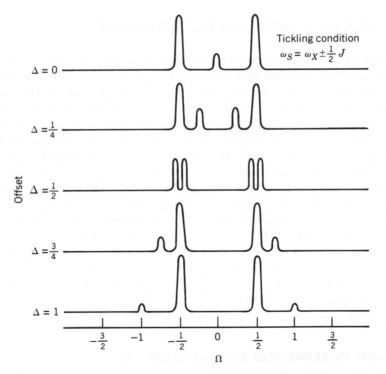

Fig. 16-10. Frequency scan spectra obtained with 5 different offsets Δ for $\gamma_X H_1/J = 0.2$. The corresponding paths on the Ω versus Δ diagram are shown. Spin tickling ocurs for an offset of $\Delta \pm \frac{1}{2}$.

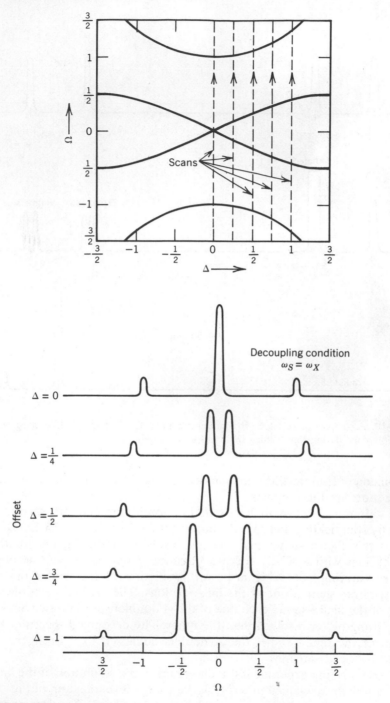

Fig. 16-11. Frequency scan spectra drawn in the manner of Fig. 16-10 for $\gamma_X H_1/J = 1$. Spin decoupling occurs for zero offset ($\Delta = 0$).

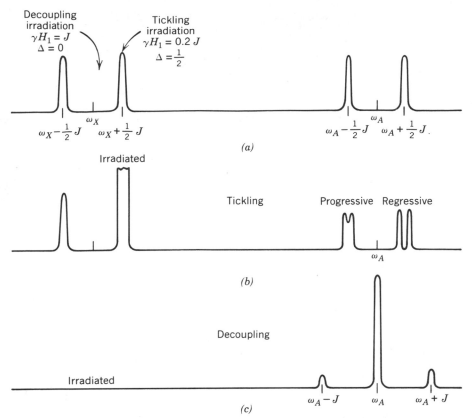

Fig. 16-12. A comparison (*a*) of the irradiation conditions for tickling and decoupling, and the spectra obtained during spin tickling (*b*) and spin decoupling (*c*).

phenomena of spin tickling and spin decoupling. In this section we elaborate a little more on these effects.[8–10]

The difference between the excitation conditions and the spectra produced by spin tickling and spin decoupling is depicted in Fig. 16-12. The top part of this figure shows that tickling involves irradiating one of the X spectral lines with a relatively lower radiofrequency power and decoupling involves carrying out a higher-power irradiation at the center of the X doublet rather than at one of the line positions. The tickled spectrum in the center of the figure shows each line of the A doublet split into two lines with one splitting better resolved than the other. The decoupled spectrum at the bottom of the figure shows the A doublet collapsed to a stronger singlet flanked by weak satellites.

In a spin tickling experiment the difference in the resolution of the lines of the A doublet is explained in terms of the energy level diagram of Fig. 16-13 which was constructed with the aid of Figs. 4-1 and 4.2. The argument that follows applies to both the homonuclear and the heteronuclear cases, so we will not distinguish them.

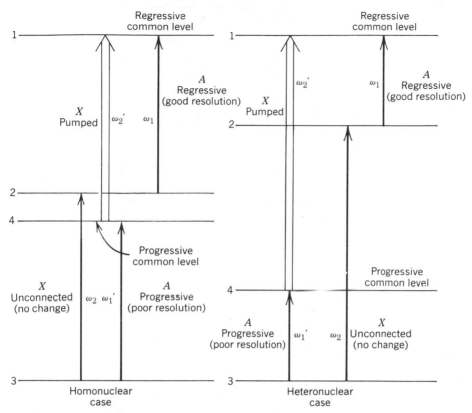

Fig. 16-13. Progressive, regressive, and unconnected transitions for the pumped $4 \rightarrow 1$ transition of a homonuclear (a) and a heteronuclear (b) AX spin system. The relative degrees of resolution for spin tickling are indicated. The transitions (ω_2', ω_2, ω_1, ω_1') are labeled using the conventions of Chapter 4.

We note from Fig. 16-13 that the two X doublet lines have no energy levels in common, and the same is true of the two A doublet lines. Such transitions are said to be unconnected and under tickling conditions a monitored transition is not influenced by the pumping of another which is unconnected to it. We also note from the figure that a pumped X transition is connected to both A type transitions, and so they are both split by the tickling. One of the A transitions has its common energy level between itself and the pumped transition, and this case is called progressive. The doublet of a progressive transition is relatively poorly resolved. The other A type transition and the pumped one have their common energy level above them and this case is called regressive. The doublet of a regressive transition is relatively well resolved. When the common level is below the two the case is also regressive with good resolution of the doublet. These rules for the relative degree of resolution of spin tickled lines are summarized in Fig. 16-13. The experimental spectra of Freeman and Anderson[22], presented in Fig. 16-14, illustrate these phenomena.

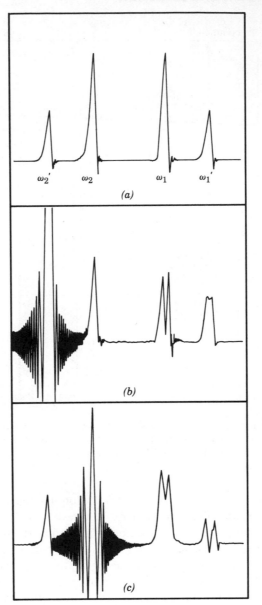

Fig. 16-14. Tickling of an AB spin system 2-bromo-5-chlorothiophene with the parameters $\omega_A - \omega_B = 9$ Hz and $J_{AB} = 3.8$ Hz showing (a) the conventional frequency swept spectrum, (b) the double resonance spectrum obtained when the highest frequency line ω_2' is irradiated with a tickling field, (c) the same with line ω_2 irradiated. The convention of Fig. 4-3 is used in labeling the transitions (from Ref. 22).

In the previous section we demonstrated decoupling for the simple AX system. When additional types of coupled nuclei are present the effect of spin decoupling is even more dramatic. Saturating a transition of one of the nuclei eliminates its effect on the remaining ones and thereby reduces the number of lines in all the multiplets to which this nucleus contributes. Spectral lines from the decoupled nucleus simply do not contribute to the spectrum.

REFERENCES

1. J. W. Emsley, J. Feeney, and L. H. Sutcliffe, *High Resolution Nuclear Magnetic Resonance Spectroscopy*, Pergamon, New York, 1965, Vol. 1.

2. U. Haeberlen, *High Resolution NMR in Solids*, Academic Press, New York, 1976.

3. R. Harris and B. Mann, eds., *NMR and the Periodic Table*, Academic Press, New York, 1978.

4. P. Lasclo, ed., *NMR of Newly Accessible Nuclei*, Vol. 1: *Chemical and Biochemical Applications*, Vol. 2: *Chemically and Biochemically Important Elements*, Academic Press, New York, 1983.

5. R. H. Webb, *Am. J. Phys.*, **29**, 429 (1961).

6. J. H. Noggle and R. E. Schirmer, *The Nuclear Overhauser Effect*, Academic Press, New York, 1971.

7. K. Wüthrich, *NMR in Biological Research*: *Peptides and Proteins*, North-Holland, Amsterdam, 1976.

8. G. Hoehler, ed., *Nuclear Magnetic Double Resonance, Principles and Applications in Solid State Physics*, Springer-Verlag, Berlin, 1973.

9. W. A. Anderson and R. Freeman, *J. Chem. Phys.*, **37**, 85 (1962).

10. R. A. Hoffman and S. Forsén, *Prog. NMR Spectrosc.*, **1**, 15 (1966).

11. M. H. Levitt, R. Freeman, and T. Frenkel, *Adv. Magn. Reson.*, **11**, 48 (1983).

12. L. G. Werbelow and D. M. Grant, *Adv. Magn. Reson.*, **9**, 189 (1977).

13. W. P. Aue, E. Bartholdi, and R. R. Ernst, *J. Chem. Phys.*, **64**, 2229 (1976).

14. R. Freeman and G. A. Morris, *Bull. Magn. Reson.*, **1**, 5 (1979).

15. A. Kumar, R. R. Ernst, and K. Wüthrich, *Biochem. Biophys. Res. Comm.*, **95**, 1 (1980).

16. R. R. Ernst, W. P. Aue, P. Bachmann, A. Höhner, M. Linder, B. Meier, L. Müller, A. Wokaun, K. Nagayama, and K. Wüthrich, *Proc. XXth Coloque Ampere, Tallinn*, 15 (1978).

17. A. Kalk and H. J. C. Berendsen, *J. Magn. Reson.*, **24**, 343 (1976).

18. W. E. Hull and B. D. Sykes, *J. Chem. Phys.*, **63**, 867 (1975).

19. G. Wagner and K. Wüthrich, *J. Magn. Reson.*, **3**, 675 (1979).

20. A. A. Bothnerby and J. H. Noggle, *J. Am. Chem. Soc.*, **101**, 5152 (1979).

21. A. Abragam, *The Principles of Nuclear Magnetism*, Oxford University Press, London, 1961.

22. R. Freeman and W. A. Anderson, *J. Chem. Phys.*, **37**, 2053 (1962).

17

ACOUSTIC, MUON, AND OPTICAL MAGNETIC RESONANCE

17-1 INTRODUCTION

The present chapter will examine the topics of acoustic resonance, muon spin resonance, and optical double magnetic resonance which are not very closely related to each other. The latter two fields have been the subject of a great deal of recent research interest.

17-2 ACOUSTIC MAGNETIC RESONANCE

In Section 11-10 we explained how the radiofrequency power that is absorbed by the spins is passed on to the surroundings such as the lattice vibrations or the Brownian motion through various relaxation mechanisms. The reverse process can also occur, whereby energy is introduced into the spin system through the excitation of lattice vibrations and its presence is detected by its effect on the resonant absorption line. This is referred to as acoustic magnetic resonance or ultrasonic magnetic resonance. The effect of ultrasonic energy on a magnetic resonance absorption was first observed indirectly by its disturbing influence on an NMR spectrum[1] and the first direct acoustic excitation of nuclear spins was carried out shortly thereafter,[2] followed by the subsequent detection of acoustic ESR.[3] We have reviewed the subject elsewhere.[4-6]

In an acoustic magnetic resonance experiment[4] ultrasonic energy is introduced into the crystal by means of a transducer and this causes the lattice to vibrate more strongly at the exciting frequency ν_u. A plot of the lattice vibrational mode amplitude versus the frequency exhibits a peak at the exciting frequency, as shown in Fig. 17-1. The lattice vibrational energy is quantized, and from a quantum-mechanical point of view we can say that the lattice has a high density of sound particles called phonons at the exciting frequency ν_u. The background curve of vibrational amplitudes shown in Fig. 17-1 corresponds to a lattice at a particular temperature, and the spike at the frequency ν_u corresponds to a heating up of the lattice at this particular frequency ν_u to a temperature T_u which is appreciably higher than the ambient temperature T of the lattice at the remaining frequencies, as indicated in Fig. 17-2.

We showed in Section 12-8 that the various spin levels of an $S = \frac{1}{2}$, $I = \frac{1}{2}$ spin system placed in a magnetic field H acquire equilibrium population distributions in accordance with the Boltzmann factor for $g\beta H \leqslant kT$

$$\exp\left(-\frac{g\beta H}{kT}\right) = 1 - 2\epsilon \tag{17-1}$$

where

$$\epsilon = \frac{g\beta H}{2kT} \tag{17-2}$$

and the nuclear Boltzmann factor $\exp(-g_N\beta_N H/kT)$ has been neglected. These equilibrium populations are indicated on the left-hand side of Fig. 17-3. Since the spins communicate with the lattice through interactions with phonons at the Larmor precession frequency ν_0, the Boltzmann factor that determines the population distribution of the spins contains the temperature of the lattice at that frequency. This means that when the ultrasonic excitation frequency ν_u is the same as the Larmor frequency ν_0 the spin system acquires the population distribution corresponding to the temperature T_u, and therefore T_u replaces T in Eqs. 17-1 and 17-2

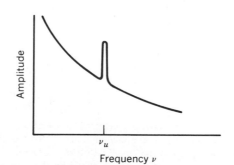

Fig. 17-1. Amplitude of vibrational modes in a crystal as a function of the frequency. The peak at the frequency ν_u is from the ultrasonic energy introduced into the crystal.

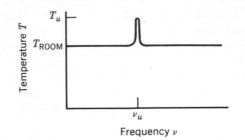

Fig. 17-2. Effective temperature of the lattice vibrational modes corresponding to the case of Fig. 17-1.

$$\exp\left(-\frac{g\beta H}{kT_u}\right) = 1 - 2\epsilon_u \qquad (17\text{-}3)$$

where

$$\epsilon_u = \frac{g\beta H}{2kT_u} \qquad (17\text{-}4)$$

Since T_u is larger than T the populations of the levels become closer together, as shown in Fig. 17-3, a result which is equivalent to a partial saturation of the spin system. If an ESR experiment is carried out while the ultrasonic energy is being applied, the intensity of the transition, which is proportional to the population difference, is reduced by the ratio ϵ_u/ϵ.

In drawing Fig. 17-3 we assumed that the spread in frequencies of the

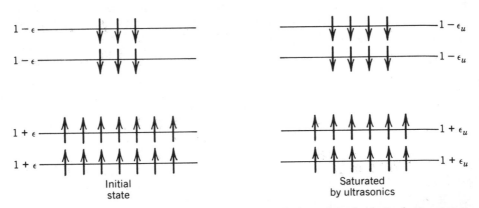

Fig. 17-3. Effect of ultrasonics on the energy level populations of the $S = \frac{1}{2}$, $I = \frac{1}{2}$ spin system. The initial state shown on the left has the thermal equilibrium Boltzmann factor ϵ and the final partly saturated state on the right has the smaller Boltzmann factor ϵ_u corresponding to the higher effective temperature T_u. The effect of the nuclei on the Boltzmann factors has been neglected. The orientations of the electron spins are shown. The difference between the number of spins in the ground and excited states is exaggerated.

ultrasonic source is sufficiently broad so that it saturates both of the allowed ESR transitions. If it is narrow enough to saturate only one of them then the two ESR spectral lines will have different intensities. A number of experimental methods have been developed for studying acoustic absorption by spin systems, and these have been reviewed elsewhere.[5-8]

17-3 MUON SPIN RESONANCE

The negative muon μ^- is a spin-$\frac{1}{2}$ lepton particle that acts in all respects like an electron except that it has a mass 206.77 times larger than that of an electron. Its electric charge is identical with that of an electron. There are also positive muons μ^+ that have the same mass and opposite charge of a negative muon and that behave like positive electrons called positrons. A muon has a magnetic moment of 3.183 nuclear magnetons. Its gyromagnetic ratio

$$\frac{\gamma_\mu}{2\pi} = 13.55 \, \text{MHz/kG} \tag{17-5}$$

is about three times that of a proton (4.2577 MHz/kG) and much smaller than that of an electron (2.8026 GHz/kG).

One dramatic difference between a muon and an electron is the fact that a muon is unstable with a lifetime of 2.199 μsec while an electron is a stable particle. A positive muon decays into a positron (e^+) and two neutrinos

$$\mu^+ \rightarrow e^+ + \nu_e + \bar{\nu}_\mu \tag{17-6}$$

with the positron emitted in a preferential direction relative to the muon magnetic moment. If P is the initial polarization of the muon beam, then the number of positrons emitted at an angle θ relative to this direction is given by

$$N_{(\theta)} \propto 3 + P \cos \theta \tag{17-7}$$

A muon may be employed as a probe of the magnetic environment of a solid on an atomic scale. Sometimes it remains as a free particle as in metallic conductors, and sometimes it combines with an electron to form muonium as in various insulators and semiconductors. Muonium is an exotic "hydrogen atom" containing an electron in orbit around a positive muon that plays the role of the nucleus.

In a solid the muons precess about the local magnetic field, and the precession is monitored by measuring the angular correlation pattern given by Eq. 17-7. This measurement of muon spin rotation (μSR) gives important information on the distribution of local magnetic fields in the solid, and it has become a useful tool in solid-state physics.

The μSR technique is limited to a few meson factories that provide large enough fluxes of muons. The main sites for μSR work are the Los Alamos

Meson Physics Facility (LAMPF) in New Mexico; the Tri University Meson Facility (TRIUMF) in Vancouver, British Columbia; the Schweizerisches Institut für Nuklearforschung (SIN) in Villigen, Switzerland; and the Russian Institute for Nuclear Studies at Dubna. The subject has been reviewed by Brewer and Crowe[9] and Denison et al.[10]

17-4 OPTICAL DOUBLE MAGNETIC RESONANCE

The magnetic resonance experiments that we have discussed until now have involved Zeeman splittings of ground state electronic energy levels such as the ground state crystal field levels treated in Chapter 10. We showed in Sections 10-2 to 10-5 how the optical transitions of paramagnetic atoms and molecules in a magnetic field take place between particular magnetic sublevels (cf. Fig. 10-2). If a scan is made through an ESR transition while monitoring the light from an optical absorption or emission line and a change occurs in the optical intensity during the passage through the ESR resonance point, this provides a clue to the identity of the particular magnetic sublevels that are involved in the optical transition. This change in the optical intensity that occurs during the ESR scan is referred to as the optical double magnetic resonance or ODMR signal.

Figure 17-4 presents the energy level diagram of a prototype ODMR experiment. On it we show how a high-power optical pump signal raises the electrons to an upper energy level from which they jump to or are

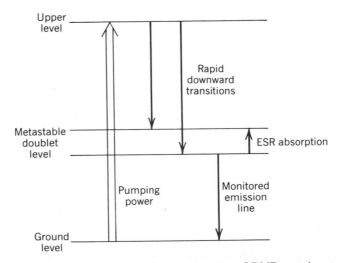

Fig. 17-4. Energy level diagram corresponding to a prototype ODMR experiment showing the excitation of the upper level by the pumping power, the rapid decay to the metastable doublet level, the final decay via the monitored emission line, and the ESR absorption transition between the metastable sublevels.

transferred to a relatively long-lived or metastable level before they return to the ground state via an optical emission signal which is monitored by a detector. The metastable level is magnetic and is shown as a doublet split by the magnetic field. Therefore microwaves applied at the Larmor precession frequency can induce transitions between the doublet sublevels, and this disturbs the level populations and hence the intensity of the downward transition from the m_S level that is being monitored. An experiment of this type can provide the spin Hamiltonian parameters of the metastable state.

The optical pumping power that populates the excited optical level can bring along a spin memory of the ground state and thereby preferentially populate particular magnetic sublevels. This can occur when the pumping signal is linearly or circularly polarized. In the case illustrated in Fig. 17-5 each level is represented as a doublet and all transitions involve the $m_S = \frac{1}{2}$ sublevels and obey the selection rule $\Delta m_S = 0$. The applied microwaves deplete the population of the lower metastable sublevel and hence attenuate the intensity of the monitored emission line. If circularly polarized microwaves are employed the ordering of the metastable doublet levels can be deduced.

Optical spectroscopy is intrinsically much more sensitive than microwave spectroscopy because the energy difference ΔE which enters into the optical Boltzmann factor $\exp(-\Delta E/kT)$ is much greater than kT so that virtually all the atoms initially populate the ground state. Therefore the use of optical detection can bring about a large increase in sensitivity, and some workers[11,12] have reported that as few as 10^4 excited-state spins can be detected by monitoring the microwave-induced modulation of phosphorescent emission.

A number of different types of ODMR experiment have been carried out with paramagnetic molecules such as those in triplet states with electronic spin $S = 1$.[11-15] In the Triplet Absorption Detection of Magnetic Resonance (TADMR) one observes microwave-induced changes in the triplet absorption intensity. In Fluorescence Microwave Double Resonance (FMDR) and Phosphorescence Microwave Double Resonance (PMDR) changes are observed, respectively, in the fluorescent and phosphorescent emission from the specimen, where phosphorescence is a delayed form of fluorescence.[16-21] A Microwave Optical Magnetic Resonance Induced by Electrons (MOMRIE) experiment involves the use of an electron beam to raise atoms to excited optical levels and the monitoring of the intensity or polarization of the light emitted by the atoms to detect the magnetic resonance absorption induced by the microwaves.[22,23] Microwave–infrared double resonance has also been reported.[24] Hills[25] recently reviewed laser magnetic resonance spectroscopy.

A technique related to ODMR is Spin-Dependent Recombination (SDR) which involves the measurement of the change in the photoinduced conductivity of silicon single crystals that occurs when one simultaneously scans through the electron spin resonance absorption line.[26-28] This experimental technique is 100 times more sensitive than ordinary ESR. The SDR results

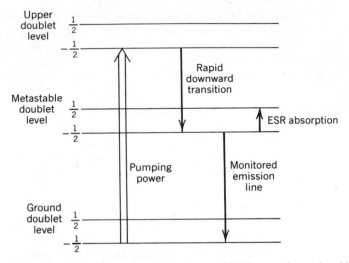

Fig. 17-5. Energy level diagram corresponding to an ODMR experiment in which all three levels are magnetic doublets so that the pumping power can bring along a spin memory of the optical ground state which is reflected in the intensity of the monitored emission line.

sometimes correlate with photoinduced ESR signals that have also been referred to as Photoconductive Resonance (PCR).[29-32]

REFERENCES

1. W. G. Proctor and W. H. Tanttila, *Phys. Rev.*, **98**, 1854 (1955).

2. M. Manes and D. I. Bolef, *Phys. Rev.*, **109**, 218 (1958).

3. E. H. Jacobsen, N. S. Shiren, and E. B. Tucker, *Phys. Rev. Lett.*, **3**, 81 (1951).

4. C. P. Poole, Jr. and H. A. Farach, *Relaxation in Magnetic Resonance*, Academic Press, New York, 1971.

5. C. P. Poole, Jr. and H. A. Farach, *Appl. Spectrosc. Rev.*, **19**, 167 (1983).

6. C. P. Poole, Jr., *Electron Spin Resonance*, 2nd ed., Wiley-Interscience, New York, 1983.

7. S. D. Devine and W. H. Robinson, *Adv. Magn. Reson.*, **10**, 54 (1982).

8. D. I. Bolef and J. G. Miller, *Phys. Acoust.*, **8** (1971).

9. J. H. Brewer and K. M. Crowe, *Annu. Rev. Nucl. Part. Sci.*, **28**, 239 (1978).

10. A. B. Denison, H. Graf, W. Kundig, and P. F. Meier, Physik-Institute der Universität Zurich, 1981.

11. C. B. Harris, R. L. Schlupp, and H. Schuch, *Phys. Rev. Lett.*, **30**, 1019 (1973).

12. W. G. Breiland, C. B. Harris, and A. Pines, *Phys. Rev. Lett.*, **30**, 158 (1973).

13. A. B. Denison, *Magn. Reson. Rev.*, **2**, 1 (1973).

14. R. H. Clark, ed., *Triplet State ODMR Spectroscopy*, Wiley-Interscience, New York, 1982.

15. K. H. Hausser and H. C. Wolf, *Adv. Magn. Reson.*, **8**, 85 (1976).

16. H. M. Van Noort, P. J. Vergragt, J. Herbich, and J. H. van der Waals, *Chem. Phys. Lett.*, **71**, 5 (1980).

17. A. M. Ponte Goncalves and R. Gillies, *Chem. Phys. Lett.*, **69**, 164 (1980).

18. H. C. Brenner, *J. Chem. Phys.*, **67**, 4719 (1977).

19. W. G. Breiland, H. C. Brenner, and C. B. Harris, *J. Chem. Phys.*, **62**, 3459 (1975).

20. R. H. Clarke and R. H. Hofeldt, *J. Chem. Phys.*, **61**, 4582 (1974).

21. T. A. Miller and R. S. Freund, *Adv. Magn. Reson.*, **9**, 50 (1977).

22. R. S. Freund and T. A. Miller, *J. Chem. Phys.*, **56**, 2211 (1972).

23. T. A. Miller, *J. Chem. Phys.*, **58**, 2358 (1973).

24. K. M. Evanson, *Faraday Disc. Chem. Soc.*, **71**, 7 (1981).

25. G. W. Hills, *Magn. Reson. Rev.*, **9**, 15 (1984).

26. G. Mendz and D. Haneman, *J. Phys. C*, **11**, 1197 (1978).

27. K. Morigaki, N. Kishimoto, and D. Lepine, *Solid State Commun.*, **17**, 1017 (1975).

28. R. Wosinski, T. Figielski, and A. Makosa, *Phys. Status Solidi A*, **37**, K57 (1976).

29. P. J. Caplan, J. N. Helbert, B. E. Wagner, and E. H. Poindexter, *Surf. Sci.*, **54**, 33 (1976).

30. V. Kurylev and S. Karyagin, *Phys. Status Solidi A*, **21**, K127 (1974).

31. D. Lepine, *Phys. Rev. B*, **6**, 436 (1972).

32. J. Ruzyllo, I. Shiota, N. Miyamoto, and J. Nishizawa, *J. Electrochem.*, **123**, 26 (1976).

18

SPIN LABELS

18-1 INTRODUCTION

When the Hamiltonian parameters of a paramagnetic molecule of free radical are anisotropic, the positions of the resonant lines of the ESR spectrum depend on the orientation of the molecule in the magnetic field, as was explained in Chapter 6. If a solid sample contains a large number of randomly oriented free radicals then each individual radical produces a spectral line at a position that depends on its orientation and the observed spectrum is a broad asymmetric powder pattern of the type described in Section 11-9, produced by the superposition of lines from all free radicals in the sample. In a low-viscosity liquid the radicals undergo rapid tumbling motions which average out the anisotropic terms in the Hamiltonian and produce a symmetric spectrum of narrow lines. For the intermediate case of high-viscosity solutions the molecular motion is sufficiently slow so that the anisotropic interactions are only partially averaged and the appearance of the spectrum is between that of a broad asymmetric powder pattern and a symmetric set of narrow lines. Thus the shape of the spectrum depends critically on the molecular motion of the individual paramagnetic centers,[1-3] and in some cases the use of partially saturated spectra can enhance this shape dependence.[4-6] This chapter describes several methods for measuring the correlation times of the molecular motion.

In biological applications this spectroscopic technique is a means of probing the correlation time τ_c of the motion that occurs near particular functional groups on macromolecules. This is accomplished by attaching paramagnetic nitroxide molecules called spin labels to particular sites of the macromolecules and deducing the degree of localized motion at the attachment positions from the lineshapes of the spectra.

18-2 CORRELATION TIMES AND LINESHAPES IN LIQUIDS

High-viscosity liquids constitute a transition region where the spectral features are intermediate between those of a liquid and a rigid lattice, and these features vary continuously from one limiting case to the other as the viscosity increases. At high temperatures the individual molecules of the liquid are in rapid, relatively unhindered motion relative to each other. Lowering the temperature increases the viscosity and slows down the molecules. This random molecular movement is the Brownian motion which is responsible for averaging out the anisotropies of Hamiltonian parameters.

Brownian motion is characterized by a correlation time τ_c which provides a timescale to the motion. It may be looked on as the average time required for a molecule to reorient through an angle of 1 radian or to translate through a distance of one molecular diameter. Only rotational reorientations are taken into account in this chapter, and the correlation time can result either from the accumulation of many successive small-angle reorientations or from individual large-angle realignments. In classical dielectric experiments the molecules are treated as spheres of radius a undergoing isotropic, small-angle rotational reorientation in a liquid of viscosity η, and the resulting Debye correlation time τ_D varies inversely with the temperature in accordance with the Einstein–Stokes relationship[7]

$$\tau_D = \frac{1}{2D}$$
$$= \frac{4\pi\eta a^3}{kT} \qquad (18\text{-}1)$$

where D is the rotational diffusion constant. The correlation time τ_c for the motion of the magnetic moments in an ESR experiment is given by

$$\tau_c = \frac{\tau_D}{f} \qquad (18\text{-}2)$$

where f varies between 1 and 3 and assumes the limiting value of 3 which appears in Eq. 11-103 based on the isotropic small-angle Brownian motion model mentioned above. If the random motion is described by a jump model wherein the molecules reorient via large angular shifts of random length then $f = 1$ so τ_c and τ_D are equal.

To clarify the manner in which the value of the correlation time τ_c affects the spectral lineshapes we examine the case of a fluctuating time-dependent axially symmetric Zeeman Hamiltonian[8] obtained from Eq. 6-19

$$\mathcal{H}(t) = \beta[g_\perp(S_x H_x + S_y H_y) + g_\parallel S_z H_z] \qquad (18\text{-}3)$$

$$= \beta[g_0 \vec{S} \cdot \vec{H}] + \tfrac{1}{3}\beta\delta g[(S_x H_x + S_y H_y) - 2S_z H_z] \qquad (18\text{-}4)$$

where

$$g_0 = \tfrac{1}{3}(2g_\perp + g_\parallel)$$
$$\delta g = g_\perp - g_\parallel$$
$$g_\parallel = g_0 - \tfrac{2}{3}\delta g \qquad\qquad (18\text{-}5)$$
$$g_\perp = g_0 + \tfrac{1}{3}\delta g$$

In the limit of small anisotropies ($|\delta g| \ll g_0$) Sillescu[8] wrote the Hamiltonian 18-4 for a radical oriented at an angle θ relative to the magnetic field direction as shown in Fig. 18-1 in the form

$$\mathcal{H}(t) = g_0\beta HS_z - \tfrac{1}{3}\delta g[3\cos^2\theta(t) - 1]\beta HS_z \qquad (18\text{-}6)$$

and the time dependence arises from random fluctuations of the polar angle $\theta(t)$ at a rate determined by the correlation time τ_c. It is assumed that the spin system has an intrinsic linewidth $\Delta\omega_{1/2}$ that is independent of the g-factor anisotropy. The lineshape depends critically on the product of the correlation time and the linewidth. If $\tau_c\Delta\omega_{1/2}$ greatly exceeds unity, the motion is so slow that very little averaging occurs and the observed spectrum is a powder pattern of the type illustrated in Fig. 18-2. In the opposite limit when $\tau_c\Delta\omega_{1/2}$ is much less than unity the molecular motion is so rapid that the factor $[3\cos^2\theta(t) - 1]$ almost completely averages out to produce the slightly asymmetric singlet illustrated in Fig. 18-3. The intermediate case where $\tau_c\Delta\omega_{1/2} \simeq 4$ corresponds to the partially averaged but still recognizable powder pattern presented in Fig. 18-4. These figures show the effect that the random fluctuation mechanism has on the lineshape by comparing curves for the Brownian motion model ($\tau_c = \tfrac{1}{3}\tau_D$) and the jump model

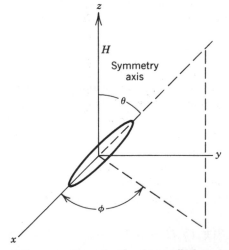

Fig. 18-1. Orientation of the symmetry axis of a free radical at an angle θ relative to the magnetic field direction.

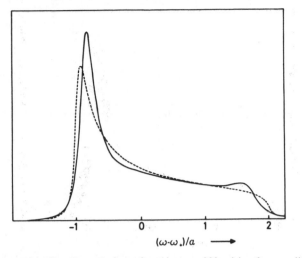

Fig. 18-2. ESR spectrum for slow reorientation $\tau\Delta\omega_{1/2} = 200$ arising from a slightly averaged, fluctuating, axially symmetric Zeeman Hamiltonian 18-6 for the Brownian motion (—) and jump (---) models. The notation $\omega_0 = g_0\beta H_0/\hbar$ and $a = (g_\parallel - g_\perp)\beta H_0/3\hbar$ is used (from Ref. 8).

$(\tau_c = \tau_D)$ mentioned above. We see from the figures that for the same value of $\tau_c\Delta\omega_{1/2}$ the Brownian motion model leaves the powder pattern features somewhat more prominent.

For very short times $(\tau_c < 10^{-9}$ sec) a single narrow line is observed while for very long times $(\tau_c > 10^{-4}$ sec) rigid lattice powder patterns are recorded, and in these limits the lineshape does not provide quantitative estimates of

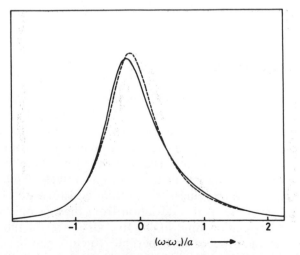

Fig. 18-3. ESR spectrum for rapid reorientation $\tau\Delta\omega_{1/2} = 0.6$ and almost complete averaging of the Hamiltonian 18-6 used in Fig. 18-2 (from Ref. 8).

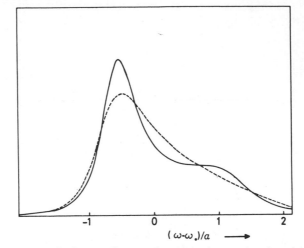

Fig. 18-4. ESR spectrum for intermediate reorientation $\tau\Delta\omega_{1/2} = 4$ and partial averaging of the Hamiltonian 18-6 of Fig. 18-2 (from Ref. 8).

the correlation times. Throughout the range of correlation times 10^{-9}–10^{-4} sec partially averaged spectra are observed which do provide numerical estimates of τ_c. In the lower limit of this range, perhaps near $\tau_c = 10^{-8}$ sec, the incompletely averaged g-factor tensor broadens the line, and the extent of this broadening can be employed to estimate the correlation time. For longer times, $10^{-6} < \tau_c < 10^{-4}$ sec, partially saturated spectra are more sensitive to the motion than low-power spectra, and saturation transfer techniques become useful for measuring the correlation time. The mechanisms which broaden and distort the line in these cases will be examined in the next two sections.

18-3 LINE BROADENING

When anisotropies are present in the Hamiltonian they produce orientation-dependent spectra in solids and they are largely averaged out by the rapid Brownian fluctuations in low-viscosity liquids. For intermediate fluctuation rates the incomplete averaging of the anisotropies can constitute the principal line-broadening mechanism. The perturbation Hamiltonian for this situation involves the deviation of the principal values of the Hamiltonian tensors from their average values. In the axial g-factor case of Eq. 18-6 the quantity δg is a measure of this deviation from the isotropic value g_0.

At the short correlation time limit (10^{-9}–10^{-8} sec) of the intermediate motion region in the absence of hyperfine structure the anisotropy in the g-factor broadens the resonant line through the following time-dependent interaction Hamiltonian[9,10] $V(t)$:

$$V(t) = \beta H_0[(g_{zx} - g_0)S_x + (g_{zy} - g_0)S_y + (g_{zz} - g_0)S_z] \quad (18\text{-}7)$$

and we will discuss how this occurs for the axially symmetric case of Eq. 18-6 where $g_{zx} = g_{zy}$. Using the direct product expansion the interaction Hamiltonian for $S = \frac{1}{2}$ becomes

$$V(t) = \frac{1}{2}\beta H_0 \begin{pmatrix} (g_{zz} - g_0) & (g_{zx} - g_0) - i(g_{zy} - g_0) \\ (g_{zx} - g_0) + i(g_{zy} - g_0) & -(g_{zz} - g_0) \end{pmatrix}$$

$$= \begin{pmatrix} V_{zz} & V_{xy} \\ V_{xy} & -V_{zz} \end{pmatrix} \quad (18\text{-}8)$$

We explained elsewhere[7] that the linewidth is related to the correlation time τ_c and the matrix elements of $V(t)$ through the expression[9]

$$\Delta H_{1/2} = \frac{8\tau_c}{\gamma^2 \hbar^2} \left(\overline{V_{zz}^2} + \frac{\frac{1}{2}\overline{V_{xy}^2}}{1 + \omega_0^2 \tau_c^2} \right) \quad (18\text{-}9)$$

where a bar over a matrix element indicates an average over the time and ω_0 is the resonant frequency. Carrington and McLaghlan[10] worked out the required matrix elements

$$\overline{(g_{zx} - g_0)^2} = \overline{(g_{zy} - g_0)^2} = \frac{1}{10}\Delta g^2$$

$$\overline{(g_{zz} - g_0)^2} = \frac{2}{15}\Delta g^2 \quad (18\text{-}10)$$

where the results are expressed in terms of the overall g-factor anisotropy parameter Δg

$$\Delta g^2 = (g_\parallel - g_0)^2 + 2(g_\perp - g_0)^2 \quad (18\text{-}11)$$

which is related to δg of Eq. 18-6 as follows:

$$\Delta g = \sqrt{\frac{2}{3}}\, \delta g \quad (18\text{-}12)$$

In the correlation time range 10^{-9}–10^{-4} sec at X band ($\omega_0/2\pi = 10^{10}$ Hz) we have $\omega_0 \tau_c \gg 1$ and the linewidth from Eq. 18-9 becomes

$$\Delta H_{1/2} = \frac{4}{15} \frac{\Delta g^2 \beta^2 H^2}{\hbar^2} \tau_c \quad (18\text{-}13)$$

Thus we see that in the intermediate motion region the contribution of the anisotropies to the width is proportional to the correlation time and to the square of the g-factor anisotropy. When first derivative spectra are recorded it is more convenient to make use of the peak-to-peak linewidth ΔH_{pp} which has the value

$$\Delta H_{pp} = \frac{\Delta H_{1/2}}{\sqrt{3}} \tag{11-109}$$

for a Lorentzian lineshape, and is given by Eq. 11-110 for a Gaussian lineshape.

18-4 SATURATION TRANSFER

This section is concerned with long correlation times (10^{-6}–10^{-4} sec), where the lineshape is close to that of a powder, small fluctuations in τ_c produce greater changes in the shape of partly saturated spectra than they do in the shape of spectra recorded at low power levels.[4-6]

We showed in Section 11-9 that an axially symmetric radical oriented at an angle θ relative to the magnetic field direction, as shown in Fig. 18-1, contributes to the resonance absorption of the powder lineshape at the magnetic field position H given by Eq. 11-95

$$H = \frac{g_0 H_0}{(g_\perp^2 \sin^2 \theta + g_\parallel^2 \cos^2 \theta)^{1/2}} \tag{18-14}$$

The probability that a molecule will be oriented at a particular angle θ determines the intensity at the corresponding magnetic field value. This probability is proportional to the quantity $\partial N / \partial H$ given by Eq. 11-96

$$\frac{\partial N}{\partial H} = \frac{N g_0^2 H_0^2}{2H^2} \left(\frac{g_\perp^2 - g_\parallel^2}{g_\perp^2 H^2 - g_0^2 H_0^2} \right)^{1/2} \tag{18-15}$$

and it provides the lineshape sketched in Figs. 18-5 and 18-6a.

All the radicals are undergoing continual random fluctuations due to the Brownian motion, and the correlation time τ_c provides the time scale for this random motion. Each such fluctuation produces a change $\delta\theta$ in the angular orientation of the g-factor axis relative to the magnetic field direction, and this changes the resonant magnetic field of the radical. For some orientations of the radical a fluctuation in angle $\delta\theta$ produces a relatively small change in the resonant field while for other orientations the same change in angle $\delta\theta$ produces a relatively large change in the resonant magnetic field. If the radical is on resonance at a high microwave power level it will be saturated, and a fluctuation which produces a very small change in the resonant field will leave it within the same region of resonance, with no change in the level of saturation. In contrast to this, fluctuations which produce large changes in the resonant field can move radicals into and out of the region of resonance, causing them to transfer their saturation to other parts of the line. This transfer of saturation will be reflected in changes in the level of saturation and hence in the observed intensity.

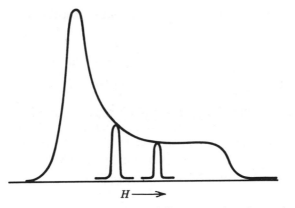

Fig. 18-5. Powder pattern lineshape for an axially symmetric g-factor showing two of its component lines.

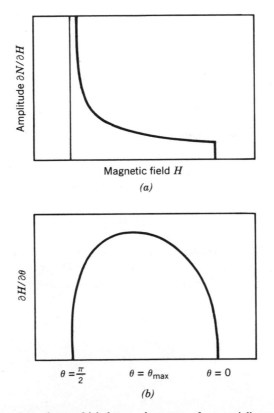

Fig. 18-6. Angular dependence of (*a*) the powder pattern for an axially symmetric g-factor, and (*b*) the derivative $\partial H/\partial \theta$ which measures the probability of transferring saturation during a random molecular fluctuation.

The rate of change of resonant field with respect to angle $\partial H/\partial\theta$ is given by Eq. 11-97:

$$\frac{\partial H}{\partial\theta} = \frac{g_0 H_0 (g_\perp^2 - g_\parallel^2)\sin\theta\cos\theta}{(g_\perp^2\sin^2\theta + g_\parallel^2\cos^2\theta)^{3/2}} \tag{8-16}$$

We see from this expression that $\partial H/\partial\theta$ is zero for radicals oriented along the magnetic field direction where $\theta = 0$ and also for radicals oriented in the plane perpendicular to the magnetic field direction where $\theta = \pi/2$. From Eq. 11-98 the derivative $\partial H/\partial\theta$ is a maximum when $\partial^2 H/\partial\theta^2 = 0$ at the intermediate value of θ denoted by θ_{max} where

$$\tan\theta_{max} = \left(\frac{g_\parallel}{g_\perp}\right)^{1/2} \tag{18-17}$$

and this is the orientation angle of the radicals for which the saturation transfer effect is greatest. The graph of $\partial H/\partial\theta$, presented in Fig. 18-6$b$, shows that the change in resonant field is most pronounced near the center of the line where Eq. 18-17 is satisfied.

The shape of the resonant line is especially sensitive to molecular motions when the azimuthal angle θ is close to the value of θ_{max}, and this sensitivity provides a means for estimating the correlation time associated with this motion.

18-5 NITROXIDE SPIN LABELS

Nitroxides and other stable free radicals called spin labels are widely used in biological studies as probes of local molecular motion. They attach themselves to or occupy positions near active sites in proteins or other macromolecules and produce spectra whose linewidths and lineshapes reflect the motions that they undergo.[1-3,11,12]

Nitroxide spin labels produce a three-line hyperfine pattern due to the unpaired electron on the nitrogen atom of the NO group. An example of such a spin label is 2,2,6,6-tetramethyl-4-piperidone-1-oxyl:

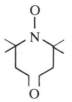

In general, both g and A are anisotropic with the following typical principal values:

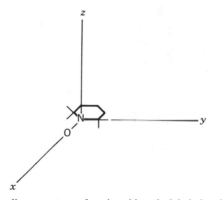

Fig. 18-7. Molecular coordinate system of a nitroxide spin label showing the x axis along the N–O bond and the z axis perpendicular to the plane of the heterocyclic ring. The z axis is the symmetry axis of the spin Hamiltonian.

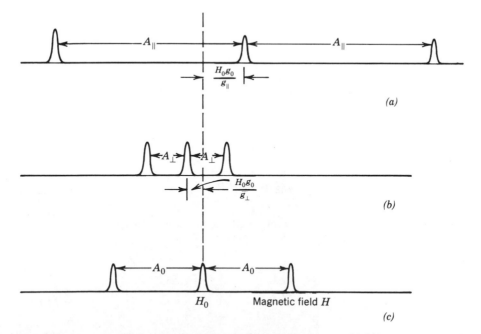

Fig. 18-8. Hyperfine triplet of a nitroxide spin label oriented in a solid (*a*) with the magnetic field along the symmetry axis z of Fig. 18-7, (*b*) with the magnetic field in the xy plane perpendicular to the symmetry axis, and (*c*) averaged isotropic spectrum obtained in a low-viscosity liquid with the averaged hyperfine splitting $A_0 = \frac{1}{3}(A_{\parallel} + 2A_{\perp})$.

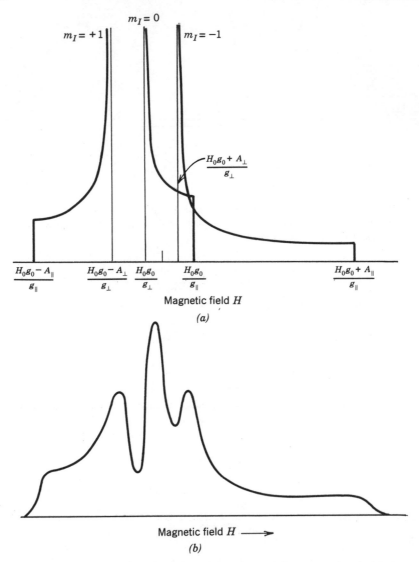

Fig. 18-9. Powder pattern of the hyperfine triplet of an axially symmetric spin label with (*a*) zero component linewidth and (*b*) a finite component width.

$$
\begin{aligned}
g_{xx} &= 2.008 & A_{xx} &= 5.5 \text{ G} \\
g_{yy} &= 2.006 & A_{yy} &= 4.0 \text{ G} \\
g_{zz} &= 2.003 & A_{zz} &= 30.0 \text{ G}
\end{aligned}
\tag{18-18}
$$

along the three molecular principal axes defined in Fig. 18-7. We note from this figure that the *z* axis is in the direction of the *p* orbital containing the unpaired electron, which is perpendicular to the plane of the piperidone ring, and the x axis is along the N–O bond direction.

Expressions for the lineshape become quite complex when both g and A are completely anisotropic so we will analyze the simpler case when each tensor has the same symmetry axis using the following values for the parameters:

$$g_\perp = 2.008 \qquad A_\perp = 7\,G$$
$$g_\parallel = 2.002 \qquad A_\parallel = 34\,G \qquad\qquad (18\text{-}19)$$
$$g_0 = 2.006 \qquad A_0 = 16\,G$$

using the notation of Eqs. 18-5.

Figure 18-8 shows the hypothetical single crystal spectrum for the magnetic field oriented along the parallel and perpendicular directions of the g-factor and hyperfine tensors. The hyperfine splittings and the g-factor shifts from the isotropic (g_0) position are shown. These two spectra provide the ranges over which the powder patterns extend. Also given in this same figure is the isotropic spectrum of three equally spaced lines centered at g_0 which is obtained in a low-viscosity solvent. Griffith and Waggoner[13] and Morrisett[14] observed spectra similar to those presented in Fig. 18-8 along the three principal axes of the completely anisotropic radical di-t-butyl nitroxide oriented in a host crystal. Appendix 2 of Ref. 2, Vol. 1, gives references to other works involving the determination of the principal values of the g-factor and hyperfine tensors of nitroxides.

The powder pattern for the case of Fig. 18-8 is the superposition of three individual patterns of the type illustrated in Fig. 11-13, and they are sketched in Fig. 18-9. We see from this figure that the center and high-field patterns overlap each other. At the lower part of the figure we sketch the absorption signal obtained by adding a finite linewidth to the patterns.

18-6 SPIN LABEL LINESHAPES DUE TO MOLECULAR MOTION

In Section 18-3 we discussed the line broadening of a free radical with an axially symmetric g-factor in a low-viscosity solvent in the absence of hyperfine structure. When hyperfine structure is present the same type of broadening occurs, as the spectra in Fig. 18-10 indicate.[15] This sequence of spectra shows the transformation of the pattern of a spin label from a well-resolved triplet at a correlation time of 10^{-11} sec to a poorly resolved powder spectrum at the longest correlation times.

For the case of axial symmetry there are two additional line-broadening terms due to the anisotropy in A and one extra term due to the combination of the g-factor and hyperfine anisotropies.[9,16] These are expressed in terms of the following parameters[9,10]

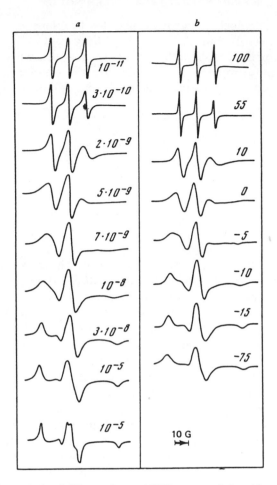

Fig. 18-10. (*a*) Theoretical and (*b*) experimental ESR spectra of nitroxide radicals for different rotational correlation times τ_c and different temperatures (in °C). The solvent is glycerol (from Ref. 11).

$$(\Delta A)^2 = (A_{\parallel} - A_0)^2 + 2(A_{\perp} - A_0)^2$$
$$\Delta g \Delta A = (g_{\parallel} - g_0)(A_{\parallel} - A_0) + 2(g_{\perp} - g_0)(A_{\perp} - A_0) \quad (18\text{-}20)$$

and Δg^2 which is given by Eq. 18-11.

The linewidth depends on the correlation time and contains terms analogous to those in Eq. 18-9. In the limit of rapid motions, $\omega_0 \tau_c \ll 1$, the width is given by

$$\Delta H_{1/2} = \Delta H_{1/2\,0} + (\alpha + \beta m_I + \gamma m_I^2)\tau_c \quad (18\text{-}21)$$

where the coefficients have the values

$$\alpha = \frac{2}{15} (\Delta g)^2 \left(\frac{\beta H_0}{\hbar} \right)^2 + \frac{3}{80} (\Delta A)^2$$

$$\beta = \frac{4}{15} \left(\frac{\beta H_0}{\hbar} \right) (\Delta g)(\Delta A) \tag{18-22}$$

$$\gamma = \frac{1}{12} (\Delta A)^2$$

and $\Delta H_{1/2\,0}$ is the width of the line in the absence of anisotropies.

Since the term βm_I carries the sign of m_I it differs for the high- and low-field components and causes the spectrum to be asymmetric in appearance. Because the area under each hyperfine absorption line does not change, the product $(Amp) (\Delta H_{pp})^2$ of the first derivative spectrum will be the same for each component.[17,18] This means that a doubling of the linewidth reduces the amplitude by a factor of 4, and so the relative line amplitudes are sensitive parameters for monitoring the line-broadening process. The broadening is least for the central ($m_I = 0$) hyperfine component and greatest for the high-field ($m_I = -1$) line. And as a result, the three component amplitudes h_m increase in the order $h_- < h_+ < h_0$ as indicated in Fig. 18-11.

As the hyperfine component lines broaden the outermost peaks of the two outer lines shift further away from each other in the manner indicated in the middle spectrum of Fig. 18-11. This movement away from the center is partly due to the line broadening and partly due to the shift towards the $A_{\parallel}$ field positions shown in the top spectrum of Fig. 18-8. The extent of this shift may be measured relative to the unshifted positions in the limit of $\tau_c \to 0$ corresponding to the top spectrum of Fig. 18-11, or they may be measured relative to the outermost peaks of the $\tau_c \to \infty$ powder pattern sketched at the bottom of this figure. Both types of parameter provide a measure of the correlation time τ_c. The first four methods for determining correlation times that are described in Section 18-7 are based on the various parameters defined in Fig. 18-11.

At high viscosities saturation transfer effects occur, as was explained in Section 18-4. In Chapter 11 we obtained the following expression (11-92)

$$\frac{\partial H}{\partial \theta} = \frac{(T_{\parallel}^2 - T_{\perp}^2) \sin \theta \cos \theta}{(T_{\perp}^2 \sin^2 \theta + T_{\parallel}^2 \cos^2 \theta)^{1/2}} \tag{18-23}$$

for the derivative $\partial H/\partial \theta$ that determines the variation of the resonant field with the angle θ. This expression is for the simpler case of an isotropic g-factor and an anisotropic hyperfine constant. Nevertheless it provides the general qualitative behavior for the present case. Figure 18-12 shows a plot of $\partial H/\partial \theta$ for an axially symmetric spin label.[19] We see from this plot that the variation of the resonant field is strongest at the center of each hyperfine

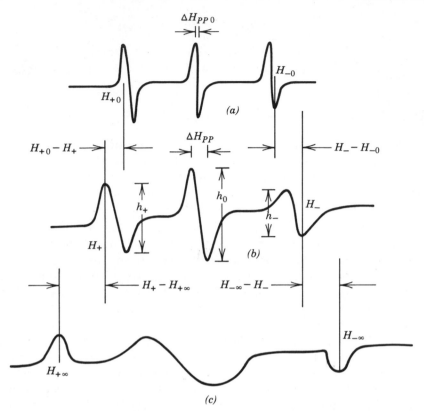

Fig. 18-11. Comparison of spin label spectra (a) in a low-viscosity solution, (b) in the slow motion region, and (c) in a rigid lattice. The amplitudes and the shifts in position of the outer hyperfine peaks are indicated.

pattern. This causes the change in the level of saturation in the center of a hyperfine line to depend strongly on the correlation time, and hence the lineshapes of partly saturated spectra provide a quantitative measure of τ_c. The last two methods for measuring correlation times that are described in the next section are based on this fact.

A comparison of experimental spectra, computer simulated spectra, and independent correlation time measurements have shown that the most sensitive methods for measuring correlation times with spin labels involve the use of the amplitude, position, and width changes of individual hyperfine lines for relatively short values of τ_c, and they involve the use of saturation transfer techniques for relatively long values of τ_c in the intermediate motion region where $10^{-9} < \tau_c < 10^{-4}$ sec. Several of these methods will be discussed in the next section.

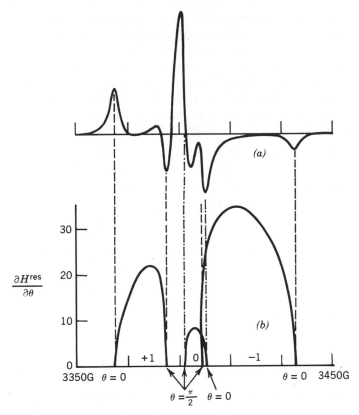

$\dfrac{\partial H^{\text{res}}}{\partial \theta}$

Fig. 8-12. (*a*) Simulated first derivative powder pattern absorption spectrum of magnetically dilute isotropically oriented nitroxide spin labels with axially symmetric Zeeman and hyperfine interactions. (*b*) Plot of the derivative $\partial H/\partial \theta$ of the three component hyperfine powder patterns labeled with their m_I values. At the turning points corresponding to $\theta = 0$ and $\theta = \pi/2$ small changes in θ have negligible effect on the spectrum. This figure should be compared with Figs. 18-6 and 18-9 (from Ref. 19).

18-7 MEASURING MOLECULAR MOTION WITH SPIN LABELS

We see from an examination of Fig. 18-10 that increasing correlation times cause the three hyperfine lines to broaden, move apart, and distort in a rather irregular manner. Various practical methods have been developed for determining the correlation time for molecular motion based on changes in the amplitudes, positions, and widths of the lines for short values of τ_c (10^{-9}–10^{-7} sec) and employing saturation transfer techniques for long values of τ_c (10^{-7}–10^{-4} sec) over the range of spectral lineshape changes that are illustrated in this figure. To some extent these methods are semiempirical in their application because in many uses they depend on the comparison of calculated spectra with those recorded under standard conditions such as using solutions with known viscosities. Computer simulations[20-22] can be a great aid in the determination of the correlation time. We will summarize several of the particular methods which have been proposed.

In this section we use the notation τ_R for values of the rotational correlation time that are determined from spectra, and τ_c for the actual correlation times of the solution. Some of the parameters which are used by these various methods are defined in Fig. 18-11.

Several methods for determining τ_R are as follows:

(1) Amplitude and width method for the shortest correlation times (10^{-9}–5×10^{-9} sec).[11,23,24] This method is based on Eq. 18-21, which may be written as follows for the $m_I = +1$ and -1 cases, respectively:

$$\tau_R[\alpha + \beta + \gamma] = \Delta H_{pp+} - \Delta H_{pp\,0} \tag{18-24}$$

$$\tau_R[\alpha - \beta + \gamma] = \Delta H_{pp-} - \Delta H_{pp\,0} \tag{18-25}$$

where peak-to-peak linewidths are now used. The sum of these two equations is

$$2\tau_R[\alpha + \gamma] = \Delta H_{pp+} + \Delta H_{pp-} - 2\Delta H_{pp\,0} \tag{18-26}$$

and their difference is

$$2\tau_R[-\beta] = \Delta H_{pp-} - \Delta H_{pp+} \tag{18-27}$$

where from Eqs. 18-19 and 18-22 the quantity β is negative. Recalling that the product of the amplitude h_i and the square of the linewidth ΔH_{ppi}^2 is the same for each hyperfine component[17,18] we can write

$$\Delta H_{ppi} = \frac{\text{const}}{h_i^{1/2}} \tag{18-28}$$

as was mentioned in the previous section. Using the numerical values of Stone et al.[25] for the g-factor and hyperfine principal components we obtain from Eqs. 11-24 and 11-25, respectively, plus Eq. 11-28 the following two expressions for the correlation time τ_R:

$$\tau_R = 5 \times 10^{-9}\left(\sqrt{\frac{h_0}{h_+}} - 1\right)\Delta H_{pp\,0}$$

$$\tau_R = 2.7 \times 10^{-10}\left(\sqrt{\frac{h_0}{h_-}} - 1\right)\Delta H_{pp\,0} \tag{18-29}$$

where the limiting peak-to-peak linewidth (11-54) $\Delta H_{pp\,0}$ of the central hyperfine component and the three hyperfine amplitudes h_-, h_0, and h_+ defined on Fig. 18-11 are used. The difference equation 18-27 gives the expression

$$\tau_R = 8.3 \times 10^{-11}\left(\sqrt{\frac{h_+}{h_-}} - 1\right)\Delta H_{pp+} \tag{18-30}$$

TABLE 18-1   **Parameters^a for fitting $\tau_R = a(1 - S)^b$ (from Ref. 16)**

Diffusion Model^b	ΔH_{pp}, G	a	b	$\tau_R{}^c$, sec	
Brownian diffusion	0.3	2.57×10^{-10}	-1.78	9×10^{-7}	A
	3.0	5.40×10^{-10}	-1.36	3×10^{-7}	B
	5.0	8.52×10^{-10}	-1.16	2×10^{-7}	
	8.0	1.09×10^{-9}	-1.05	12×10^{-7}	
Free diffusion	0.3	6.99×10^{-10}	-1.20	3×10^{-7}	C
	3.0	1.10×10^{-9}	-1.01	1×10^{-7}	D
Strong diffusion (jump)	0.3	2.46×10^{-9}	-0.589	4×10^{-8}	E
	3.0	2.55×10^{-9}	-0.615	5×10^{-8}	F

^{a} These values are calculated for an axial nitroxide with $A_{\parallel} = 32\,G$, $A_{\perp} = 6\,G$, $g_{\perp} - g_{\parallel} = 0.0041$, and isotropic reorientation (from Ref. 20).
^{b} These models are discussed in Ref. 16.
^{c} τ_R values compared for $S = 0.99$.

(5) Progressive saturation method for intermediate correlation times $(10^{-7}\text{–}10^{-6}\,\text{sec})$.[5,30,31] This method, which is based on an article by Goldman et al.,[31] is recommended by Hyde and Dalton[5] as most sensitive in the correlation time range from 10^{-7} to 10^{-6} sec. They measured T_1 by the saturation method described in Section 11-11 by plotting the peak-to-peak linewidth squared $(\Delta H_{pp})^2$ versus the microwave power and using Eqs.

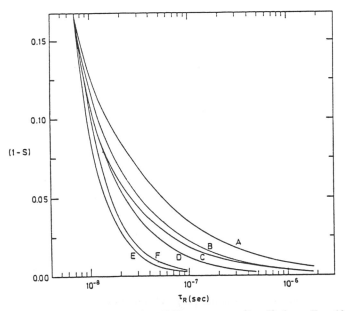

Fig. 18-17. Dependence of the hyperfine shift parameter $(1 - S)$ from Eq. 18-35 on the correlation time. Table 18-1 gives the linewidth and reorientation model corresponding to each of the curves A–F (from Ref. 21).

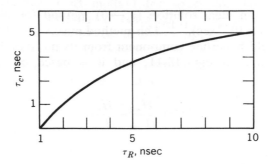

Fig. 18-13. Comparison of the rotational correlation time τ_R calculated from Eq. 18-30 with τ_c (from Ref. 15).

which is plotted in Fig. 18-13. We see from this graph that the method becomes insensitive for correlation times longer than 5×10^{-9} sec. Hoffman et al.[24] made use of the sum equation (18-26) to obtain the expression[26]

$$\tau_R = 5.47 \times 10^{-10} \left[\sqrt{\frac{h_0}{h_+}} + \sqrt{\frac{h_0}{h_-}} - 2 \right] \Delta H_{pp\,0} \qquad (18\text{-}31)$$

for the correlation time.

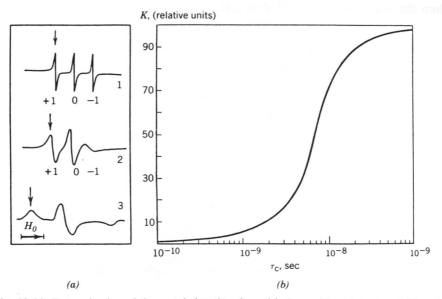

(a)   (b)

Fig. 18-14. Determination of the correlation time from (a) the position of the low-field peak marked with the arrow on the rapid motion region ($\tau_c = 0$) spectrum (1), the slow motion region spectrum (2), and the rigid lattice spectrum (3). (b) Plot of parameter K from Eq. 18-32 versus correlation time τ_c (from Ref. 15).

11-124 and 11-127 to determine the ratio T_1/T_2 from the slope and T_2 from the intercept of the plot, as described in Sections 3-3 of Ref. 7 and 13C of Ref. 18. Care was taken to evaluate the constant k of Eq. 11-120 which gives the ratio of the H_1 field to the square root of the microwave power.

(6) First harmonic out of phase dispersion method for long correlation times (10^{-6}–10^{-5} sec).

(7) Second harmonic out of phase absorption method for long correlation times (10^{-5}–10^{-4} sec).

The last two methods, which will be treated together, are based on the saturation transfer behavior of the slow motion regime spectrum.[4-6] The phenomenon of saturation transfer was described in Section 18-4. Signals recorded with the lock-in detector set 90° out of phase are more sensitive to saturation transfer phenomena than are the usual in-phase signals. Figure 18-18 presents out of phase experimental and computer simulated spectra for maleimide spin labeled II hemoglobin over the correlation time range from 10^{-7} sec to infinity.[19,32] We see that both the first harmonic dispersion

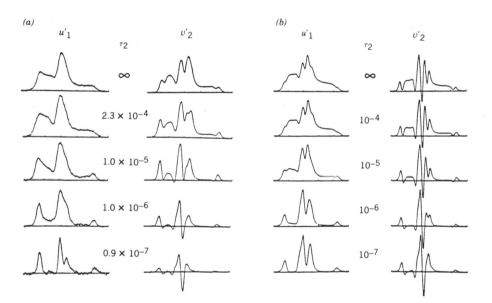

Fig. 18-18. (*a*) Representative saturation transfer spectra obtained from maleimide spin labeled hemoglobin. From top to bottom the samples are ammonium sulfate precipitated Hb, Hb in 90% glycerol at $-12°C$, Hb 80% at 5°C, Hb 60% at 5°C, and Hb 40% at 5°C. Correlation times were calculated from Eq. 18-2 using measured solvent viscosities η and an effective radius $a = 29$ Å for the hydrated protein. The displayed signals are first harmonic dispersion out of phase (U_1') modulated at 100 kHz and second harmonic absorption out of phase (V_2') modulated at 50 kHz, both detected at 100 kHz. A 0.25 G microwave field H_1 and a 5 G modulation amplitude were employed using scans from 3350 to 2450 G. (*b*) Simulated spectra were calculated using the parameters $T_1 = 6.6 \times 10^{-6}$ sec, $T_2 = 2.4 \times 10^{-8}$ sec, $A_{\parallel} = 35$ G, $A_{\perp} = 7$ G, $g_{\parallel} = 2.00241$, and $g_{\perp} = 2.00741$ (from Ref. 19).

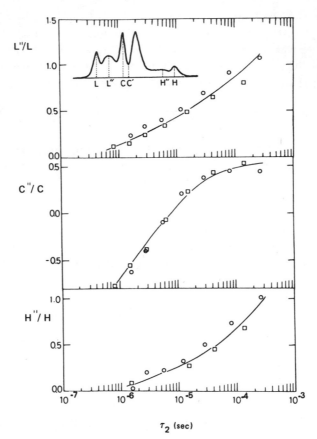

Fig. 18-19. Dependence of the parameter ratios L''/L, C''/C, and H''/H defined at the upper left on the correlation time. The data are for hemoglobin ($\bigcirc$) and bovine serum albumin ($\square$) labeled with maleimide spin labels and dissolved in glycerol water mixtures (from Ref. 32).

and the second harmonic absorption signals exhibit pronounced variations in shape over the range of correlation times presented. Hyde and Dalton[5] assert that the former provides greater sensitivity for the shorter correlation times, with the latter being more sensitive for the longer times in the range 10^{-6}–10^{-4} sec. Other articles[33,34] have presented series of spectra similar to those in Fig. 18-18.

To determine the correlation time for a system, empirical parameters are selected from the spectra, and calibration curves are made to ascertain the variation of the parameters with τ_c. An example of this procedure is shown in Fig. 18-19 in which three pairs of amplitude parameters provide a greater range of sensitivity to τ_c than any one parameter alone. The use of amplitude ratios, in the manner illustrated in Fig. 18-19, provides for internal consistency in the determination.

REFERENCES

1. H. M. Schwartz, J. R. Bolton, and D. D. Borg, *Biological Applications of Electron Spin Resonance*, Wiley, New York, 1972.

2. L. S. Berlinger, ed., *Spin Labeling*: *Theory and Practice*, Academic Press, New York, 1976, Vol. 1; 1979, Vol. 2.

3. P. F. Knowles, D. Marsh, and H. W. E. Rattle, *Magnetic Resonance of Biomolecules*: *An Introduction to the Theory and Practice of NMR and ESR in Biological Systems*, Wiley-Interscience, New York, 1976

4. L. R. Dalton, B. H. Robinson, L. A. Dalton, and P. Coffey, *Adv. Magn. Reson.*, **8**, 149 (1976).

5. J. S. Hyde and L. R. Dalton, in *Spin Labeling*: *Theory and Practice*, Vol. 2, L. S. Berlinger, ed., Academic Press, New York, 1979, Chap. 1.

6. J. S. Hyde and D. D. Thomas, *Ann. Rev. Phys. Chem.*, **31**, 293 (1980).

7. C. P. Poole, Jr. and H. A. Farach, *Relaxation in Magnetic Resonance*, Academic Press, New York, 1971.

8. H. Sillescu, *J. Chem. Phys.*, **54**, 2110 (1971).

9. C. P. Poole, Jr. and H. A. Farach, *Relaxation in Magnetic Resonance*, Academic Press, New York, 1971, Chap. 6.

10. A. Carrington and A. D. McLachlan, *Introduction to Magnetic Resonance*, Harper & Row, New York, 1967.

11. G. I. Likhtenshtein, *Spin Labeling Methods in Molecular Biology*, P. S. Shelnitz, translator, Wiley-Interscience, New York, 1974.

12. P. F. Devaux and J. Davoust, "Physical Aspects of the Spin Labelling Technique," in *ESR and NMR of Paramagnetic Species in Biological and Related Systems*, I. Bertini and R. S. Drago, eds., Reidel, Dordrecht, 1979, p. 381.

13. O. H. Griffith and A. S. Waggoner, *Acc. Chem Res.*, **2**, 17 (1969).

14. J. D. Morrisett, in: *Spin Labeling*: *Theory and Practice*, Vol. 1, L. S. Berlinger, ed., Academic Press, New York, 1976, Chap. 8.

15. A. N. Kuznetsov, A. M. Wasserman, A. Yu. Volkov, and N. N. Korst, *Chem. Phys. Lett.*, **12**, 103 (1971).

16. J. H. Freed, in *Spin Labeling*: *Thoery and Practice*, Vol. 1, L. S. Berlinger, ed., Academic Press, New York, 1976, Chap. 3.

17. C. P. Poole, Jr. and H. A. Farach, "Electron Spin Resonance," in *Handbook of Spectroscopy*, Vol. 2, J. W. Robinson, ed., CRC Press, Cleveland, 1974.

18. C. P. Poole, Jr., *Electron Spin Resonance*, Wiley-Interscience, New York, 1983.

19. D. D. Thomas, L. R. Dalton, and J. S. Hyde, *J. Chem. Phys.*, **65**, 3006 (1976).

20. J. H. Freed, G. V. Bruno, and C. F. Polnazek, *J. Phys. Chem.*, **75**, 3385 (1971).

21. S. A. Goldman, G. V. Bruno, and J. H. Freed, *J. Phys. Chem.*, **76**, 1858 (1972).

22. A. M. Bobst, T. K. Sinha, and Y. C. E. Pan, *Science*, **188**, 153 (1975).

23. J. H. Freed and G. H. Frenkel, *J. Chem. Phys.*, **39**, 326 (1963).

24. M. M. Hoffman, P. Schofield, and A. Rich, *Proc. Natl. Acad. Sci. U.S.A.*, **62**, 1195 (1969).

25. T. J. Stone, T. Buchman, P. L. Nordio, and H. M. McConnell, *Proc. Natl. Acad. Sci. U.S.A.*, **54**, 1010 (1965).

26. A. M. Bobst, in *Spin Labeling*: *Theory and Practice*, Vol. 2, L. S. Berlinger, ed., Academic Press, New York, 1979, Chap. 7.

27. R. C. McCalley, E. J. Shimshick, and H. M. McConnell, *Chem. Phys. Lett.*, **13**, 115 (1972).

28. E. J. Shimshick and H. M. McConnell, *Biochem. Biophys. Res. Commun.*, **46**, 321 (1972).

29. R. Mason and J. H. Freed, *J. Phys. Chem.*, **78**, 1321 (1974).

30. J. S. Hyde and L. Dalton, *Chem. Phys. Lett.*, **16**, 568 (1972).

31. S. A. Goldman, G. V. Bruno, and J. S. Freed, *J. Chem. Phys.*, **59**, 3071 (1973).

32. A. Kusumi, T. Sakaki, T. Yoshizawa, and S. Ohsishi, *J. Biochem.*, **88**, 1103 (1980).

33. K. Balasurbramanian, L. R. Dalton, K. D. Schmalbein, and A. H. Heiss, *Chem. Phys.*, **29**, 163 (1968).

34. R. C. Perkins, Jr., T. Lionel, B. H. Robinson, L. A. Dalton, and L. R. Dalton, *Chem. Phys.*, **16**, 393 (1976).

19

FOURIER TRANSFORM NUCLEAR MAGNETIC RESONANCE

19-1 INTRODUCTION

Traditionally, magnetic resonance spectra are recorded by varying either the frequency or the magnetic field strength through the region of resonance while holding the other at a constant value. When searching for unknown spectra this technique has the disadvantage that useful information is only obtained while one is actually recording a resonant line. Most of the time is spent scanning between resonances when only noise appears.

An alternate type of spectroscopy consists in irradiating the sample with a pulse that contains a range of frequencies covering the entire spectrum and monitoring the absorption from all the superimposed resonant lines as the signal decays with time due to the various relaxation processes.[1-4] The result is a time domain spectrum of the type illustrated in Fig. 19-1a. The corresponding frequency domain spectrum obtained by a Fourier transformation of the time domain spectrum is presented in Fig. 19-1b. Since useful information is being obtained during the entire period that the time domain spectrum is being recorded a considerable increase in sensitivity can result.

In high-resolution NMR experimental measurements are generally made in the time domain and theoretical calculations of Hamiltonian parameters readily provide frequency domain spectra. As a result, the Fourier transform procedure is the link between experimentation and theory. In ESR, on the other hand, both theoretical and experimental aspects of the spectroscopy are ordinarily carried out in the frequency domain.

In this chapter we discuss spectra obtained by the use of this Fourier transformation method. Before doing so, however, we summarize some of the properties of Fourier transforms.

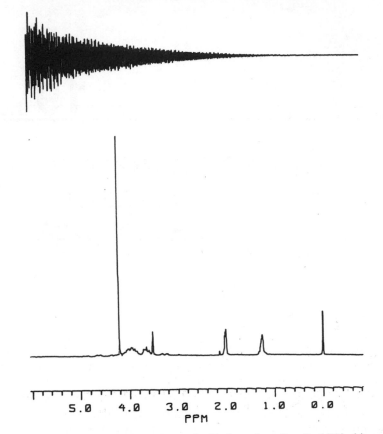

Fig. 19-1. Proton NMR spectrum of a liposaccharide from the cell wall of Rhizobium showing (*a*) time domain and (*b*) frequency domain scans (P. T. Ellis, private communication).

19-2 FOURIER TRANSFORMS

Fourier transforms[5-7] are mathematical functions that involve pairs of conjugate variables, and in our case these variables are the frequency ω and the time t. A function of the frequency $Y(\omega)$ is related to a function of the time $G(t)$ through the following two Fourier transformation integrals:

$$Y(\omega) = \int_{-\infty}^{\infty} G(t) \exp(i\omega t)\,dt \qquad (19\text{-}1)$$

$$G(t) = \frac{1}{2\pi} \int_{-\infty}^{\infty} Y(\omega) \exp(-i\omega t)\,d\omega \qquad (19\text{-}2)$$

The functions $Y(\omega)$ and $G(t)$ are referred to as Fourier transformation pairs. In our case $Y(\omega)$ corresponds, for example, to the amplitude of a spectrum as a function of the frequency and $G(t)$ provides the decay of a signal with time.

In practice the measured signal $G(t)$ is real, and for this case we see from Eq. 19-1 and the relation $\exp(i\omega t) = \cos \omega t + i \sin \omega t$ that $Y(\omega)$ has imaginary and real parts $Y'(\omega)$ and $Y''(\omega)$, respectively, given by

$$Y'(\omega) = \int_{-\infty}^{\infty} G(t) \sin \omega t \, dt \qquad (19\text{-}3)$$

$$Y''(\omega) = \int_{-\infty}^{\infty} G(t) \cos \omega t \, dt \qquad (19\text{-}4)$$

where $Y'(\omega)$ corresponds to a dispersion signal and $Y''(\omega)$ corresponds to an absorption signal. Gade and Lowe[8] calculated a large number of Fourier transforms for the free induction decay of nuclear spins $I = \frac{1}{2}, \frac{3}{2}$, and infinity in various lattices.

We see from Eqs. 19-1 and 19-2, respectively, that the response function $G(t)$ must be known at all times to calculate the value of $Y(\omega)$ at a single frequency, and likewise one must have a complete knowledge of $Y(\omega)$ at all frequencies to determine the value of $G(t)$ at a single instant of time. Sometimes experimental conditions or computational limitations require one to truncate the integration in evaluating a Fourier transform. Experimentally, the exciting rf pulse will have a finite range of frequencies and at long times the measured resonance response will become quite weak and eventually drop below the noise level. Computationally, one makes the approximation

$$Y(\omega) = \int_{0}^{t_f} G(t) \exp(i\omega t) dt \qquad (19\text{-}5)$$

and the validity of this assumption depends on the extent to which the experimentally determined response function $G(t)$ is distorted by the pulsing conditions and has an appreciable magnitude at times before zero and beyond t_f.

It is a general property of Fourier transform pairs that the width of the distribution in the frequency domain is the reciprocal of the width in the time domain

$$\left(\begin{array}{c} \text{Width of } Y(\omega) \\ \text{in frequency domain} \end{array} \right) \left(\begin{array}{c} \text{Width of } G(t) \\ \text{in time domain} \end{array} \right) \approx 1 \qquad (19\text{-}6)$$

independent of the shapes of the functions that are involved. In more conventional magnetic resonance nomenclature this means that the half-width of the resonant line $\frac{1}{2}\Delta\omega_{1/2}$ in the frequency spectrum is the reciprocal of the time constant τ of the decay of the response with time

$$(\tfrac{1}{2}\Delta\omega_{1/2})\tau \approx 1 \qquad (19\text{-}7)$$

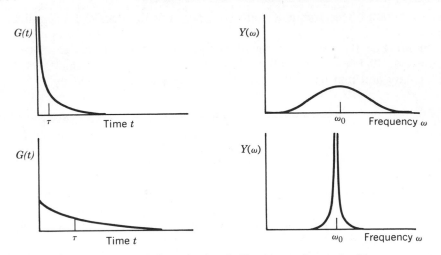

Fig. 19-2. The top of the figure shows graphs of a Fourier transform pair with a narrow spread in the time domain and a broad spread in the frequency domain; the bottom of the figure shows the opposite case.

This reciprocal relationship is illustrated in Fig. 19-2.

19-3 LORENTZIAN LINESHAPE AND EXPONENTIAL DECAY

When the relaxation processes cause the signal to decay exponentially with time in the manner illustrated in Fig. 19-3a after the removal of the rf power at the time $t = 0$, then the time domain spectrum, called a free induction decay, is given by

$$G(t) = 0 \qquad\qquad t < 0$$

$$G(t) = \exp\left(-\frac{t}{\tau}\right) \qquad t > 0 \tag{19-8}$$

Inserting $G(t)$ from these expressions into Eq. 19-1 gives the following frequency domain signal:

$$Y(\omega) = \int_0^\infty \exp\left(-\frac{t}{\tau}\right) \exp(i\omega t)\, dt \tag{19-9}$$

The integration is easily carried out to give

$$Y(\omega) = \tau \, \frac{1 + i\omega\tau}{1 + \omega^2\tau^2} \tag{19-10}$$

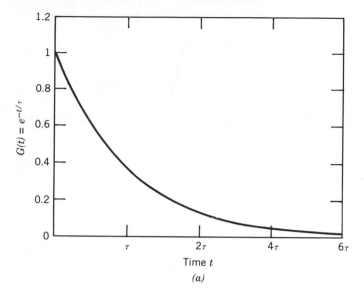

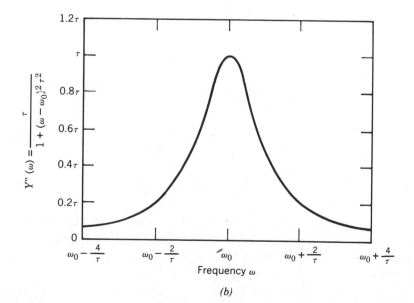

Fig. 19-3. Lorentzian lineshapes showing, from left to right, (*a*) free induction exponential decay in the time domain, (*b*) absorption, and (*c*) dispersion in the frequency domain.

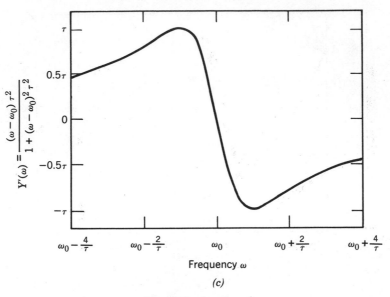

(c)

Fig. 19-3   (continued).

The imaginary and real parts $Y'(\omega)$ and $Y''(\omega)$ are easily recognized as the dispersion and absorption (Eq. 11-49) Lorentzian lineshapes, respectively,[9-11]

$$Y'(\omega) = \frac{\omega\tau^2}{1 + \omega^2\tau^2} \tag{19-11}$$

$$Y''(\omega) = \frac{\tau}{1 + \omega^2\tau^2} \tag{19-12}$$

centered around the frequency $\omega = 0$. They are easily rewritten in terms of the center frequency ω_0 as follows:

$$Y'(\omega) = \frac{(\omega - \omega_0)\tau^2}{1 + (\omega - \omega_0)^2\tau^2} \tag{19-13}$$

$$Y''(\omega) = \frac{\tau}{1 + (\omega - \omega_0)^2\tau^2} \tag{19-14}$$

These dispersion and absorption lineshapes are plotted in Figs. 19-3b and 19-3c, respectively. The quantity τ in these expressions is equal to twice the reciprocal of the full linewidth at half-amplitude $\Delta\omega_{1/2}$ defined in Fig. 11-7 which gives

$$\tfrac{1}{2}\Delta\omega_{1/2} = \frac{1}{\tau} \tag{19-15}$$

in agreement with Eq. 19-7 and also with Eq. 11-49.

19-4 GAUSSIAN SELF-TRANSFORM

A Gaussian function is its own Fourier transform. To show this we substitute the following Gaussian time decay function

$$G(t) = \exp\left(-\frac{t^2}{2\tau^2}\right) \tag{19-16}$$

into Eq. 19-1 to obtain the frequency domain signal $Y(\omega)$

$$Y(\omega) = \int_0^\infty \exp\left(-\frac{t^2}{2\tau^2}\right) \exp(i\omega t) dt \tag{19-17}$$

The change of variable

$$t = \tau x + i\omega\tau^2 \tag{19-18}$$

converts Eq. 19-17 to the form

$$Y(\omega) = \tau \exp(-\tfrac{1}{2}\omega^2\tau^2)\left[\int_0^\infty \exp(-\tfrac{1}{2}x^2) dx + \int_0^{i\omega\tau} \exp(-\tfrac{1}{2}x^2) dx\right] \tag{19-19}$$

The first term of this expression integrates directly to $\sqrt{\tfrac{1}{2}\pi}$ and the second may be converted to a real integral by the change of variable

$$x = iz \tag{19-20}$$

to give

$$Y(\omega) = \tau\sqrt{\frac{\pi}{2}} \exp(-\tfrac{1}{2}\omega^2\tau^2)\left[1 + i\sqrt{\frac{2}{\pi}} \int_0^{\omega\tau} \exp(\tfrac{1}{2}z^2) dz\right] \tag{19-21}$$

The real or absorption part $Y''(\omega)$ of the Gaussian lineshape is a true Gaussian (cf. Eq. 11-48),

$$Y''(\omega) = \exp[-\tfrac{1}{2}(\omega - \omega_0)^2\tau^2] \tag{19-22}$$

where we have dropped the factor $\tau\sqrt{\pi/2}$ and added the center frequency ω_0 to the lineshape. In like manner, we have for the dispersion mode

$$Y'(\omega) = \sqrt{\frac{2}{\pi}} \exp[-\tfrac{1}{2}(\omega - \omega_0)^2\tau^2] \int_0^{\omega\tau} \exp(\tfrac{1}{2}z^2) dz \tag{19-23}$$

where part of the frequency dependence is in the upper limit of the integral. Unfortunately, this integral can not be evaluated in closed form.

Figure 19-4 compares the Lorentzian and Gaussian lineshapes in both the time and the frequency domains. We see from these figures that the Gaussian shapes are more peaked in the center and the Lorentzian ones are

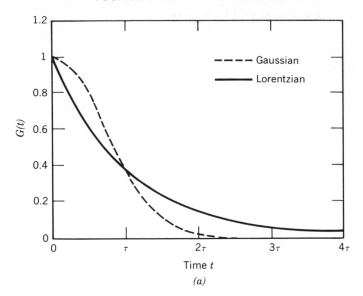

(a)

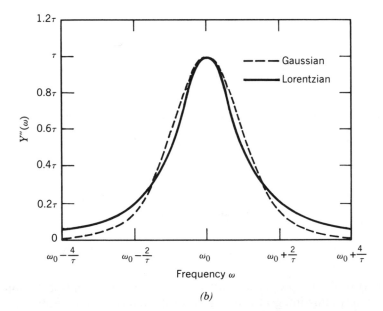

(b)

Fig. 19-4. Comparison of Gaussian and Lorentzian lineshapes (a) time domain and (b) absorption part of frequency domain.

more spread out in the wings. In practical applications the Gaussian and Lorentzian absorption lines are equally common, but in the time domain NMR of low viscosity liquids Lorentzian shapes are encountered more often.

19-5 FREE INDUCTION DECAY

Fourier transform NMR spectroscopy is ordinarily carried out with pulse sequences of particular types, as was mentioned at the end of Section 11-11, so before describing time domain spectra we make some remarks about these pulses.[1-4,12-16]

Consider an ensemble of nuclear spins with a net magnetization M_z along the direction of an external magnetic field H_0. If a radiofrequency pulse of field strength H_1 in the x direction is applied at a frequency that satisfies the resonance condition, then the spins will experience a torque $\vec{M} \times \vec{H}_1$ which tends to tip the magnetization vector away from the z direction. This will induce a transverse magnetization which precesses around H_0 and, after the cessation of the pulse, decays with time in the xy plane as $\exp(-t/T_m)$. In a perfectly homogeneous magnetic field the time constant T_m for the decay is the spin–spin relaxation time T_2 introduced in Section 11-10. Ordinarily, magnetic field inhomogeneities broaden the line through the factor $2/T_2^*$ and this adds to the intrinsic width $\gamma\Delta H_{1/2} = 2/T_2$ of Eq. 11-109 to give for the total width

$$\Delta H_{1/2} = \frac{2}{\gamma T_m} \tag{19-24}$$

where the decay constant T_m is given by

$$\frac{1}{T_m} = \frac{1}{T_2} + \frac{1}{T_2^*} \tag{19-25}$$

The exponential decay of the transverse magnetization, which is shown in Fig. 19-5, is called a free induction decay.

The pulses used to produce the free induction decay must be longer than several rf periods $2\pi/\omega_0$ to be well defined and shorter than the decay time constant T_m otherwise the magnetization vector will not follow the pulse

$$T_m > t_w > \frac{2\pi}{\omega_0} \tag{19-26}$$

Typical NMR values are 2×10^{-8} sec for $2\pi/\omega_0$ and 10^{-2} sec for T_m. The requirements of Eq. 19-26 are more difficult to satisfy at microwave frequencies, so the spin echo experiments that are described in the next section are much less common in ESR.

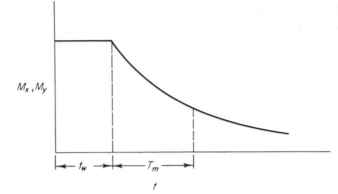

Fig. 19-5. The application of a single high-power rf pulse of width t_w and the subsequent free induction decay with the time constant T_m following the cessation of the pulse.

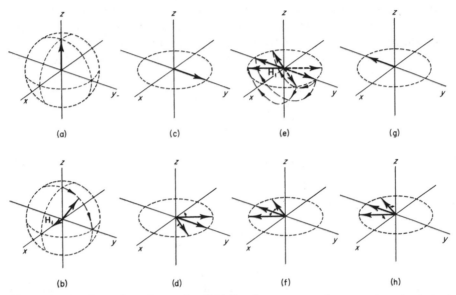

Fig. 19-6. The formation of an echo. Initially, the net magnetic moment vector is in its equilibrium position (*a*) parallel to the direction of the strong external field. The rf field H_1 is then applied. As viewed from above the rotating frame of reference, the net magnetic moment appears (*b*) to rotate quickly about H_1. At the end of a 90° pulse this net magnetic moment vector is in the equatorial plane (*c*). During the relatively long period of time following the removal of H_1 the incremental moment vectors begin to fan out slowly (*d*). This is caused by the variations in H_1 over the sample. At the time $t = \tau$ the rf field H_1 is again applied and the moments (*e*) are rotated quickly about the direction of H_1. This time H_1 is applied just long enough to satisfy the 180° pulse condition which implies that at the end of the pulse (*f*) all the incremental moment vectors slowly begin to recluster. Because of the inverted relative positions following the 180° pulse and because each incremental moment vector continues to precess with its former frequency, these vectors will be perfectly reclustered (*g*) at $t = 2\tau$ when a maximum signal is induced in the pickup coil. This maximum signal, or echo, then begins to decay as the incremental vectors again fan out (*h*) (from Ref. 17).

19-6 SPIN ECHOES

When an rf pulse of duration t_w and rf magnetic field strength H_1 is applied to the spin system it causes the magnetization vector to tilt away from the applied magnetic field direction through an angle equal to $\gamma H_1 t_w$ in radians. A 90° rf pulse

$$\gamma H_1 t_w = \tfrac{1}{2}\pi \qquad (19\text{-}27)$$

rotates the magnetization into the xy plane perpendicular to the magnetic field direction,[17] as shown in Fig. 19-6. When the pulse is turned off the spins in this plane precess around H_0, and the magnetic field inhomogeneities cause various spin packets to have slightly different Larmor frequencies and hence to precess at slightly different rates. Therefore the magnetization vectors of the individual spin packets spread out in the xy plane, as shown in Fig. 19-6d. The spreading out corresponds to a decrease in the magnitude of the overall magnetization vector in the plane, and this produces the decrease in signal after the initial pulse that is shown in Fig. 19-7. The application of a 180° pulse of length $2t_w$

$$\gamma H_1(2t_w) = \pi \qquad (19\text{-}28)$$

reverses the spin directions as indicated in Fig. 19-6e, and they begin to come together to produce an echo when they coalesce at the point of Fig. 19-6g. After this they spread out again, always with an amplitude that decreases exponentially with time. Successive 180° pulses can be employed to produce spin echoes which sample the magnetization while it decays with time, as indicated in Fig. 19-7. A number of different spin echo schemes

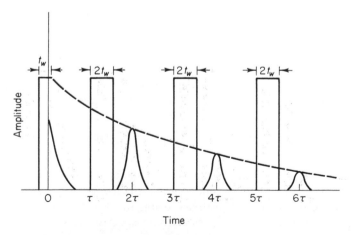

Fig. 19-7. Application of a 90° pulse ($t_w = \pi/2\gamma H_1$) followed by successive 180° pulses, and the resultant exponential decay (dotted line) of the echos.

have been devised which make use of various combinations of 90°, 180°, and other pulse lengths.[2,4,14,18]

19-7 TIME DOMAIN NMR SPECTRA

In time domain NMR spectroscopy a broadband rf pulse bends the magnetization away from the H_0 direction and its return to the z direction is monitored by the detector. In the early days of NMR a series of 90° rf pulses was used with a waiting time between them long enough for the system to approach equilibrium, which in practice amounts to waiting much longer than one spin lattice relaxation time T_1. The signals from successive pulses were added to provide a stronger detected signal. Repetitive pulsing without waiting for equilibration to occur is faster but it reduces the signals from lines with long T_1 values, and this can be partly compensated for by employing spin flipping angles less than 90°. Several special pulse sequences are currently utilized to circumvent this difficulty and acquire data more efficiently.

Multiplet structure is ordinarily not evident on a time domain spectrum which typically has the appearance of exponentially decaying noise, as illustrated in Fig. 19-1. The structure appears when a computer is employed to Fourier transform this spectrum and thereby convert it to the more usual frequency domain type shown in Fig. 7-7. Thus a time domain spectrum is mainly of interest as the raw data to be processed by a Fourier transform algorithm which converts it to a conventional spectrum and sometimes calculates line positions, line intensities, chemical shifts, and perhaps spin–spin coupling constants.

Ernst and Anderson[10] compared the efficiency of the Fourier transform method with that of a conventional cw frequency domain scan made under slow passage conditions so that the scanning time satisfied the inequality

$$t_s \gg \frac{1}{\gamma H_1} \tag{19-29}$$

They found the following signal to noise ratios S/N obtained using time domain (TD) and frequency domain (FD) procedures

$$\frac{(S/N)_{TD}}{(S/N)_{FT}} = 0.8K \left(\frac{\text{Scan width}}{\text{Linewidth}} \right)^{1/2} \tag{19-30}$$

where the factor K is a function of T_1 and the spacing between the pulses, and it is generally close to 1. Since linewidths are typically much less than scanning times the gain in sensitivity using the Fourier transform method can be quite large. Most frequency domain spectra are scanned too fast to satisfy the slow passage criterion so in practice the improvement in sensitivity is less.

Fourier transform NMR spectroscopy has become a routine analytical tool, and the literature is quite extensive.[1-4,18] Fourier transform techniques have also been applied to ESR,[19-21] but thus far only to a somewhat limited extent.

19-8 TWO-DIMENSIONAL SPECTROSCOPY

Until now we have been discussing one-dimensional Fourier transforms which have one frequency variable and one time variable. If the detected NMR signal $G(t_1, t_2)$ is a function of two times t_1 and t_2, then a double Fourier transformation

$$Y(\omega_1, \omega_2) = \int_{-\infty}^{\infty} e^{i\omega_1 t_1} dt_1 \int_{-\infty}^{\infty} G(t_1, t_2) e^{i\omega_2 t_2} dt_2 \qquad (19\text{-}31)$$

will provide a spectral lineshape function $Y(\omega_1, \omega_2)$ that depends on two frequencies ω_1 and ω_2 which are conjugate to their respective time counterparts. It is customary to label the two frequency axes of the resulting spectrum with the symbols F_1 and F_2 instead of ω_1 and ω_2.

A general two-dimensional (2D) experiment[22,23] has the four time periods shown in Fig. 19-8. During the first period, called the preparation, thermal equilibrium is established followed by one or more rf pulses. In the second period the spin system evolves for a variable time t_1 under the action of these and perhaps additional pulses. The mixing period is one with fixed pulses and delays and is sometimes absent. During the final detection period the signal is recorded for successive times t_2, just as in a one-dimensional experiment.

A simple example of a two-dimensional experiment is a free induction decay recorded as a function of the pulse length t_1 and the time t_2 after the termination of the pulse, as indicated in Fig. 19-9. There is no mixing period in this case. The signal $G(t_1, t_2)$ will be strongest for a $\pi/2$ pulse (19-27) corresponding to

$$t_1 = \frac{\pi}{2\gamma H_1} \qquad (19\text{-}32)$$

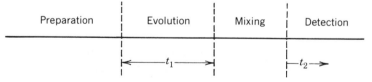

Fig. 19-8. The four time periods of a general two-dimensional experiment. The preparation period establishes thermal equilibrium and ends with one or more pulses, the variable length evolution period provides for the development of the spin system under the influence of pulses, the fixed time mixing period provides for further development under additional pulses and delays, and the detection period involves signal acquisition as a function of the time t_2. The mixing period is sometimes omitted (from Ref. 23).

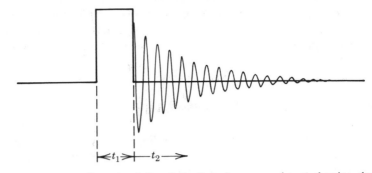

Fig. 19-9. Simple two-dimensional free induction decay experiment showing the pulse of variable length t_1 and the signal acquisition during the time t_2 (from Ref. 23).

which rotates the magnetization into the xy plane, as indicated in Fig. 19-6c. A π pulse (19-28) produces no signal because no x, y component exists after it, and a $3\pi/2$ pulse rotates the magnetization around to the $-x$ direction and produces a signal with the opposite phase of the $\pi/2$ pulse. A cosine Fourier integration (19-4) of the detected signal $G(t_1, t_2)$ over t_2

$$J(t_1, \omega_2) = \int_{-\infty}^{\infty} G(t_1, t_2) \cos(\omega_2 t_2) dt_2 \qquad (19\text{-}33)$$

will give the usual absorption spectrum as a function of the pulse length t_1, and Fig. 19-10a shows plots of the spectrum $J(t_1, \omega_2)$ along the F_2 axis for successive values of t_1. On the figure we see the maximum signal appearing at the $\pi/2$ and $3\pi/2$ points and we also observe the change in phase after π. A second cosine Fourier transform over the time t_1

$$Y(\omega_1, \omega_2) = \int_{-\infty}^{\infty} J(t_1, \omega_2) \cos(\omega_1 t_1) dt_1 \qquad (19\text{-}34)$$

provides the two-dimensional absorption spectrum shown in Fig. 19-10b.

We now illustrate these expressions for the simple case of an exponential time decay of the magnetization for which the various detected and Fourier transformed functions can be written explicitly in terms of the frequency Ω_1

$$\Omega_1 = \gamma H_1 \qquad (19\text{-}35)$$

at which the pulse rotates the magnetization around the x axis, and the deviation Ω_2 of the rotating frame frequency Ω_R from the resonant frequency $\Omega_0 = \gamma H_0$ is

$$\Omega_2 = \Omega_0 - \Omega_R \qquad (19\text{-}36)$$

The detected signal is given by

$$G(t_1, t_2) = G_m[\sin(\Omega_1 t_1)e^{-t_1/\tau_1}](e^{i\Omega_2 t_2}e^{-t_2/\tau_2}) \qquad (19\text{-}37)$$

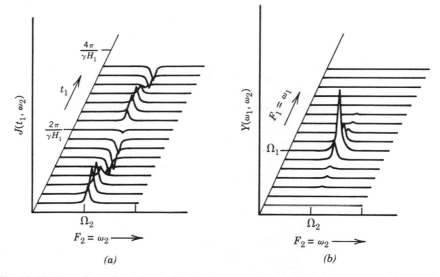

Fig. 19-10. Two-dimensional free induction decay results showing (*a*) a sequence of Lorentzian shaped absorption spectra $J(t_1, \omega_2)$ with a damped sine wave amplitude dependence for successive pulse lengths t_1 obtained from a single Fourier integration and (*b*) the corresponding two-dimensional spectrum $\gamma(\omega_1, \omega_2)$ obtained from two Fourier integrations showing the Lorentzian lineshapes along both axes centered at $\omega_1 = \omega_0$, $\omega_2 = \omega_0$ (from Ref. 23).

where τ_1 and τ_2 are the time decay constants during the evolution and detection periods, respectively, and G_m is the maximum value of $G(t_1, t_2)$. The first cosine Fourier integration (19-33) over t_2 gives the Lorentzian-type absorption signal

$$J(t_1, \omega_2) = J_m \sin(\Omega_1 t_1) e^{-t_1/\tau_1} \left[\frac{\tau_2}{1 + (\omega_2 - \Omega_2)^2 \tau_2^2} \right] \qquad (19\text{-}38)$$

of maximum amplitude $J_m \tau_2$ that is plotted in Fig. 19-10*a*. We see from this figure how the $\sin \Omega_1 t_1$ dependence is damped in amplitude by the $\exp(-t_1/\tau_1)$ term since the signal beyond $t_1 = 2\pi/\gamma H_1$ is weaker than that for earlier times. The second cosine Fourier integration (19-34) gives the double Lorentzian function

$$Y(\omega_1, \omega_2) = Y_m \left[\frac{\tau_1}{1 + (\omega_1 - \Omega_1)^2 \tau_1^2} \right] \left[\frac{\tau_2}{1 + (\omega_2 - \Omega_2)^2 \tau_2^2} \right] \qquad (19\text{-}39)$$

of maximum value $Y_m \tau_1 \tau_2$ which is plotted in Fig. 19-10*b*.

Bax[23] mentions three main categories of two-dimensional experiments:

1. Shift correlation spectroscopy in which, for example, chemical shifts of protons might constitute the ordinate F_1 and chemical shifts of ^{13}C nuclei might be plotted along the abscissa axis F_2.

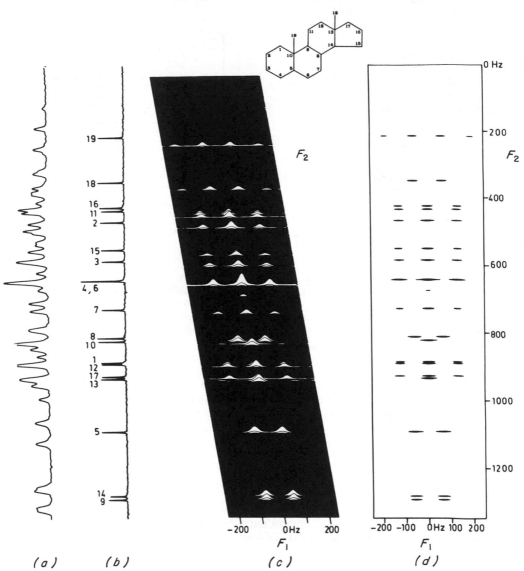

Fig. 19-11. Two-dimensional J spectroscopy showing (a) conventional one-dimensional ^{13}C spectrum of 5α-androstane recorded at 50 MHz, (b) the same spectrum with proton decoupling, (c) two-dimensional J spectrum in three-dimensional relief, and (d) contour plot of the relief spectrum (from Ref. 23).

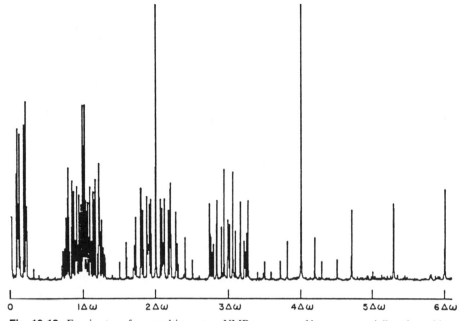

Fig. 19-12. Fourier transform multiquantum NMR spectrum of benzene, partially oriented in a nematic liquid crystal, exhibiting a great multiplicity of lines. The transitions appear in groups according to the change in Zeeman quantum number, as indicated. The complexity of the line groupings decreases from left to right until the highest-order one, labeled $6\Delta\omega$, comprises only a single transition (from Ref. 24).

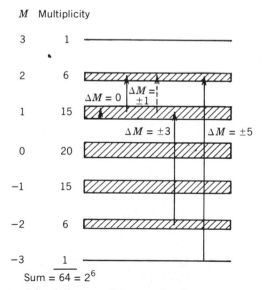

Fig. 19-13. NMR energy level diagram of benzene showing examples from the various groups of multiquantum transitions. The multiplicity or number of sublevel states for each total M quantum number is indicated.

2. J spectroscopy in which chemical shifts are plotted along the F_1 ordinate axis and scalar or spin–spin couplings are plotted along the F_2 abscissa axis. Figure 19-11 gives an example of this. We see from this figure that adding the intensity along horizontal or constant F_2 lines gives the conventional spectrum on the left, and for a particular chemical shift or F_2 value the F_1 axis displays spin–spin type couplings.

3. Multiple quantum spectroscopy[24] in which two-dimensional methods facilitate the detection of multiquantum transitions. Figure 19-12 shows the multiquantum spectrum of benzene[24,25] which exhibits all the transitions of the types indicated in Fig. 19-13.

Recent reviews[23,26–29] of two-dimensional NMR spectroscopy may be consulted for further information on this rapidly growing field.

REFERENCES

1. D. Shaw, *Fourier Transform NMR Spectroscopy*, Elsevier, Amsterdam, 1976.

2. T. C. Ferrar and E. D. Becker, *Pulse and Fourier Transform NMR*, Academic Press, New York, 1973.

3. K. Mullin and P. S. Pregosin, *Fourier Transform Nuclear Magnetic Resonance Techniques: A Practical Approach*, Academic Press, New York, 1977.

4. J. W. Akitt, *NMR and Chemistry: An Introduction to the Fourier Transform–Multinuclear Era*, 2nd ed., Chapman & Hall, London, 1983.

5. C. S. Rees, S. M. Shah, and C. V. Štanojevič, eds., *Theory and Applications of Fourier Analysis*, Dekker, Amsterdam, 1980.

6. J. H. Weaver, *Applications of Discrete and Continuous Fourier Analysis*, Wiley-Interscience, New York, 1983.

7. M. R. Spiegel, *Fourier Analysis*, McGraw-Hill, New York, 1974.

8. S. Gade and I. J. Lowe, *Phys. Rev.*, **148**, 382 (1966).

9. I. J. Lowe and R. E. Norberg, *Phys. Rev.*, **107**, 46 (1957).

10. R. R. Ernst and W. A. Anderson, *Rev. Sci. Instrum.*, **37**, 93 (1966).

11. R. C. Ernst, *Adv. Magn. Reson.*, **2**, 1 (1966).

12. C. P. Poole, Jr. and H. A. Farach, *Relaxation in Magnetic Resonance*, Academic Press, New York, 1981.

13. E. Fukushima and S. B. W. Roeder, *Experimental Pulse NMR: A Nuts and Bolts Approach*, Addison-Wesley, Reading, MA, 1981.

14. U. Haeberlen, Advances in Magnetic Resonance Supplement I, *High Resolution NMR in Solids, Selective Averaging*, Academic Press, New York, 1976.

15. A. G. Redfield and R. K. Gupta, *Adv. Magn. Reson.*, **5**, 82 (1971).

16. J. D. Ellet, Jr., M. G. Gibby, U. Haeberlen, L. M. Mehring, A. Pines, and J. S. Waugh, *Adv. Magn. Reson.*, **5**, 117 (1971).

17. H. Y. Carr and E. M. Purcell, *Phys. Rev.*, **94**, 630 (1954).

18. R. A. Komoroski and G. C. Levy, *Magn. Reson. Rev.*, **3**, 289 (1974).

19. P. A. Narayana and L. Kevan, *Magn. Reson. Rev.*, **7**, 239 (1983).

20. W. B. Mims, in *Electron Paramagnetic Resonance*, S. Geschwind, ed., Plenum, New York, 1972, Chap. 4.

21. W. B. Mims, *The Linear Electric Field Effect in Paramagnetic Resonance*, Clarendon Press, Oxford, 1976.

22. J. Jeener, B. H. Meier, P. Bachman, and R. R. Ernst, *J. Chem. Phys.*, **71**, 4546 (1979).

23. A. Bax, *Two-Dimensional Nuclear Magnetic Resonance in Liquids*, Reidel, Dordrecht, 1982.

24. D. P. Weitekamp, *Adv. Magn. Reson.*, **11**, 111 (1983)

25. S. Sinton, *NMR Studies of Oriented Molecules*, Ph.D. Dissertation, University of California, Berkeley, 1981.

26. B. E. Mann and H. Günther, *Angew. Chem. (Int. Ed. Engl.)*, **22**, 350 (1983).

27. K. Wüthrich, *Biochem. Soc. Symp.*, **46**, 17 (1981).

28. R. R. Ernst, *ACS Symp. Ser.*, 191 (1982).

29. H. C. E. McFarlane and W. McFarlane, *Specialist Periodical Report NMR*, **13**, 174 (1983).

PHYSICAL CONSTANTS AND ENERGY CONVERSION FACTORS

PHYSICAL CONSTANTS

Planck constant	h	6.62618×10^{-34}	$\mathrm{J\,Hz^{-1}}$
h-Bar	$\hbar$	1.05459×10^{-34}	$\mathrm{J\,s}$
Boltzmann constant	k	1.38066×10^{-23}	$\mathrm{J\,K^{-1}}$
Velocity of light	c	2.997924×10^{8}	$\mathrm{m\,s^{-1}}$
Avogadro number	N	6.022045×10^{23}	$\mathrm{mole^{-1}}$
Electronic charge	e	1.60219×10^{-19}	C
Electron mass	m_e	9.10953×10^{-30}	kg
Proton mass	m_p	1.67265×10^{-27}	kg
Neutron mass	m_n	1.67495×10^{-27}	kg
Bohr magneton	β	9.27408×10^{-24}	$\mathrm{J\,T^{-1}}$
Nuclear magneton	β_N	5.05082×10^{-27}	$\mathrm{J\,T^{-1}}$
Proton gyromagnetic ratio	γ_p	2.67520×10	$\mathrm{rad\,G^{-1}\,s^{-1}}$
	$\gamma_p/2\pi$	4.25771×10^{3}	$\mathrm{G^{-1}\,s^{-1}}$
Free electron g-factor	g_e	2.002319	
Proton magnetic moment	μ_p	1.41062×10^{-26}	$\mathrm{J\,T^{-1}}$
Electron magnetic moment	μ_e	9.28483×10^{-24}	$\mathrm{J\,T^{-1}}$

ESR $\qquad 1/\lambda = 0.0466860gH$ (λ in cm, H in kG)

$\qquad\qquad \nu = 1.39961gH$ (ν in GHz, H in kG)

Proton
NMR $\qquad \left\{ \nu = 4.25771\ H \right.$ (ν in MHz, H in kG)

$\beta/\beta_N = 1836.15$

ENERGY CONVERSION FACTORS

The energy in a unit at the left in Table A-1 is multiplied by the appropriate factor to convert it to the energy unit given at the top of the corresponding column. For example, 1 electron volt (eV) is equal to $8065.47\ \mathrm{cm}^{-1}$ and corresponds to a temperature of $1.16049\ 10^4$ K.

TABLE A-1 Energy Conversion Factors

	eV	cal/mole	cm^{-1}	erg	K	sec^{-1}
eV	1	23055	8065.47	1.60219×10^{-12}	1.16049×10^{4}	2.41797×10^{14}
cal/mole	4.3374×10^{-5}	1	0.3499	0.6949×10^{-16}	0.50336	1.0489×10^{10}
cm^{-1}	1.2398×10^{-4}	2.8583	1	1.9862×10^{-16}	1.4387	2.998×10^{10}
erg	0.6242×10^{12}	1.439×10^{16}	0.5035×10^{16}	1	0.7244×10^{16}	1.5092×10^{26}
K	8.6170×10^{-5}	1.9867	6.9501×10^{-1}	1.3805×10^{-16}	1	2.0837×10^{10}
sec^{-1}	4.1357×10^{-15}	9.534×10^{-11}	3.3356×10^{-11}	6.625×10^{-27}	4.799×10^{-11}	1

INDEX